Groundwater Hydrology: Issues, Challenges and Management

Groundwater Hydrology: Issues, Challenges and Management

Edited by **William Sobol**

SYRAWOOD
PUBLISHING HOUSE

New York

Published by Syrawood Publishing House,
750 Third Avenue, 9th Floor,
New York, NY 10017, USA
www.syrawoodpublishinghouse.com

Groundwater Hydrology: Issues, Challenges and Management
Edited by William Sobol

© 2016 Syrawood Publishing House

International Standard Book Number: 978-1-68286-044-1 (Hardback)

Printed in the United States of America.

Contents

Preface

Groundwater hydrology is an emerging field of study that focuses upon movement, quality as well as availability of groundwater. The prominent topics such as groundwater modelling, availability of groundwater resources, quality and assessment of groundwater from different regions, etc. have been covered in this text. Researches and case studies by eminent experts and scientists have been incorporated in this advanced text. Students, researchers, and all associated with hydrology, water resource management and allied fields will benefit alike from this book.

Significant researches are present in this book. Intensive efforts have been employed by authors to make this book an outstanding discourse. This book contains the enlightening chapters which have been written on the basis of significant researches done by the experts.

Finally, I would also like to thank all the members involved in this book for being a team and meeting all the deadlines for the submission of their respective works. I would also like to thank my friends and family for being supportive in my efforts.

Editor

Heavy metals assessment of private drinking water supplies in Ibadan, Nigeria and associated health risks

B. A. Adelekan and Oguntoso N.

Department of Agricultural Engineering, Federal College of Agriculture, Moor Plantation, Ibadan, Oyo State, Nigeria.

The aim was to measure the concentrations of heavy metals in groundwater obtained from 30 randomly selected domestic wells and 10 stream locations all in Ibadan, Nigeria, compare the results with the World Health Organization guidelines, draw conclusions and make recommendations. Water samples were obtained and analysed for Pb, As, Cd, Zn, Cu, Cr, Fe and Mn. Overall, the minimum concentrations of Pb, As, Cd, Cr and Fe in the well water samples were below detection limit (BDL). The maximum values were 0.02, 0.45, 0.01, 0.445, 0.135, 0.09, 0.245 and 0.155 mg/l respectively. In the surface water samples, the minimum concentrations of Pb, As, Cd and Cr were below detection level, while the maximum concentrations of Pb, As, Cd, Zn, Cu, Cr, Fe and Mn were respectively 0.075, 0.05, 0.001, 0.445, 0.120, 0.065, 0.45 and 0.16 mg/l. No evidence of contamination of these water supply sources with heavy metals was found going by the fact that the values obtained were lower than the guideline values established by the World Health Organization. A possible exception is As which in some samples had higher concentrations than the WHO guideline. The recommendations of the study include continuous and close monitoring of these private drinking water supplies. There must also be strict compliance to regulatory limits in sludge and wastes to be released into the environment, and enforcement of other environmental protection regulations. Findings from this study will be of immense help to the general public as well as researchers and environmental regulators working in this area of interest in developing countries.

Key words: Heavy metals, drinking water, Ibadan

INTRODUCTION

Begun et al. (2009) observed that large quantities of pollutants have continuously been introduced into ecosystems as a consequence of urbanization and industrial processes. Metals are persistent pollutants that can be biomagnified in the food chains, and natural waters, becoming increasingly dangerous to human beings and wildlife. The use of private wells and surface sources to provide water for domestic purposes in both urban areas and rural communities is a common practice in Nigeria and communities all over Africa (Adelekan and

Ogunde, 2012; Adetunji and Odetokun, 2011; Adelekan, 2010). Therefore, assessing the concentrations of pollutants in different components of the ecosystem has become an important task in preventing risk to natural life and public health. Heavy metals enter into the environment mainly via three routes namely: (i) deposition of atmospheric particulate; (ii) disposal of metal enriched sewage sludges, sewage effluents and solid wastes, as well as (iii) by-products from metal mining process. Corrosion of metallic installations, ash amendment to manures for odor reduction and lime amendment for disinfection purposes are other possible sources of heavy metals, e.g. lead, cadmium, chrome, zinc and nickel (Eckel, 2005). Air pollution caused by the emissions of toxic metals is one of the main types of environmental pollution, and it has been recognized as a potential threat to both environment and human health. Ingestion, dermal contact absorption and inhalation are the main routes of air particle metals entering human bodies (Shi et al., 2011). All these readily happen in urban environments. Soil is one of the repositories for anthropogenic wastes. Biochemical processes can mobilize its content of wastes to pollute water supplies and impact food chains. Heavy metals such as Pb, As, Cd, Zn, Cu, Cr, Fe, and Mn are potential soil and water pollutants.

Globally, the problem of environmental pollution due to heavy metals has begun to cause concern in most large cities since this may lead to geoaccumulation, bioaccumulation and biomagnifications in ecosystems. Ibadan (7°23'47" N 3°55'0" E) is one of the three largest metropolises in Nigeria. It occupies an area of 828 km^2 and has a population of approximately 2.6 million according to the 2006 census (NPC, 2009). Reported researches in respect of water quality in the city of Ibadan, Nigeria include Adelekan and Abegunde (2011) which investigated the heavy metals contamination of soils and groundwater at automobile mechanic villages in the city and Adelekan and Alawode (2011) which assessed the contributions of municipal refuse dumps to heavy metals concentrations in soil profile and groundwater on the dump sites in the city. The two papers reported that the concentrations of certain heavy metals are steadily increasing due to industrial activities and waste dumping at those locations studied. With that position established, the objective of this research is to investigate the concentrations of these heavy metals in domestic water sources at residences in the city in order to establish the presence or otherwise of any health risks to the city residents. From observations, appropriate remedies can then be proposed for the protection of human health and the environment. It is with that objective in mind that this research work was undertaken.

MATERIALS AND METHODS

Water samples were obtained from 10 locations on Odo Ona and Ogunpa streams which are important surface water sources used for domestic purposes by Ibadan residents. Samples were also obtained from 30 wells on private residences scattered throughout the city following standard water sampling procedure. The wells ranged from 8 to 10 m in depth, and were lined to ensure recharge from the bottom only. Three replicates were sampled at each location. Each sample was directly collected into a factory-fresh 1.5 L plastic bottle, with cap securely tightened. This ensured there was no prior contamination, and there was no possibility of contamination after sampling that could affect the samples and invalidate results obtained. After collection the bottles were placed inside ice coolers for transportation to the laboratory where they

were then transferred to the refrigerator. This ensured that any form of microbial activity due for example to iron bacteria, occurring after sampling, would be stopped or severely curtailed. Laboratory analysis commenced the same day. The methods used are described in APHA et al. (1998). Readings were made on the Atomic Absorption Spectrophotometer (AAS). These investigations were conducted in November 2011.

RESULTS

Results obtained are presented in the tables and figure. From Table 1, it is noticed that the concentration of each of the heavy metals in most of the surface water samples met the guidelines of the World Health Organization (2004a). From Table 2, it is noticed that the maximum and minimum concentrations of heavy metals occurred variously in different surface water samples. From Table 3, it is noticed that the concentration of each of the heavy metals in most of the well water samples met the guidelines of the World Health Organization (2004a). Measurements for well water samples obtained from the tables are plotted in the following figures.

Figure 1 shows that Pb concentration in the well samples ranged from 0 (that is, below detection limit) to 0.02 mg/l. The trendline started around 0.005 mg/l and dipped slightly as well number changed from 1 to 30.

Figure 2 shows that As concentration in the well samples ranged from 0 (that is, below detection limit) to 0.055 mg/l (well number 18). The trendline started at 0.03 mg/l and dipped strongly to about 0.02 mg/l as well number changed from 1 to 30.

Figure 3 shows that Zn concentration in the well samples ranged from 0.12 (well 18) to 0.445 mg/l (well 26). The trendline started around 0.35 mg/l and climbed slightly above that value as the well number changed from 1 to 30.

Figure 4 shows that Cu concentration in the well samples ranged from 0.07 mg/l (well 14) to 0.135 mg/l (wells 12, 13, 17 and 25). The trendline started around 0.104 mg/l and remained practically on the same level as well number changed from 1 to 30.

Figure 5 shows that Cr concentration in the well samples ranged from 0 (that is, below detection limit) to 0.15 mg/l. The trendline started around 0.04 mg/l and dipped slightly as well number changed from 1 to 30.

Figure 6 shows that Fe concentration in the well samples ranged from 0 (1 sample) (that is, below detection limit) to 0.025 mg/l. The trendline started around 0.16 mg/l and dipped slightly as well number changed from 1 to 30.

Figure 7 shows that Mn concentration in the well samples ranged from 0.07 to 0.16 mg/l. The trendline started around 0.12 mg/l and dipped slightly as well number changed from 1 to 30.

Figure 8 shows that Zn concentration in the well samples ranged from 0.13 to 0.45 mg/l. The trendline started around 0.34 mg/l and rose slightly as well number changed from 1 to 30. Cu concentration ranged

Table 1. Heavy metals concentrations in surface water samples (mg/l).

Sample	Pb	As	Cd	Zn	Cu	Cr	Fe	Mn
S1	BDL	0.050	BDL	0.280	0.090	0.050	0.225	0.135
S2	0.050	BDL	0.001	0.280	0.105	BDL	0.175	0.105
S3	BDL	0.035	0.001	0.435	0.085	0.050	0.180	0.145
S4	0.010	BDL	BDL	0.295	0.105	0.065	0.450	0.135
S5	BDL	BDL	BDL	0.355	0.110	BDL	0.225	0.105
S6	0.075	0.050	BDL	0.295	0.105	BDL	0.185	0.125
S7	0.007	0.035	0.001	0.385	0.075	0.035	0.205	0.160
S8	0.010	0.030	BDL	0.405	0.085	BDL	0.205	0.100
S9	BDL	0.040	BDL	0.335	0.120	BDL	0.250	0.105
S10	BDL	0.040	BDL	0.445	0.105	0.060	0.270	0.135
Average	0.015	0.028	0.0003	0.351	0.099	0.026	0.233	0.125
WHO guideline	0.01	0.01	0.003	3	1	0.050	0.30	0.200

Values are means of 3 measurements; BDL, below detection limit in the sample analysed.

Table 2. Maximum and minimum measured concentrations of heavy metals in surface water samples (mg/l).

Heavy metal	Measured limits		Sample of occurrence
Pb	Maximum	0.075	S6
	Minimum	BDL	S1, S3, S5, S9, S10
As	Maximum	0.050	S1
	Minimum	BDL	S2, S4, S5
Cd	Maximum	0.001	S2, S3, S7
	Minimum	BDL	S1, S4, S5, S6, S8, S9, S10
Zn	Maximum	0.445	S10
	Minimum	0.280	S1, S2
Cu	Maximum	0.120	S9
	Minimum	0.085	S3, S8
Cr	Maximum	0.065	S4
	Minimum	BDL	S2, S5, S6, S8, S9
Fe	Maximum	0.450	S4
	Minimum	0.175	S2
Mn	Maximum	0.160	S7
	Minimum	0.100	S8

BDL, Below detection limit in the sample analysed.

from 0.07 to 0.15 mg/l and its trendline starting around 0.1 mg/l remained level. Concentration of Zn was more than that of Cu in all the samples. Figure 9 shows that Fe concentration in the well samples ranged from 0 (that is, below detection limit) to 0.25 mg/l. The trendline started around 0.16 mg/l and dipped slightly as well number changed from 1 to 30. Cr concentration ranged from 0 (below detection limit) and peaked at 0.12 mg/l. Its

trendline dipped slightly from 0.04 mg/l. For the majority of samples, the concentration of Fe was higher than that of Cr.

Figure 10 shows that Mn concentration in the well samples ranged from 0.08 to 0.15 mg/l. The trendline started around 0.12 mg/l and dipped slightly as well number changed from 1 to 30. As ranged from 0 (below detection limit) to 0.06 mg/l, while its trendline started

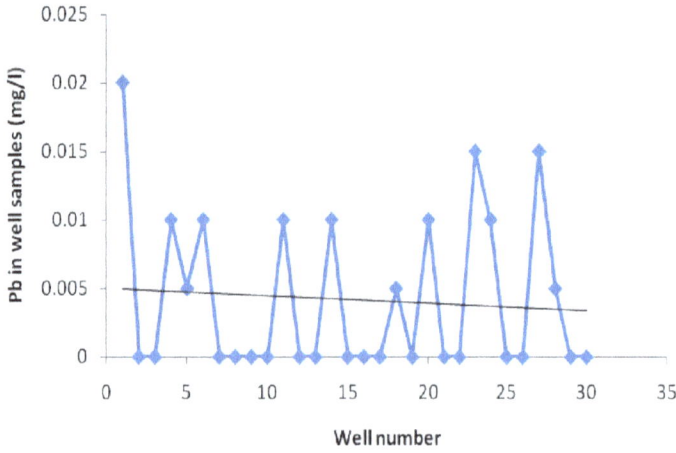

Figure 1. Lead in well water samples.

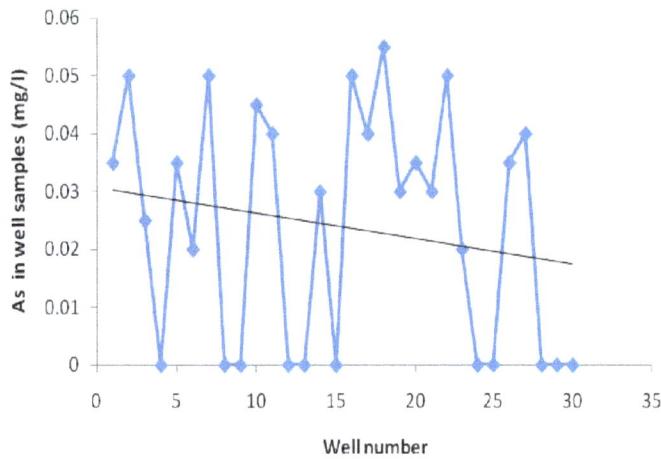

Figure 2. Arsenic in well water samples.

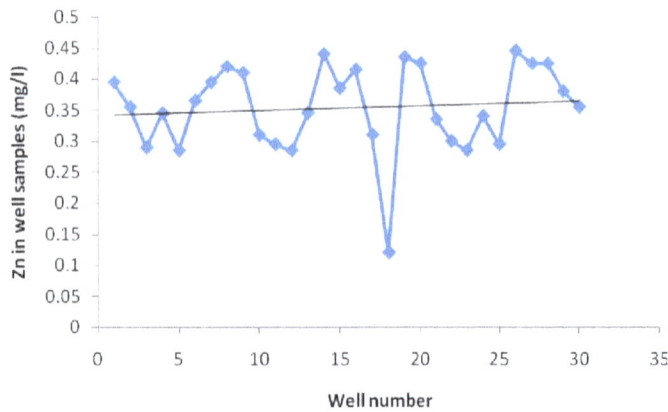

Figure 3. Zinc concentration in well water samples.

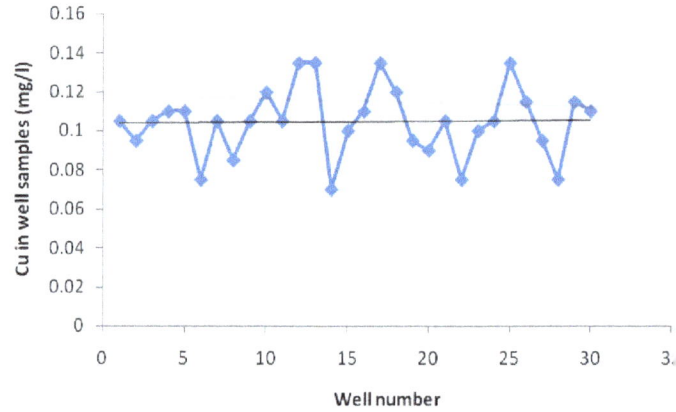

Figure 4. Copper in well water samples.

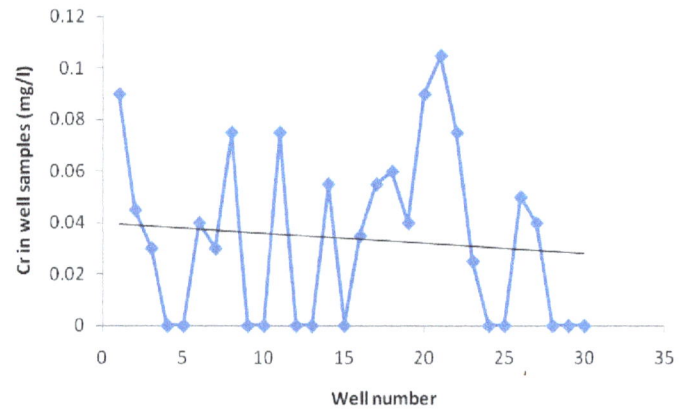

Figure 5. Chromium in well water samples.

DISCUSSION

Table 1 shows the concentrations of heavy metals in surface water samples obtained from 10 locations, the average values as well as the WHO (2004a) guidelines. With the exception of Cr and Fe, concentrations of heavy metals measured were lower than the WHO guideline values, for all samples. In fact, values were below detection limits (BDL) in 5, 3, 7, and 5 locations for Pb, As, Cd, and Cr respectively. For Zn and Cu in particular, concentrations measured were far below the WHO guideline values. For Cr, concentration above WHO guideline value was measured in two out of the ten locations, while for Fe just one location had a concentration higher than the WHO guideline value.

Table 2 shows the maximum and minimum concentrations of the heavy metals as well as their samples of occurrence. For Pb, the maximum concentration of 0.075 mg/l occurred in sample S6, and this sample also had the minimum concentrations for Cd and Cr, in both cases, below detection limits. The maximum values of 0.065 and 0.45 mg/l measured for Cr and Fe respectively shown in Table 2 were higher than

around 0.03 mg/l and dipped slightly. Concentrations of manganese were clearly higher than those of As in all the samples.

Table 3. Concentrations of heavy metals measured in well water samples (mg/l).

Sample	Pb	As	Cd	Zn	Cu	Cr	Fe	Mn
W1	0.020	0.035	BDL	0.395	0.105	0.090	0.140	0.155
W2	BDL	0.050	BDL	0.355	0.095	0.045	0.160	0.090
W3	BDL	0.025	0.010	0.290	0.105	0.030	0.090	0.095
W4	0.010	BDL	BDL	0.345	0.110	BDL	0.130	0.120
W5	0.005	0.035	BDL	0.285	0.110	BDL	0.145	0.125
W6	0.010	0.020	BDL	0.365	0.075	0.040	0.185	0.145
W7	BDL	0.050	BDL	0.395	0.105	0.030	0.195	0.105
W8	BDL	BDL	BDL	0.420	0.085	0.075	0.220	0.105
W9	BDL	BDL	BDL	0.410	0.105	BDL	0.150	0.085
W10	BDL	0.045	BDL	0.310	0.120	BDL	0.205	0.085
W11	0.010	0.040	BDL	0.295	0.105	0.075	0.155	0.115
W12	BDL	BDL	BDL	0.285	0.135	BDL	0.195	0.110
W13	BDL	BDL	BDL	0.345	0.135	BDL	0.215	0.115
W14	0.010	0.030	BDL	0.440	0.070	0.055	0.245	0.135
W15	BDL	BDL	0.001	0.385	0.100	BDL	BDL	0.140
W16	BDL	0.050	BDL	0.415	0.110	0.035	0.090	0.150
W17	BDL	0.040	BDL	0.310	0.135	0.055	0.105	0.085
W18	0.005	0.055	BDL	0.120	0.120	0.060	0.100	0.110
W19	ND	0.030	BDL	0.435	0.095	0.040	0.105	0.130
W20	0.010	0.035	0.005	0.425	0.090	0.030	0.145	0.085
W21	BDL	0.030	BDL	0.335	0.105	0.035	0.105	0.100
W22	BDL	0.050	BDL	0.300	0.075	0.025	0.150	0.100
W23	0.015	0.020	BDL	0.285	0.100	0.025	0.095	0.140
W24	0.010	BDL	0.001	0.340	0.105	BDL	0.075	0.115
W25	BDL	BDL	0.001	0.295	0.135	BDL	0.135	0.135
W26	BDL	0.035	0.001	0.445	0.115	0.050	0.155	0.135
W27	0.015	0.040	BDL	0.425	0.095	0.040	0.220	0.075
W28	0.005	BDL	BDL	0.425	0.075	BDL	0.125	0.100
W29	BDL	BDL	BDL	0.380	0.115	BDL	0.170	0.090
W30	BDL	BDL	BDL	0.355	0.110	BDL	0.160	0.105
Average	0.006	0.02	0.001	0.354	0.105	0.03	0.15	0.1
WHO guideline	0.01	0.01	0.003	3	1	0.05	0.3	0.2

Values are means of 3 measurements; BDL, below detection limit in the sample analysed.

the WHO (2004a) guidelines of 0.05 and 0.3 mg/l (Table 1). All the other values measured are certainly within the WHO guidelines, and consumers do not appear to be exposed to heavy metal contamination through these water samples. Table 3 shows the concentrations of heavy metals in water samples obtained from 30 private water wells, their average values as well as the WHO (2004a) guideline values. Cd appears to be the rarest heavy metal in these samples. Its concentration was below detection limit in 77% of the wells. In 20% of the well samples, the concentration of Cd measured was lower than the WHO (2004a) guideline, while it was higher in 3% of the samples. Pb, As and Cr were measured to be below detection limit in 60, 37 and 40% of the well samples respectively. Although, Zn, Cu and Mn were found in all the water samples, their concentrations were however below the WHO guidelines.

For the majority of samples, the average concentrations for all the heavy metals were lower than the WHO (2004a) guideline values. The only exception to this was sample W1 in which 0.09 mg/l was measured for Cr while the WHO guideline value was 0.05 mg/l. Table 4 shows average concentrations of heavy metals measured in surface and well water samples as compared to the WHO (2004a) guidelines. Regarding the well samples, all the average values are lower than the guidelines. Regarding the surface water samples, average concentrations measured for Cd, Zn, Cu, Cr, Fe, and Mn were lower than the WHO guidelines. Average concentrations of Pb and As were slightly higher than the guidelines. Table 5 shows the maximum and minimum concentrations of heavy metals measured in the well water samples. Comparing Table 5 and Table 2, it is noticed that for Pb, Fe and Mn, the maximum concentrations measured in

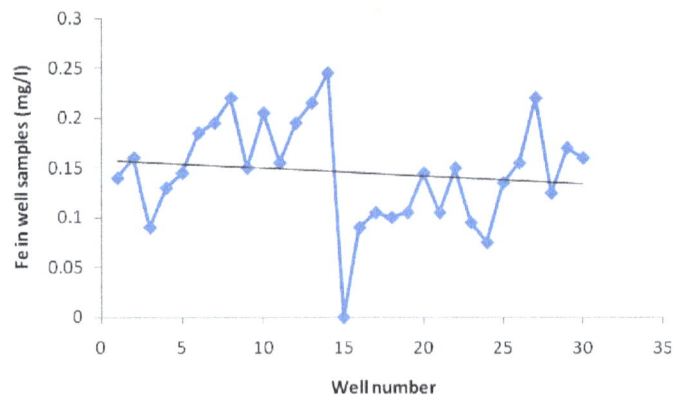

Figure 6. Iron in well water samples.

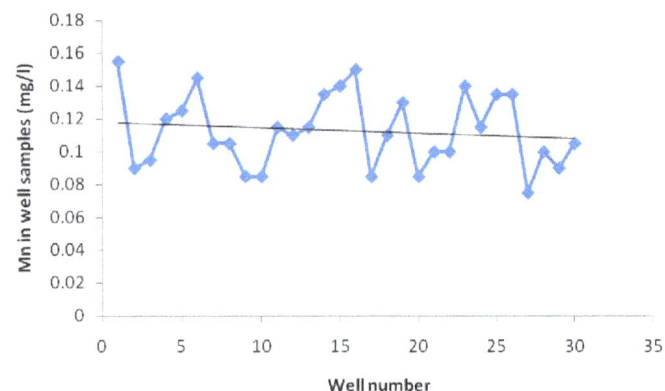

Figure 7. Manganese in well water samples.

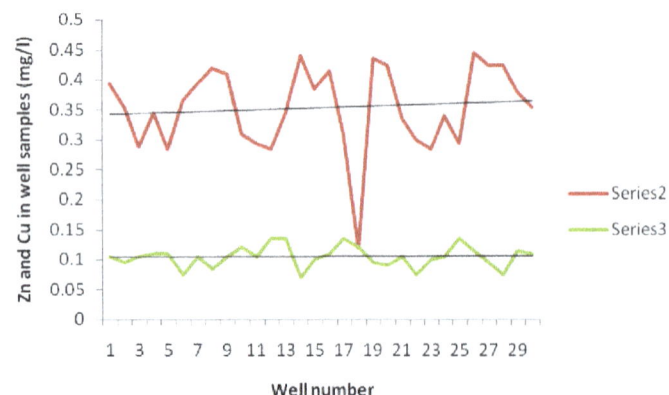

Figure 8. Concentrations of Zn (series 2) and Cu (series 3) in well water samples.

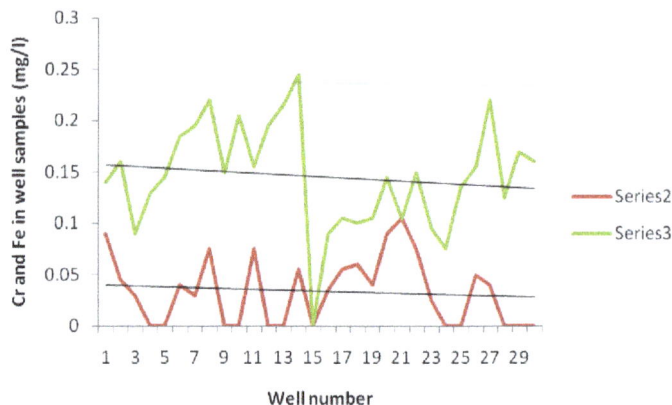

Figure 9. Concentrations of Cr (series 2) and Fe (series 3) in well water samples.

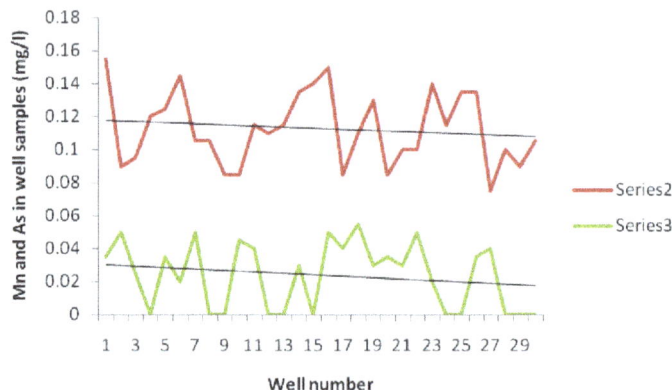

Figure 10. Concentrations of Mn (series 2) and As (series 3) in well water samples.

well water samples were less than those measured for surface water samples. The reverse is the case for As, Cd, Cu and Cr; while for Zn, it was the same value for both well water and surface water samples.

According to USDA (2000), acute (immediate) poisoning from heavy metals is rare through ingestion or dermal contact, but it is possible. Chronic problems associated with long-term heavy metal exposures are mental lapse (lead); toxicological effects on kidney, liver and gastrointestinal tract (cadmium); skin poisoning and harmful effects on kidneys and the central nervous system (arsenic). There is a link between long term exposure to copper and decline of intelligence in young adolescents (Lenntech, 2009). Chronic cadmium exposures result in kidney damage, bone deformities, and cardiovascular problems (Goyer and Clarkson, 2001). Human diseases have resulted from consumption of cadmium contaminated foods (Kobayashi, 1978; Nogawa et al., 1987). The threat that heavy metals pose to human and animal health is aggravated by their low environmental mobility, even under high precipitations, and their long term persistence in the environment (Mench et al., 1994; Chirenje et al., 2004).

USEPA (2004) noted that manganese is an essential element for many living organisms including human beings. It is necessary for proper functioning of some enzymes (manganese superoxide dismutase) and for the activation of others (notably kinases, decarboxylases). Adverse health effects can be caused by inadequate intake or overexposure. The average concentration of manganese

Table 4. Average concentrations of heavy metals measured in surface water and well water samples (mg/l).

Description	Pb	As	Cd	Zn	Cu	Cr	Fe	Mn
Surface water	0.015	0.028	0.0003	0.351	0.099	0.026	0.233	0.125
Well water	0.006	0.02	0.001	0.354	0.105	0.03	0.15	0.1
WHO guideline	0.01	0.01	0.003	3	1	0.05	0.3	0.2

Table 5. Maximum and minimum measured concentrations of heavy metals in well water samples (mg/l).

Heavy metal	Measured limits		Sample of occurrence
Pb	Maximum	0.020	W1
	Minimum	BDL	18 different samples
As	Maximum	0.450	W10
	Minimum	BDL	11 different samples
Cd	Maximum	0.010	W3
	Minimum	BDL	24 different samples
Zn	Maximum	0.445	W26
	Minimum	0.120	W18
Cu	Maximum	0.135	W12, W13, W17, W25
	Minimum	0.070	W14
Cr	Maximum	0.090	W1
	Minimum	BDL	12 different samples
Fe	Maximum	0.245	W14
	Minimum	BDL	W15
Mn	Maximum	0.155	W1
	Minimum	0.075	W27

BDL, Below detection limit in the sample analysed.

measured in all the well samples was 0.1 mg/l while the surface samples averaged 0.125 mg/l. These low values however should not be of much concern since manganese deficiency in human beings appears to be rare because manganese is present in many common foods. According to WHO (2004b), a health-based guideline value of 0.4 mg/l should be adequate to protect public health. Excessive levels of Mn in the brain produces extra pyramidal symptoms similar to those in patients with Parkinson's disease (Stredrick et al., 2002), decreased learning ability in school age children and increased propensity for violence in adults (Finley, 2004). According to the Wisconsin Department of Health and Family Services (2007), manganese levels below 300 µg/l are generally not a health concern. Infants should not drink water that is above the health advisory level of 300 µg/l. Many years of exposure to high levels of manganese can cause harm to the nervous system. A disorder similar to Parkinson's disease can result.

Frequently found in water due to large deposits in the earth's crust, iron, is also an important heavy metal from the point of view of human health and aesthetics. In the presence of hydrogen sulfide, iron causes sediment to form that may give the water a blackish color. WHO (2004a) has a guideline value of 0.3 mg/l of iron in drinking water. A range of 0 to 0.22 mg/l was measured for the well samples in this study (Table 3), while the range measured in the surface samples was 0.175 to 0.450 mg/l (Table 1). In only one sample was Fe measured to be above 0.3 mg/l which is the WHO (2004a) guideline value. All the other samples contained iron at concentrations lower than the WHO guideline. The Illinois Environmental Protection Agency (IEPA) has established a maximum concentration for iron in drinking water of 1.0 mg/l and this is even a more relaxed value than the WHO guideline. According to WHO (2004c), anaerobic groundwaters may contain iron (II) at concentrations up to several milligrams per litre without discoloration or turbidity in the water when directly pumped from a well. Taste is not usually noticeable at iron concentrations below 0.3 mg/l, although turbidity and colour may develop in piped systems at levels above 0.05 to 0.1 mg/l. Laundry and sanitary ware will stain at iron concentrations above 0.3 mg/l. Iron is an essential element in human nutrition. Estimates of the minimum daily requirement for iron depend on age, sex, physiological status, and iron bioavailability and range from about 10 to 50 mg/day.

Heavy metals are regularly found in liquid pig manure (Gerber et al., 2005). Animal feeding is supposed to be the main source of occurrence for copper and zinc (Li et al., 2007; Nicholson et al., 1999). Mineral feed may also contribute to occurrence of cadmium, lead, arsenic and mercury (de La Calle Guntinas et al., 2011). Livestock rearing is a common activity in urban agriculture and this also occurs in Ibadan. Corrosion of metallic installations, ash amendment to manures for odor reduction and lime amendment for disinfection purposes are other possible sources of heavy metals, e.g. lead, cadmium, chromium, zinc and nickel (Eckel, 2005). Arsenic (As) is a well-known heavy metal which is ubiquitous in the environment (Vahidnia et al., 2007; Garelick et al., 2008). Arsenic is classified as a poison and its exposure is mainly through ingestion and inhalation of arsenic-bearing chemicals from drinking water, food, and air (Rahman et al., 2009; Aposhian et al., 2004). Once it enters the body, arsenic and its metabolites generate free radicals, which damage proteins, fatty acids, DNA, and RNA, and cause oxidative stress or death to cells (Gong and O'Bryant, 2010). Inflammatory response is one of the hallmarks of arsenic-induced toxicity (Valko et al., 2005). Clinically, arsenic is well-known as a carcinogen, causing prostate, lung, liver, bladder, and other cancers (Jomova et al., 2011; Mink et al., 2008; Benbrahim-Tallaa and Waalkes, 2008; Celik et al., 2008; Chiu et al., 2004).

The measured well water arsenic concentrations ranged from 0 to 0.05 mg/l with a mean of 0.02 mg/l (Tables 3 and 4), while the surface water arsenic concentrations ranged from 0 to 0.05 mg/l with a mean of 0.028 mg/l (Tables 1 and 4). The WHO guideline value is 0.01 mg/l. It is conclusive that as far as arsenic is concerned, concentrations in many of these samples were higher than the WHO guideline. The difference is however not presently alarming. Previous studies have shown that cancer risk increases by 100-fold among people exposed to drinking water with arsenic concentration at 50 mg/l (Smith et al., 2002). As a result, U.S. Environmental Protection Agency (USEPA) set the current standard concentration of arsenic in drinking water to10 mg/l effective in 2006 in the United States (Smith et al., 2002). Concentrations found in this present research were far lower than this. A paper by Liao et al. (2009) reported that bladder cancer risk increased significantly among Taiwanese men with long-term chronic arsenic exposure at low level and noted that a recommended safe arsenic level in drinking water is 3.4 mg/l, far below the current USEPA standard, 10 mg/l. The National Research Council (2001) has recommended examining the effect of

low-level arsenic exposure on health outcomes. Arsenic exposure not only causes cancer but also increases the risks of many other diseases. Chen et al. (1988) reported that compared with the general population, mortality rates from cardiovascular diseases, peripheral vascular diseases, as well as cancers of bladder, skin, lung, and liver were significantly higher among patients with blackfoot disease resulting from exposure to high concentrations of arsenic (from 350 to 1140 ppb or mg/l with a median of 780 mg/l) in certain areas in Taiwan. Tseng et al. (2003) reported that ischemic heart disease prevalence was significantly correlated with cumulative arsenic exposure (arsenic concentration multiplied by the number of years individuals had lived there) in arseniasis-hyperendemic villages in Taiwan. A dose–response relationship existed between ischemic heart disease mortality and long-term arsenic exposure (Chen et al., 1996). Coronary heart disease and hypertension were associated with low-level arsenic exposure, while coronary heart disease and hyperlipidemia were associated with AS3MT polymorphism in a rural cohort in Texas, USA (Gong and O'Bryant, 2012).

The concentration of Chromium in the well water samples was found to range from 0 to 0.09 mg/l with a mean of 0.03 mg/l. In the surface water samples, Cr was measured from a range of 0 to 0.065 mg/l with a mean of 0.026 mg/l. The WHO (2004a) guideline value for Cr in drinking water is 0.05 mg/l. Chromium is one of those heavy metals the environmental concentration of which is steadily increasing due to industrial growth, especially the development of metal, chemical and tanning industries. Other sources of chromium permeating the environment are air and water erosion of rocks, power plants, liquid fuels, brown and hard coal, and industrial and municipal waste. Although there is no risk of chromium contamination on a global scale, local permeation of the metal to soil, water or the atmosphere might result in excessive amounts of this pollutant in biogeochemical circulation (Wyszkowska, 2002). As observed by Ghosh and Singh (2005) non-biodegradability of chromium is responsible for its persistence in the environment; once mixed in soil, it undergoes transformation into various mobile forms before ending into the environmental sink (Bartlett and James, 1983; Bartlett, 1988). Although Cr toxicity in the environment is relatively rare, it still presents some risks to human health since chromium can be accumulated on skin, lungs, muscles fat, and it accumulates in liver, dorsal spine, hair, nails and placenta where it is traceable to various heath conditions (Reyes-Gutiérrez et al., 2007).

CONCLUSIONS AND RECOMMENDATIONS

Overall, the minimum concentrations of Pb, As, Cd, Cr and Fe in the well water samples were below detection limit (BDL), while the maximum concentrations were 0.02, 0.055, 0.01, 0.09, and 0.245 mg/l. For Zn, Cu and Mn, the minimum concentrations were 0.120, 0.135 and 0.075 mg/l and the maximum concentrations were 0.0445, 0.135 and 0.155 mg/l respectively. Similar results were also obtained for the surface water samples. There is therefore no evidence of wide-scale contamination of these water samples with heavy metals going by the fact that most of the values measured were well within the guideline limits established by the World Health Organization. A possible exception is As which showed values higher than the WHO guideline. The recommend-dations of the study include continuous and close monitoring of these private drinking water supplies in view of increasing urbanization and industrialization of the town in which they are situated. There must also be strict compliance to regulatory limits in sludge and wastes to be released into the environment as well as enforcement of other environmental protection regulations to arrest the earlier reported ongoing buildup of heavy metals in soils on those locations. Findings from this study will be of immense help to the consumers of these waters as well as researchers and environmental regulators working in this area of interest in developing countries. On one hand the consumers have a scientific basis for assessing the quality of the water which they are using. On the other hand researchers also have useful scientific knowledge on the water supply situation in Ibadan, which is a typical African city. and this they can apply in finding solutions to problems which they may encounter in similar situations existing around the African continent.

Conflict of Interests

The author(s) have not declared any conflict of interests.

REFERENCES

Adelekan BA (2010). Water quality of domestic wells in typical african communities: case studies from Nigeria. Int. J. Water Res. Environ. Eng. 2(6):137-147, September. Published online at www.academicjournals.org/IJWREE

Adelekan BA, Abegunde KD (2011). Heavy metals contamination of soil and groundwater in automobile mechanic villages in Ibadan, Nigeria. Int. J. Phy. Sci. 6(5):1045-1058. March 4. Available online at www.academicjournals.org/IJPS

Adelekan BA, Alawode AO (2011). Contributions of municipal refuse dumps to heavy metals concentrations in soil profile and groundwater in Ibadan, Nigeria. J. Appl. Biosci. 40:2727-2737. Available online at www.m.elewa.org/JABS

Adelekan BA, Ogunde OA (2012). Quality of water from dug wells and the lagoon in Lagos Nigeria and associated health risks. Sci. Res. Essays. 7(11):1195-1211. Available online at www.academicjournals.org/SRE

Adetunji VO, Odetokun IA (2011). Groundwater contamination in Agbowo community, Ibadan, Nigeria: Impact of septic tanks distances to wells. Mal. J. Mocrobiol. 7(3):159-166.

Aposhian HV, Zakharyan RA, Avram MD, Sampayo-Reyes A, Wollenberg ML (2004). A review of the enzymology of arsenic metabolism and a new potential role of hydrogen peroxide in the detoxication of the trivalent arsenic species. Toxicol. Appl. Pharmacol. 198:327-335.

Bartlett RJ (1988). Mobility and Bioavailability of Chromium in Soils. Adv. Environ. Sci. Technol. 20:267-304.

Bartlett RJ, James B (1983). Behavior of Chromium in Soils: VII. Adsorption and Reduction of Hexavalent Forms. J. Environ. Qual. 12(2):177-181.

Begun A, Ramaiah M, Harikrishna, Khan I, Veena K, (2009). Analysis of Heavy Metals Concentration in Soil and Litchens from Various Localities of Hosur Road, Bangalore, India. E-J. Chem. 6(1):13-22. www.e-journals.net

Benbrahim-Tallaa L, Waalkes MP (2008). Inorganic arsenic and human prostate cancer. Environ. Health Perspect. 116:158-164.

Celik I, Gallicchio L, Boyd K, Lam TK, Matanoski G, Tao X (2008). Arsenic in drinking water and lung cancer: A systematic review. Environ. Res. 108:48-55.

Chen CJ, Chiou HY, Chiang MH, Lin LJ, Tai TY (1996). Dose-response relationship between ischemic heart disease mortality and long-term arsenic exposure. Arterioscler. Thromb. Vasc. Biol. 16:504-510.

Chen CJ, Wu MM, Lee SS, Wang JD, Cheng SH, Wu HY (1988). Atherogenicity and carcinogenicity of high-arsenic artesian well water. Multiple risk factors and related malignant neoplasms of blackfoot disease. Arteriosclerosis 8:452-460.

Chirenje T, Ma L, Reeves M, Szulczewski M (2004). Lead Distribution in Near-Surface Soils of Two Florida Cities: Gainesville and Miami. Geoderma 119:113-120.

Chiu HF, Ho SC, Wang LY, Wu TN, Yang CY (2004). Does arsenic exposure increase the risk for liver cancer? J. Toxicol. Environ. Health A 67:1491-1500.

de La Calle Guntinas MB, Semeraro A, Wysocka I, Cordeiro F, Quetel C, Emteborg H, Charoud-Got J, Linsinger TPJ (2011). Proficiency test for the determination of heavy metals in mineral feed.The importance of correctly selecting the certified reference materials during method validation. Food Addit. Contam. Part A Chem. Anal. Control Exposure Risk Assess. 28(11):1534-1546.

Eckel H (2005). Assessment and Reduction of Heavy Metal Input Into Agro-Ecosystems: Final Report of the EU-Concerted Action AROMIS. M" unster, Westf: Landwirtschaftsverl. M" unster.

Finley JW (2004). Does Environmental Exposure to manganese pose a health risk to healthy adults? Nutr. Rev. 62:148-153.

Garelick H, Jones H, Dybowska A, Valsami-Jones E (2008). Arsenic pollution sources. Rev. Environ. Contam. Toxicol. 197:17-60.

Gerber P, Chilonda P, Franceschini G, Menzi H (2005). Geographical determinants and environmental implications of livestock production intensification in Asia. Bioresour. Technol. 96:263-276.

Ghosh M, Singh SP (2005). Comparative Uptake and Phytoextraction Study of Soil Induced Chromium by Accumulator and High Biomass weed Species. Appl. Ecol. Environ. Res. 3(2):67-79. www.ecology.kee.hu.

Gong G, O'Bryant SE (2010). The arsenic exposure hypothesis for Alzheimer disease. Alzheimer Dis. Assoc. Disord. 24:311-316.

Gong G, O'Bryant SE (2012). Low-level arsenic exposure, AS3MT gene polymorphism and cardiovascular diseases in rural Texas counties. Environ. Res. 113(2012):52-57.

Goyer RA, Clarkson TW (2001). Toxic Effects of Metals. In, Casarett and Doullis Toxicology: The basic science of poisons, Sixth Edition (C.D. Klaassen, ed.) Mc-Graw-Hill, New York, pp. 811-867.

Jomova K, Jenisova Z, Feszterova M, Baros S, Liska J, Hudecova D, Rhodes CJ, Valko M (2011). Arsenic: Toxicity, oxidative stress and human disease. J. Appl. Toxicol. 31:95-107.

Kobayashi J (1978). Pollution by cadmium and the itai-itai disease in Japan. In: Toxicity of heavy metals in the environment. Oehme FW (ed.) Marcel Dekker Inc, New York, pp. 199-260.

Lenntech WT (2009). Chemical Properties, Health and Environmental Effects of Copper. Lenntech Water Treatment and Purification Holding B.V. www.lenntech.com/periodic/elements/cu.htm.

Li Y-X, Li W, Wu J, Xu LC, Su QH, Xiong X (2007). Contribution of additives Cu to its accumulation in pig faeces: Study in Beijing and Fuxin of China. J. Environ. Sci. (China) 19:610-615.

Liao CM, Shen HH, Chen CL, Hsu LI, Lin TL, Chen SC, Chen CJ (2009). Risk assessment of arsenic-induced internal cancer at long-term low dose exposure. J. Hazard. Mater. 165:652-663.

Mench M, Didier V, Löffler M, Gomez A, Masson P (1994). A Mimicked in-situ Remediation Study of Metal-contaminated Soils with emphasis

Mink PJ, Alexander DD, Barraj LM, Kelsh MA, Tsuji JS (2008). Low level arsenic exposure in drinking water and bladder cancer: A review and meta-analysis. Regul. Toxicol. Pharmacol. 52:299-310.

National Population Commission, NPC (2009). Federal Republic of Nigeria Official Gazette. 96:2, Abuja, 2nd February.

National Research Council Subcommittee (2001). Arsenic in drinking water Report. Committee on Toxicology, Board on Environmental Studies and Toxicology, Division on Earth and Life Studies. Arsenic in Drinking Water, 2001 update. Washington, D.C. National Academy Press.

Nicholson FA, Chambers BJ, Williams JR, Unwin RJ, (1999). Heavy metal contents of livestock feeds and animal manures in England and Wales. Bioresour. Technol. 70:23-31.

Nogawa K, Honda R, Kido T, Tsuritani I, Yamanda Y (1987). Limits to protect people eating cadmium in rice, based on epidemiological studies. Trace Subst. Environ. Health 21:431-439.
on Cadmium and Lead. J. Environ. Qual. 23:58-63.

Rahman MM, Ng JC, Naidu R (2009). Chronic exposure of arsenic via drinking water and its adverse health impacts on humans. Environ. Geochem. Health 31(Suppl. 1):189-200.

Reyes-Gutiérrez LR, Romero-Guzmán ET, Cabral-Prieto A, Rodríguez-Castillo R (2007). Characterization of Chromium in Contaminated Soil Studied by SEM, EDS, XRD and Mössbauer Spectroscopy. J. Miner. Mater. Character. Eng. 7(1):59-70.

Shi G, Chen Z, Bi C, Wang L, Teng J, Li Y, Xu S (2011). A comparative study of health risk of potentially toxic metals in urban and suburban road dust in the most populated city of China. Atmos. Environ. 45:764-771.

Smith AH, Lopipero PA, Bates MN, Steinmaus CM, (2002). Arsenic epidemiology and drinking water standards. Sci. 296:2145-2146.

Stredrick DL, Stokes AH, Worst TJ, Freeman WM, Johnson EA, Lash LH, Vahter M, Berglund M, Akesson A, Liden C (2002). Metals and Women's Health. Environ. Res. 88:145-155.

Tseng CH, Chong CK, Tseng CP, Hsueh YM, Chiou HY, Tseng CC (2003). Long-term arsenic exposure and ischemic heart disease in arseniasis- hyperendemic villages in Taiwan. Toxicol. Lett. 137:15-21.

United States Department of Agriculture, USDA (2000). Heavy metals contamination. Soil Quality Urban Technical Note 3, Natural Resources Conservation Service.

United States Environmental Protection Agency, USEPA (2004). Drinking water health advisory for manganese. US Environmental Protection Agency Office of Water (4304T) Health and Ecological Criteria Division Washington, DC 20460. Available online at http://www.epa.gov/safewater/EPA-822-R-04-003 January, Accessed April 11, 2013.

Vahidnia A, vander Voet GB, deWolff FA (2007). Arsenic neurotoxicity - A review. Hum. Exp. Toxicol. 26:823-832.

Valko M, Morris H, Cronin MTD (2005). Metals, toxicity and oxidative stress. Curr. Med. Chem. 12: 1161-1208.

Wisconsin Department of Health and Family Services, WDHFS (2007). manganese in drinking water. Information for home owners and other members of the public. Human Health Hazards. Available online at www.dhfs.wisconsin.gov/eh

World Health Organization, WHO (2004a). Guidelines for drinking water quality, Third Edition. Volume 1: Recommendations. World Health Organization, Geneva.

World Health Organization, WHO (2004b). Manganese in drinking water. Background document for development of WHO Guidelines for Drinking water Quality. World Health Organization, Geneva, Switzerland.

World Health Organization, WHO (2004c). Iron in drinking water. Background document for development of WHO Guidelines for Drinking Water Quality. World Health Organization, Geneva, Switzerland.

Wyszkowska J (2002). Soil contamination by chromium and its enzymatic activity and yielding. Pol. J. Environ. Stud. 11(1):79-84.

Use of multivariate statistical and geographic information system (GIS)-based approach to evaluate ground water quality in the irrigated plain of Tadla (Morocco)

Naima EL Hammoumi[1,2]*, M. Sinan[2], B. Lekhlif[2] and M. Lakhdar[1]

[1]Laboratory of Catalysis and Environment, Faculty of Science, University Hassan II Ain Chock- km 8 Road of El Jadida P. O. BOX 5366, Casablanca Morocco.
[2]Research Team of Hydrogeology, Treatment and Purification of Water and Climate Change, Department of Hydraulics, Hassania School of Public Works, km 8 Road of El Jadida P. O. BOX 8108, Casablanca Morocco.

Twenty five chemical, physical and bacteriological features of water samples from 25 wells were analyzed by multivariate statistical tools to provide the characterization of the ground water distribution in the basin of Tadla province (Morocco). The 25 parameters determined include: temperature, pH, conductivity, IP (permanganate index), dry residue (RS), total hardness, total iron (FeT), several bacteriological residues (faecal, total coliforms and faecal streptococcus) and several cations and anions (Ca^{2+}, Na^+, K^+, Mg^{2+}, SO_4^{2-}, Mn^{2+}, NH_4^+, NO_3, HCO_3^-, NO_2^-, Cl^-). All sampling was performed in the period between December 2007 and February 2007(rain season). Principal Component Analysis together with the GIS approach (kriging methods) which provided a description of the area investigated with respect to the characteristics of the ground water was used. It is demonstrated that the water, quality in this region is critical. Nitrate levels are situated between 11.3 and 100 mg/L, with 73% of the observations exceeding the critical level of 50 mg/L fixed by the standards of the World Health Organization (WHO) for drinking water. However, bacteriological residues (faecal, total coliforms, and faecal streptococcus) are retrieved in nearly all water samples. Principal Component Analysis indicates that Bacteriological contamination is merely correlated with nitrite and ammonia amount rather than with nitrate amount, indicating a possible contribution of local pollution sources to ground water deterioration. The variability of the nitrate and bacteriological pollution is important and spatially correlated. A significant difference in water composition has been highlighted between water table of Beni Amir and water table of Beni Moussa and also a difference between ground water near cities (with a probable human polluting effect) and zones far from the built-up areas. Cluster analysis was also performed in order to evaluate the different wells similarity and to confirm the results obtained by Principal Component Analysis.

Key words: Morocco, ground water quality, Kriging methods, principal component analysis, geographic information system (GIS), multivariate analysis.

INTRODUCTION

Water resources in Morocco are currently under serious pressure. National averaged available water resources are estimated to be equal to 740 m³/habitant/ year (Second National Communication of Climate Change of Morocco, 2010). Morocco thereby ranks on the 155th position on a scale of 180 countries, measuring the pressure that is exerted on available fresh water resources (Food and Agriculture Organization (FAO), 2002). The volume of available groundwater in Morocco is estimated

*Corresponding author. E-mail: hammoumi_naima@yahoo.fr.

to be equal to approximatively 4×10^9 m³/year. Groundwater resources are distributed within 32 profound and 46 superficial major groundwater bodies (Khattabi, 2003). Groundwater is a resource that should appropriately be protected.

Irrigated agriculture is responsible for 93% of the total water demand in Morocco. Irrigated agriculture is also the principal driver for ensuring food security and economical development (Bouchama and Benchakroun, 1991). Yet, the irrigated agriculture in Morocco is also reported to have negative impacts on groundwater quality (Abouzaid, 1980; Lhadi et al., 1993). Laftouhi et al. (2003) for instance observed high nitrate contamination of groundwater in the central Moroccan Essaouira basin and explained this contamination by an intensifying agricultural activity. Berdaï et al. (2004) studied the groundwater quality in the Tadla plain for the period of August 1996 until April 1998. He reported that nitrate contamination levels vary between 3 and 228 mg/L for the Beni Amir part in the north of the plain and between 6 and 152 mg/L for the Beni Moussa part in the south of this plain.

Literature shows the permanent increasing potentiality of chemometric methods in obtaining useful information from environmental data, which could be otherwise correlated and interpreted. In particular, many examples of the application of multivariate analysis to sets of variables collected for surface and ground waters can be found. Pattern recognition methods were applied in the treatment of well waters in Vulture mount (Basilicata, Italy) to suggest the origin of mineral water samples (Caselli et al., 1998). A study based on Principal Component Analysis (PCA) and Kohonen neural networks, has been devoted to assess the quality of the waters of Mura River in Slovenia and evidenced the improvement of the quality of the waters along the 9 years of the project (Brodnjak-Vonin et al., 2002). Studies concerning the Suquia River in Cordoba (Argentina) making use of the Factor Analysis (FA) and PCA techniques, as well as cluster and discriminant analysis permitted to evaluate the sources of variability and the pattern of some variables in a graphical analysis were carried out. The results were correlated with seasonal variations, urban run-off and pollution sources (Wunderlin et al., 2001).

By robust cluster analysis and multivariate treatments freshwaters from wells in Friuli and karsic freshwaters of pollution were studied and the temporal evolution of the pattern was evaluated for the different sampling sites (Barbieri et al., 2001; Reisenhofer et al., 1998).

Vega et al. (1998) applied Principal Component Analysis and cluster analysis, to collect information from environmental data in lagoon waters and to discriminate the possible sources of spatial and temporal variations that can affect the quality of water.

Groundwater samples collected from wells located down gradient from the potential nuclear waste repository at Yucca Mountain (Nevada) were treated with PCA and QFA (Q-model factor analysis) to evaluate possible contamination (Farnham et al., 2003). Multivariate analysis techniques were employed to study the possibility that livestock slurries used as fertilizers in rural areas in Spain could influence the quality of drinkable water (Vidal et al., 2000). These techniques have been also used to investigate the possible interactions of India river waters and adjacent groundwater or mixing of different ground waters (Reghunath et al., 2002). A study based on 3-way PCA in addition to multivariate analysis techniques was used to collect information on the similarities and differences between ion amount / ion composition of underground waters with the aim to improve the criteria of water resource management in Sicily (Librando et al., 1998).

Unfortunately, most of the previous studies in this region focused only on one singular groundwater quality parameter, typical nitrate or total salinity. Previous studies did not allow to assess the overall water quality in holistic way, encompassing thereby physico-chemical, organic and microbiological quality criteria. The potential correlation which may exist between quality parameters is further rarely explored, in as much as their space-time dynamics.

The mapping of groundwater quality is therefore often subjected to a lot of uncertainty, which complicates the assessment of the overall quality of the groundwater body itself. We also observe that only few studies have been reported illustrating these problems for the aquifers in Morocco such us the management of regional aquifers using a combined procedure based on geostatistical and geographic information system (GIS) tools (Haouz groundwater of Marrakekech) (Sinan and Razack, 2005, 2008).

This paper is focused on the analysis of the distribution of ground water in the Tadla plain (Morocco). Water samples from 25 wells were characterized by 25 parameters, including: air and water temperature, pH, conductivity, permanganate Index, dry residue , TH , TAC, total hardness, Total iron (FeT), several cations and anions (NH_4^+, Ca^{2+}, Na^+, K^+, Mg^{2+}, Mn^{2+}, Ca^{2+}, Cl^-, NO_3^-, NO_2^-, SO_4^{2-}) and microbiological quality criteria (total streptococcus bacteria, Fecal streptococcus bacteria and fecal coliforms bacteria).

The analysis was devoted to illustrate the influence of possible sources (irrigated agriculture, the domestic and factories activities) on as well physico-chemical and bacteriological groundwater quality parameters for the Tadla aquifer. This aquifer is a major groundwater body of the Oum R'bia catchment, situated in central part of Morocco. The study focuses on nutrient related and bacteriological quality parameters of the groundwater body. The latter parameters are further closely linked to health and sanitation problems, and it was carried out by means of multivariate statistical tools: Principal Component Analysis, cluster analysis and kriging

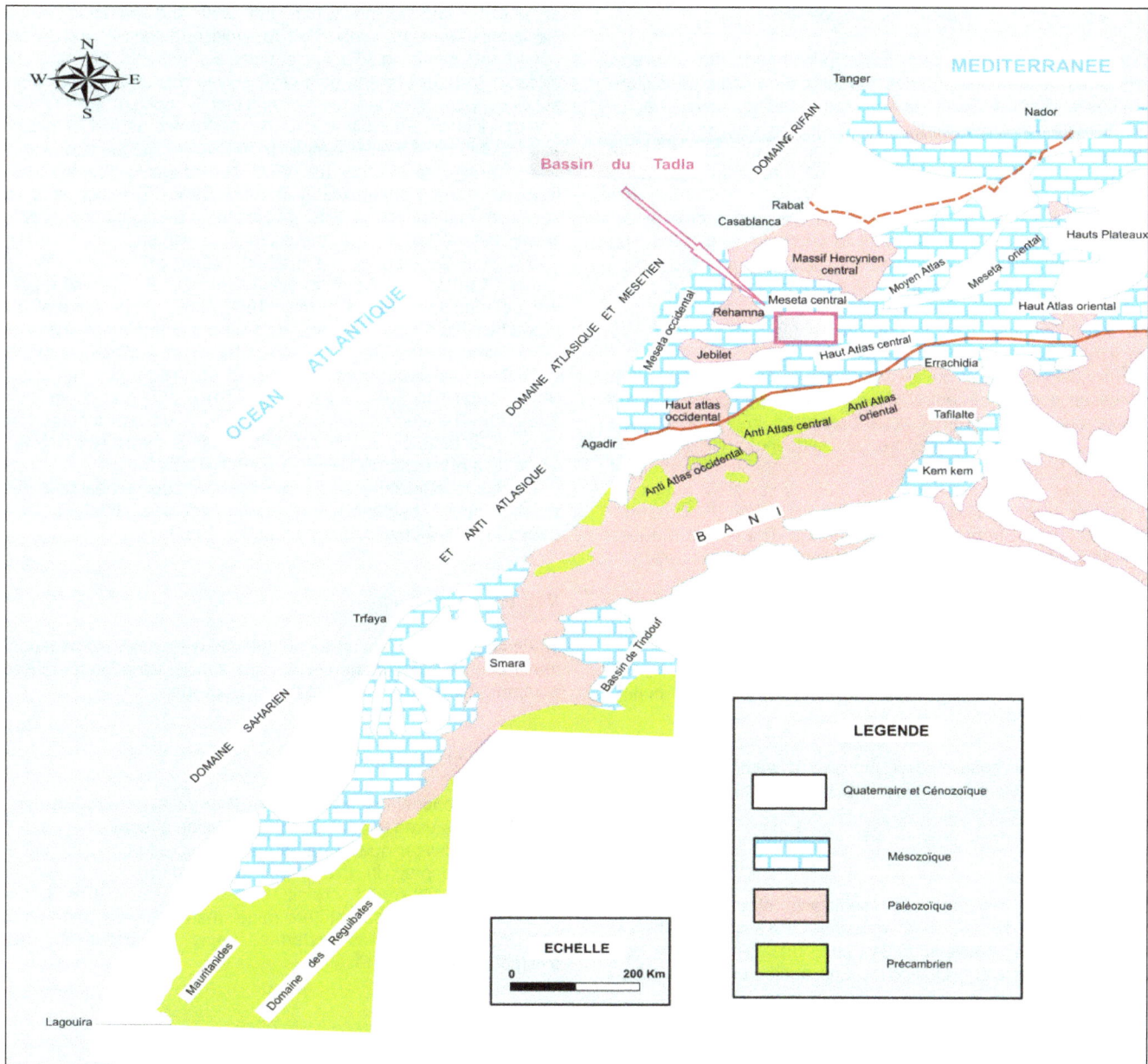

Figure 1. Map showing the location and geology of the case study area.

methods, for an improvement of management and protection of its water resources.

The research deals with both the analysis of the water samples and the chemometric treatment.

MATERIALS AND METHODS

The case study region

The plain of Tadla is a vast synclinal located in central Morocco. It spreads in ENE–WSW direction between the phosphate plateau and the High Atlas of Beni-Mellal in the basin of the Oum R'bia River approximately 270 Km of Rabat (capital of Morocco) which covers a surface of 3600 km approximately (Figure 1). It presents a slope varying from 1 to 3% and is traversed by approximately 160 Km of the Oum R'bia River which divides it into two large perimeters irrigated with different hydraulic characters: Beni Amir in the North and Beni Moussa in the South whose irrigated surfaces are respectively 33 000 ha and 69 500 ha. At the beginning of the XXe century, the population counted approximately 95 000 rural inhabitants with an average density of about 26 habitants/Km^2 practising extensive cereal crop primarily. The installation of the hydro farming infrastructure was accompanied:

(a) With a spectacular growth of the population, counting approximately 420 000 in 1982, 560 000 in 1994 including 440 000 rural and 571.000 inhabitants in 2004, of whom 65% were rural.

(b) With a development of the economic activities.

Thus the intensive agriculture is based primarily on the culture of cereals, sugar beet, market garden, fodder, cotton and arboriculture (olive-trees and citrus fruits), and on the breeding. Several agro-industrial units were installed in the perimeter, especially sugar refineries (3 units of which only one is functional), oil mills (461 units with 7 modern 195 semi-modern and 259 traditional) and three dairies.

The plain of Tadla has a semi-arid climate with a cold winter pluviometry spends an interval of 275 to 1025 mm over the period of 1935 to 1980 with an interval 175 to 625 mm over the period 1980 to 2008. It will be retained that this area knew an average degradation of the annual pluviometry of 70 mm (20%) over this period. It is thus undeniable that the zone regularly knows from now on more marked periods of dryness (PDAIRE, 2008).

From a geological point of view, the basin of Tadla is attached at the Mesetian domain of central Morocco, more particularly, at the synclinal zone "Bahira-Tadla". It is presented in the form of a synclinal basin filled by the sedimentary sequence from which the age extends from Permo-Trias to the Quaternary. In the whole of the basin, the quaternary coverings mask completely subjacent grounds, so that geological information available (FAO, 2002; BRGM, 1993) come from some major drillings. The results of these drillings reveal the existence of four great lithostratigraphic sets:

(i) A thick paleozoic base, of Cambrien to the Carboniferous one;
(ii) Triassic and Jurassic formations localised in the central edge of the basin;
(iii) Cretaceous marine transgressive deposits with Eocene;
(iv) An unmatched filling Neocene and quaternary mainly continental (Figure 2).

From hydrogeologic point of view: the ground water walks on in a Plio-Quaternary complex primarily made up of francs limestones, of marno-limestones and clays.

This aquifer shows a groundwater flow from the north to the southwest:

In Beni Amir, the transmissivity coefficients of the aquifer are situated between 1.10^{-3} and 1.10^{-1} m^2/s while the storage coefficient generally oscillate between 0.01 and 0.1. The Least low transmissivity meet along Oum R'bia and apart from the irrigated perimeters.

In Beni Moussa, the transmissivity coefficients aquifer situated between 1.10^{-3} and 5.10^{-1} m^2/s, and the storage coefficient between 0.06 and 1 (BRGM, 1993).

Sampling strategy and water analysis

Twenty-five groundwater sampling locations were selected in the study region (Figure 3). The sampling points were distributed all over the study area to ensure appropriate spatial coverage of the entire aquifer. Most of the sampling points were pumped wells.

All sampling was performed in the period between December 2007 and February 2007 (rain season), period during which contamination of groundwater is highest because of the infiltration of surface water (rain water and water of river).

The physicochemical and bacteriological analyses enter within the framework of the monitoring of the Oum R'bia basins water quality. A complete network of groundwater analyses is put by the Hydraulic Agency of Oum R'bia (ABHEOR) in collaboration with the Public Laboratory of the Studies and Tests (LPEE) in Casablanca.

Pumped groundwater samples were collected in polyethylene bottles of 1 L. Before sampling, the recipient was cleaned several times using the pumped water. Flushing allowed also to clean the adductions and tubing within the well and pumping system. Recipients were gradually filled to avoid turbulences and aeration during the sampling. To avoid sampling artifacts and analytical artifacts, in particular the gain of dissolved gas and microbiological activity, water samples were immediately cooled at 4°C using portable icebox. Analysis was further performed as fast as possible and this within 24 h after sampling.

The analytical method for NH_4^+-N dosage is the method of Solorzano and Koroleff (cited in Afnor T90-015) which is a very sensitive method for low NH_4^+-N content in the liquid phase. NO_2^--N was determined through the diazotation with sulfanilamide (Afnor T90-013), NO_3^--N with the method based on sodium salicylate (Afnor T90-012). For the determination of PO_4-P, use was made of the colorimetric method (Afnor T90-023). Other mineralogical properties have been determined following the French system of normalisation (Afnor): chloride (Afnor T90-014), sulfates (Afnor T90-009), the total toughness of water (Afnor T90-003), the calcium (Afnor T90-016), sodium (Afnor T90-019), and potassium (Afnor T90-020). The electric conductivity is measured with a conductivity meter (COND 330i, LUTRONTM). The pH is measured directly in water by a pH-meter (WTW530, LUTRONTM).

The bacteriological analysis for total streptococcus bacteria, fecal streptococcus bacteria and sulfito-reducing bacteria were determined by Rodier's method (Rodier, 1984).

Methods of analysis

The statistical and geostatistical methods were selected because of the strong capacity to characterize the special variation for different parameters from the quality of the groundwater.

Statistical modelling

Principal component analysis (PCA) (Box et al., 1978; Massart et al., 1988) was used to analyze the correlation structure between the set of groundwater quality parameters collected during the survey. The PCA adopted in this paper is based on normalized and standardized data and exploits the correlation matrix between groundwater quality parameters rather than the covariance matrix. The PCA analysis was performed using the statistical toolbox available in MATLAB7™.

Cluster analysis

Cluster analysis (CA) encompasses a number of different methods which organize objects (observations) into groups called clusters without explanation or interpretation. Objects within the same clusters are similar whereas objects in different clusters are dissimilar. This exploratory method is used to discover the data structure not only among observations, but also among variables, arranged into a tree diagram, usually called a dendrogram. The utilized methods, algorithms, and similarity/dissimilarity measures are described elsewhere in the literature (Everitt, 2001). In this study, the commonly applied average group and the Ward's clustering methods were used. The Euclidean distance was used as a similarity measure. The cluster analysis was performed using the software XLSTAT 2009.

Kriging method

Kriging methods are interpolation methods firstly proposed by Krige (Isaaks, 1990; Krige, 1978; Sinan, 2000) that allow taking into consideration the spatial continuity of the points. These methods are based on the minimisation of the variance of the estimation

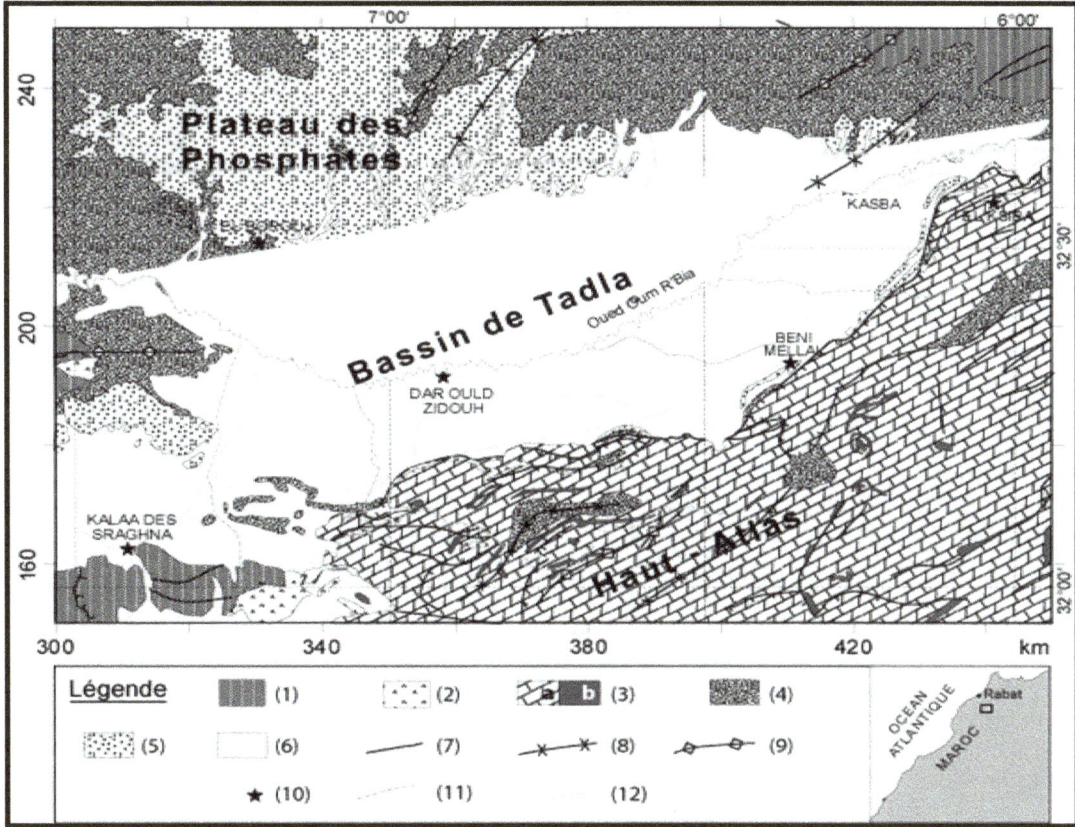

Figure 2. Geological map of the Tadla basin. (**1**) Palaeozoic (schist and quartzite). (**2**) Trias (red clay and basalt). (**3**) Jurassic (**a**, limestone and dolomites; **b**, diorite, gabbro). (**4**) Cretaceous (red detritic facies). (**5**) Tertiary (limestone, marls and phosphatic sands). (**6**) Quaternary (alluvium). (**7**) Fault. (**8**) Synclinal axis. (**9**) Anticlinal axis. (**10**) City. (**11**) River. (**12**) Road.

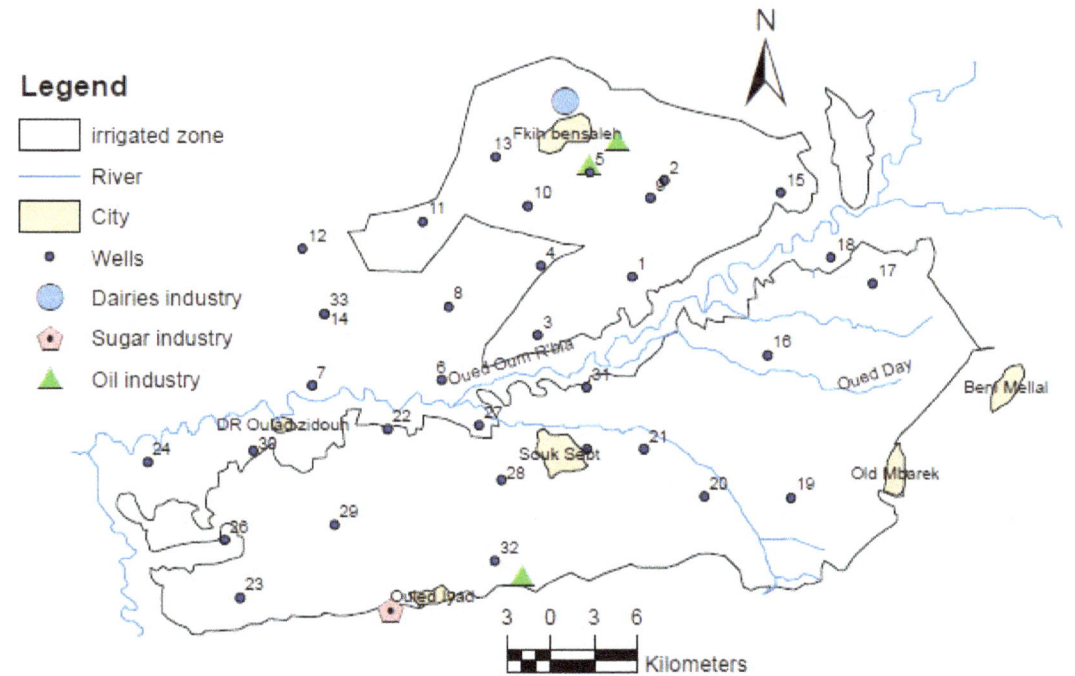

Figure 3. Map of the Tadla plain with the numbered sampling sites.

error, through a weighted linear combination of the available points. Different variants have been proposed but ordinary Kriging is certainly the most exploited method (Massart et al., 1997). The paper map of the region study was digitilized to north Morocco degree coordinate system.

The development of the map of distribution of each variable and the kriging of each axis were performed using the kriging approach integrated in the Geographic Information System (GIS) by the software ArcGIS 9.2 version.

RESULTS AND DISCUSSION

Data pre-treatment

Table 1 reports the average, minimum and maximum values calculated for each variable obtained in groundwater sample from wells in Tadla plain in comparison with WHO standards.

The pH of water varied between 3.85 and 6.08. The pH represents the intensity of acidity or alkalinity and measures the concentration in ions hydronium in water. The interval values of pH recommended by the O.M.S are 6.5 to 8.5. PH of analyzed water was in all the cases smaller than the limit recommended by the WHO. What shows the neutral character of this water.

As well anthropogenic, natural processes determine nitrogen release in soil and soil organic matter turn-over produces mineral N from the organic pool while biologically and chemically mediated processes sustain nitrogen transformation. Within the soil mineral N pool, nitrate-N receives particular attention as it is dissolved within the soil solution and therefore extremely mobile. It can easily be incorporated in the plant biomass, but also leach from the root zone of the crops and as such contaminate groundwater. Nitrate levels in the Tadla aquifer are situated between 11.4 and 99.4 mg/L.

Nitrite concentration in all wells never exceeds the drinking water norms proposed by the WHO. Observed values situate between 0 and 0.48 mg/L. This is in agreement with the study of Kholtei (2002) who analysed the nitrite quality of the Berrechid aquifer in the central part of Morocco and who observed nitrite contamination levels situated between 0.0014 and 0.066 mg/L. Ammonium is far less mobile than nitrite and nitrate.

Under normal pH conditions, ammonium will be protonated and retained in the soil by cation's exchange. Within the wells, the ammonium values never exceeded the drinking water norm proposed by the WHO.

The electric conductivity which is a measurement of the capacity of an aqueous solution to lead, it electrical current varied between 730 and 10800 µS/cm. All water of the wells had an electric conductivity higher than the limit recommended by the WHO which is of 2700 µS/cm, except the wells numbers 6, 13, 14, 15, 16, 17, 21, 23 and 25 which had an electric conductivity less than the value recommended by the WHO.

The results of the analysis indicate the calcium values range between 82.2 and 825 mg/L. This shows that the

values exceeded the desirable limits set by WHO standards except the well number 12 and 22.

The results of the analysis indicated that the sodium values range between 7.68 and 1152 mg/L and exceeded the desirable limits set by WHO standards.

The results of the analysis indicated that the chloride values ranged between 57.6 and 3763 mg/L. This shows that the values exceeded the desirable limits set by WHO standards except wells number 6, 13, 14, 15, 16, 17, 21, 23 and 25.

The total hardness of water (THt) is attached mainly to the quantity of calcium and magnesium in water. In the water samples, THt varied between 38 and 429 mg/L of $CaCO_3$ For all the points of water studied, THt was higher than the value guides WHO which is of 200 mg/L of $CaCO_3$ for the majority of water except the wells numbers 5, 7, 10, 20, 21. According to the classification of Durfor and Becker (1964), the analyzed groundwater is hard.

Human activities at the land surface may cause groundwater contamination by bacterial pathogens. In general, the microbiological quality of groundwater will be better than that of surface water, since pathogens will be filtered by the soil and subsoil during the recharge process (Maier et al., 2000). Leaking waste water evacuation systems, leaking septic tanks, slurry and organic waste deposits, buried biomass, including animal cadavers may be responsible for pathogen contamination of groundwater (EPA, 2000). Contamination pathways of groundwater include pathogen transport through the soil matrix and fractures and preferential transport along pumping wells and boreholes. Viability of bacterial pathogens in the soil–groundwater system is determined as well by the soil–groundwater properties as by the properties of the bacteria themselves (Harvey and Harmas, 2002).

The results of the bacteriological survey show that 88% of the sampled wells exhibit a total bacteriological contamination which exceeds the norms. The total coliform, the faecal coliform and faecal streptococcus content exceeds at least once the norm of respectively 10, 1 and 1 count per 100 ml of sampled water. Total coliform, faecal coliform and faecal streptococcus content varies respectively between 0 and 80 000, 0 and 1200, and 0 and 18000 counts per 100 ml (WHO, 2000) Values range between 2 and 192 counts per 20 ml. These types of bacteria are indicative of the vulnerability of the aquifer system or the poor maintenance of the pumping and distribution systems.

Principal Component Analysis

In the aim of investigating the role of each variable and to simplify the original data structure, the first step of our analysis was the extraction of principal components from the original data set. The Principal Component Analysis was applied to a matrix of 25 observations and 25 variables. Four components explaining 69.76% of the

Table 1. Summary statistics of groundwater samples.

Variable	Acronym	Unit of measurement	Min	Max	Average	STD	The maximum permissible limits prescribed by WHO for drinking purposes
Air Température	Ta	°C	10.5	22.5	18.0	3.2	-
Water Températurc	Tea	°C	14.5	22.5	19.5	1.7	25
pH	pH		6.9	7.4	7.1	0.2	9.6
Conductivity	Cond	µS/cm	730.0	10800.0	4092.4	2962.9	2700
permanganate Index (COD)	IP	mg/L	0.3	13.1	4.3	2.9	-
Ammonuim	NH_4^+	mg/L	0.0	0.3	0.1	0.1	0.5
Sodium	Na^+	mg/L	7.7	1152.0	408.9	361.3	100
Potassium	K^+	mg/L	0.6	5.0	2.3	1.2	-
Calcium	Ca^{2+}	mg/L	82.2	825.0	260.6	194.8	100
Magnésium	Mg^{2+}	mg/L	28.6	649.0	176.6	165.8	50
Manganèse	Mn^{2+}	mg/L	0.0	0.1	0.0	0.0	0.5
Chloride	Cl^-	mg/L	57.6	3967.0	1204.4	1164.3	750
Nitrites	NO_2^-	mg/L	0.0	0.5	0.1	0.1	0.5
Nitrates	NO_3^-	mg/L	11.4	99.4	44.0	24.8	50
Hydrocarbonates	HCO_3^-	mg/L	183.0	494.1	380.7	81.2	-
Sulfates	SO_4^{2-}	mg/L	15.4	1076.0	179.3	295.0	200
Complete Alkalinity Titration	TAC	(°F)	15.0	40.5	31.2	6.7	-
Titre Hydrotimétrique (hardness)	TH	°F	37.9	429.0	137.4	110.3	200
Résidu Sec à 105°C (dry residue)	RS	mg/L	542.0	7685.0	3083.9	2249.9	1500
Fer Total (total iron)	FeT	mg/L	0.0	0.3	0.1	0.1	0.3
Faecal Coliforms	CF	/100ml	0.0	12000.0	1624.4	3002.8	-
Total coliforms	CT	/100ml	0.0	80000.0	10642.0	19098.6	-
Faecal Streptococcus	SF	/100ml	0.0	18000.0	1789.0	3934.9	-

total variance were retained for further analysis. The results obtained from the principal component decomposition of the original data set show a high correlation between the variables (Table 2); the total variance being highly partitioned over many components (the first five PCs hardly exceed 76.9%).

The loading and the score plots for the first four PCs are represented in Figures 4 and 5 respectively. From the analysis of these figures it is possible to argue about some particular relationships between observations and variables, as follows:

(1) The first Principal Component mainly discriminates well waters having large conductivity, due to the higher concentration of chlorides, hardness (including calcium and magnesium), hydrocarbonates, permanganate index, larger amount of some metal ions as Na^+, Ca^{2+}, Mg^{2+}, K^+, NH^{4+}, and dry residue, (large negative loadings). In the corresponding score plot, the first PC clearly extracts wells 21, 19 and 24 from all the others, and also, even if at a minor extent, wells 16. These samples are characterised by large amount of conductivity, chlorides, hardness and large values of some metal ions.

PC1 can be considered as the salt component because it is mainly saturated with conductivity and hardness (including calcium).

(2) The second Principal Component discriminates well waters having a larger amount of nitrate, nitrite, bacteriological contamination (CF, CT, SF) and Potassium. The score plot of PC2 versus PC1 allows to point out some particular samples: well 24 showing a large amount of conductivity, chlorides, hardness, Permanganate Index and large values of some metal ions: a large amount of potassium and NH^+ and a small amount of SO_4^-. Wells 3, 11, 14, 16 and 20 show a large amount of nitrite, bacteriological contamination (CF, CT, SF), potassium and NH^+ and a small amount of SO_4 for what regards PC2, but they are characterised by a small amount of conductivity, chlorides, hardness, Permanganate Index, dry residue and large values of some metal ions (negative values on PC1).

At positive values on PC2, it is possible to point out: wells 1, 2, 4, 5, 6, 9, 10, 12, 13, 18, 21, 22, 23 and 25 characterised by large amount of nitrate and large amount of Nitrite and Bacteriological residues (CF, CT, SF) (negative values on PC2) for what regards PC2 but a small conductivity, chlorides, hardness, Permanganate Index and large values of some metal ions as Na^+, Ca^{2+}, Mg^{2+}, FeT and dry residue, (negative values on PC1).

The PCA analysis did not allow identifying a strong

Table 2. Principal component weights.

Parameter	First principal component (PC1)	Second principal component (PC2)	Third principal component (PC3)	Fourth principal component (PC4)
Ta	0.2606	-0.0327	-0.3274	0.2078
pH	0.3361	-0.2229	**0.4220**	**-0.4438**
CE	**-0.9377**	0.0131	-0.2804	-0.0306
IP	**-0.7242**	-0.0151	0.2077	-0.0128
NH	-0.5706	-0.2278	0.1095	**0.4726**
Na	-0.6330	-0.0010	**-0.5452**	0.2004
K+	-0.6534	**-0.4433**	0.1557	0.2765
Ca	**-0.9105**	-0.0226	0.0396	-0.1813
Mg	**-0.8802**	0.0766	-0.1519	-0.2335
Mn	0.0010	-0.2620	0.0906	**0.7623**
Cl	**-0.9364**	-0.0160	-0.2150	0.0278
NO$_2$	-0.1224	**-0.7950**	0.3423	-0.0869
NO$_3$	0.2183	**0.4756**	0.2281	0.1983
HCO	0.6946	-0.0603	**-0.5503**	0.2390
SO$_4$	-0.0703	0.1074	-0.3784	-0.3012
TH	**-0.9450**	0.0373	-0.0764	-0.2241
TAC	0.6944	-0.0605	**-0.5503**	0.2393
RS	**-0.9428**	0.0042	-0.2758	-0.0640
FeT	-0.3807	-0.1699	0.2417	**0.5030**
CT	0.1955	**-0.4468**	-0.5689	0.0115
CF	0.2205	**-0.7803**	-0.2691	-0.1292
SF	0.2254	**-0.8708**	0.0501	-0.2857
Eigenvalue	8.3152	2.8488	2.2765	1.9082
% Total variance	37.7963	12.9493	10.3476	8.6737
% Cumulative variance	37.7963	50.7456	61.0932	69.7668

positive correlation between nitrate amount and bacteriological quality. However, rather a strong negative correlation exists between the nitrite and bacteriological contamination. Weak positive correlations are observed between nitrite/ammonia amount on the one hand and pathogenic pressure on the other hand.

Explaining this weak but significant correlation may be indicative of recently generated groundwater pollution from a rather local pollution source, such as a leaking farmyard stock or septic tank, local contamination by human wastewater and local contamination by livestock excrements around wells used to watering animals aggravated by the absence of proper protection of the wells. Indeed, in case of a local and close by pollution source, high amounts of fresh dissolved organic matter will be released likely together with a set of pathogenic bacteria.

This release will stimulate the oxygen consumption during the degradation of the fresh organic matter and the oxidation of organic N. The degradation of organic N will further result in a release of ammonia, which oxidizes fast to nitrite. The further oxidation of nitrite to nitrate may be impeded by a limited amount of oxygen during this first stage of the mineralization process. The partial

mineralization of the fresh organic matter released from such local pollution sources may therefore explain a weak but positive correlation between bacteriological quality and nitrite/ammonia amount on the one hand, and a weak, but negative, correlation with oxygen and nitrate amount on the other hand. The PCA graph further shows that the nutrient and bacteriological related parameters are poorly correlated with the basic mineralogical parameters of the groundwater body. This confirms that external anthropogenic sources are most likely contributing to the pollution of the groundwater body. Similar observation of correlation between nutrient and bacteriological parameters of aquifers in morocco were in irrigated plain of Triffa (Fetouani et al., 2008).

(3) The third principal component mainly explains the differences to the amount of pH (positive loadings on PC3), to the alkalinity, hydrocarbonates, total coliforms and sodium (negative loadings on PC3). The score plot points out wells 1, 2, 4, 5, 6, 9, 12, and 13 at large positive values on PC3 with a small amount of Na$^+$ and a small conductivity (negative values on PC1). Wells 3, 8 and 11 are characterised by similar behaviour to wells 14 and 16 for what regards the variables weighing on PC3 but they show a small conductivity, Cl$^-$, dry residue, Mg

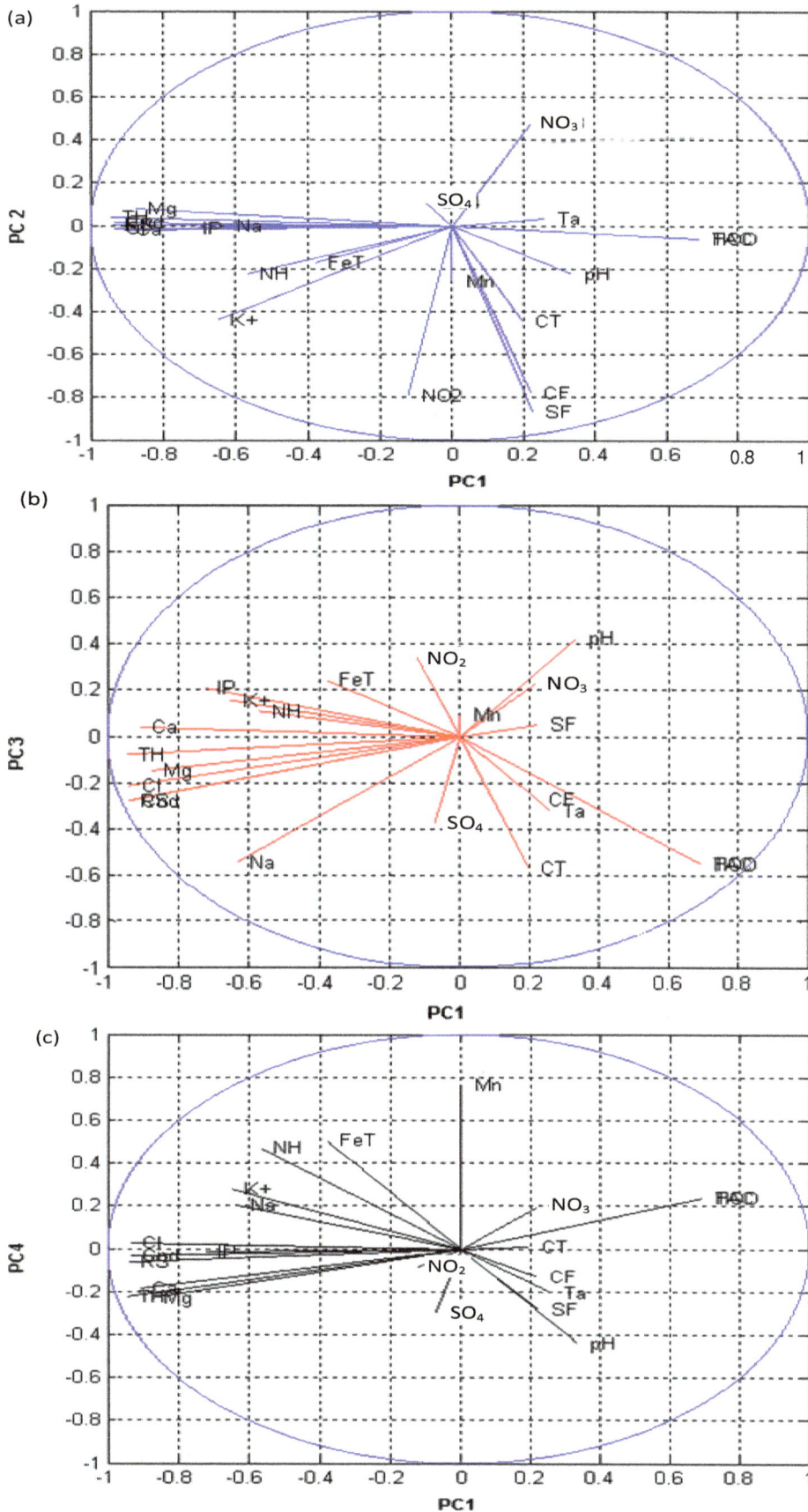

Figure 4. Loading plots: (a) PC1 versus PC2 (b) PC1 versus PC3, (c) PC1 versus PC4.

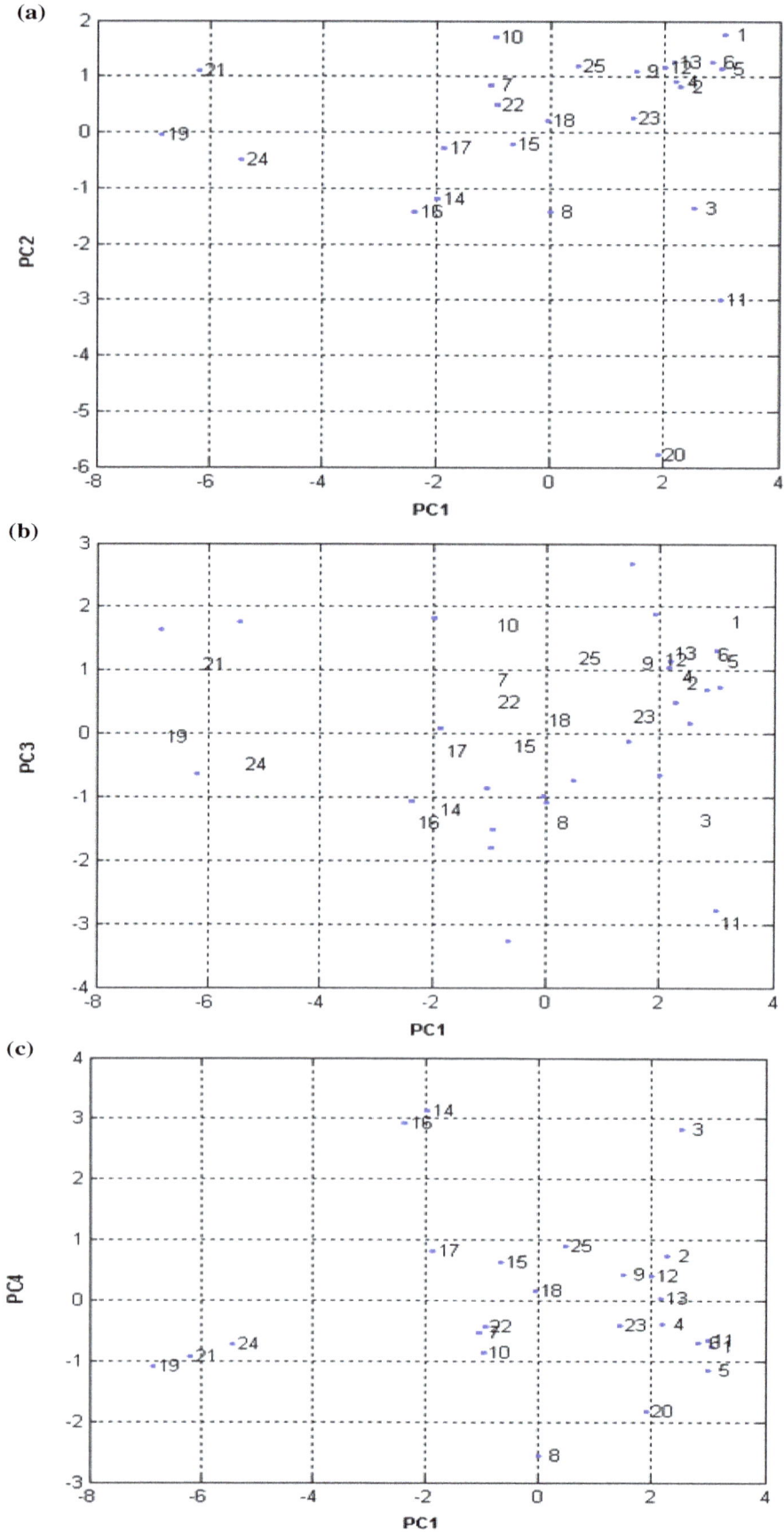

Figure 5. Score plots: (a) PC1 versus PC2, (b) PC1 versus PC3 (c) PC1 versus PC4.

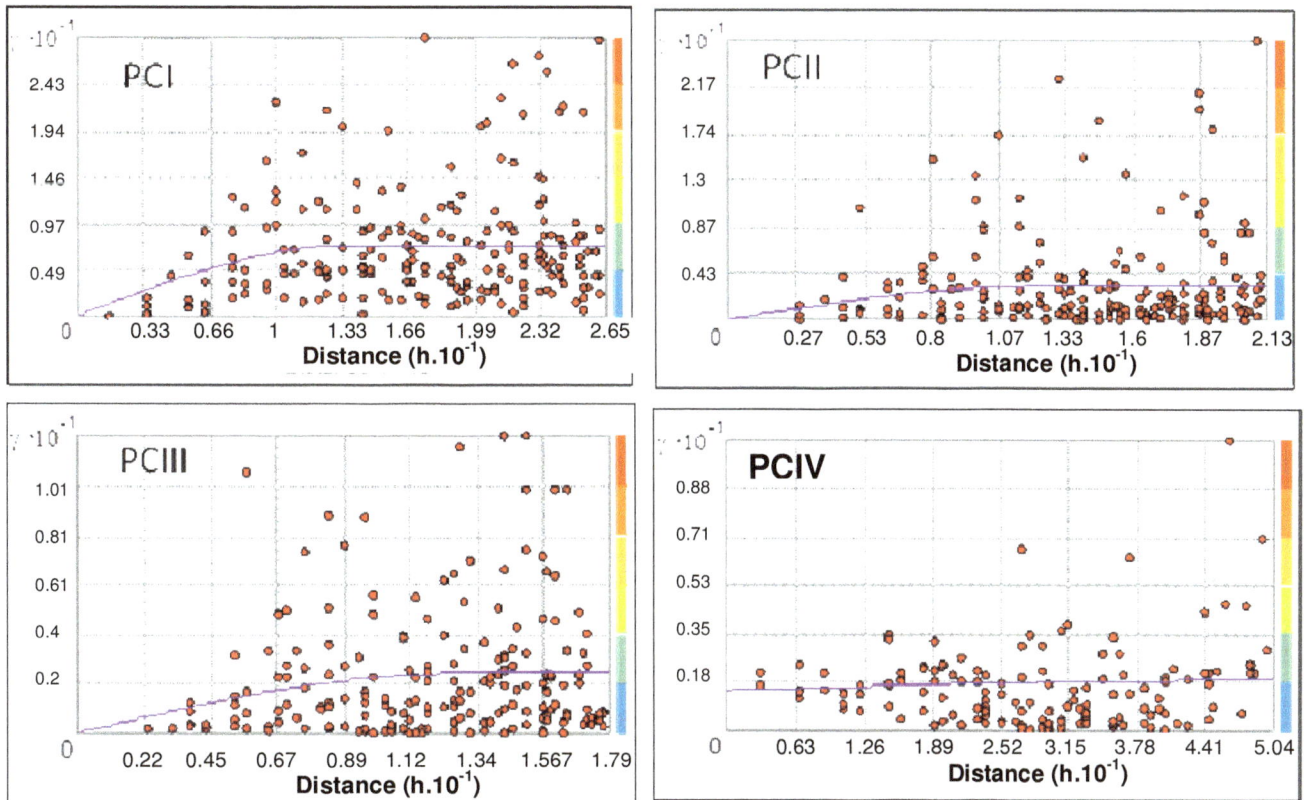

Figure 6. Observed and modelled semi-variograms of PC1 semi-variogram model is spherical, with range 13,99 km and sill 7,54. – Observed and modelled semi-variograms of PC2 semi-variogram model is spherical, with range. 12,45 km and sill 3,134 . – Observed and modelled semi-variograms of PC3 semi-variogram model is spherical, with range 12,87 km and sill 2,431. – Observed and modelled semi-variograms of PC4 semi-variogram model is spherical, with range 49,810 km and sill 0,405399.

and TH (positive values on PC1). At negative values on PC3 (large hydrocarbonates, alkalinity, total coliforms and sodium) we can distinguish well 24 and 16 (large conductivity) from wells 3, 8 and 11 (small conductivity).

(4) The fourth principal component discriminates the samples according to their amount of total iron (FeT), NH$^+$, pH and Mn. This PC discriminates wells 14, 15, 16, 17 (large amount of total iron (FeT), NH$^+$, pH and Mn) from well 7, 10, 22 (small values of the above mentioned variables) and in a minor way well 8. These samples are further separated according to their conductivity, chlorides, hardness, Permanganate Index and large values of some metal ions as Na$^+$, Ca^{2+}, Mg^{2+} and dry residue,: well 19, 21, 24 being characterised by large values of these variables and wells 8 by small values.

The geostatistical study

The geostatistical study includes two main phases:

i) The first is the characterization of the spatial structure of the regionalized variable; development of variogramme.

ii) The second is to estimate this variable using an interpolation tool by kriging method.

Variogram modeling

Once the experimental variogram established, he must calibrate a model that suits best, it is not always easy to find the theoretical model that corresponds to it. four models are proposed, namely linear, spherical, Gaussian and exponential.

The empirical semi-variogram for de PC4 illustrates that observed parameters concentrations are characterized by a significant spatial dependence up to a distance of 49.81 Km.

The nugget effect is relatively high and may be explained by the coarse sampling network, the lack of short distance data pairs within the experimental data set, and the expected short range variability of nitrate contamination within the groundwater body. In our case, the adjustment is done graphically. We opted for the spherical variogram for the four principal components (Figure 6).

Kriging method

Ordinary kriging was used to make the spatial predictions. It assumes a constant but unknown mean.

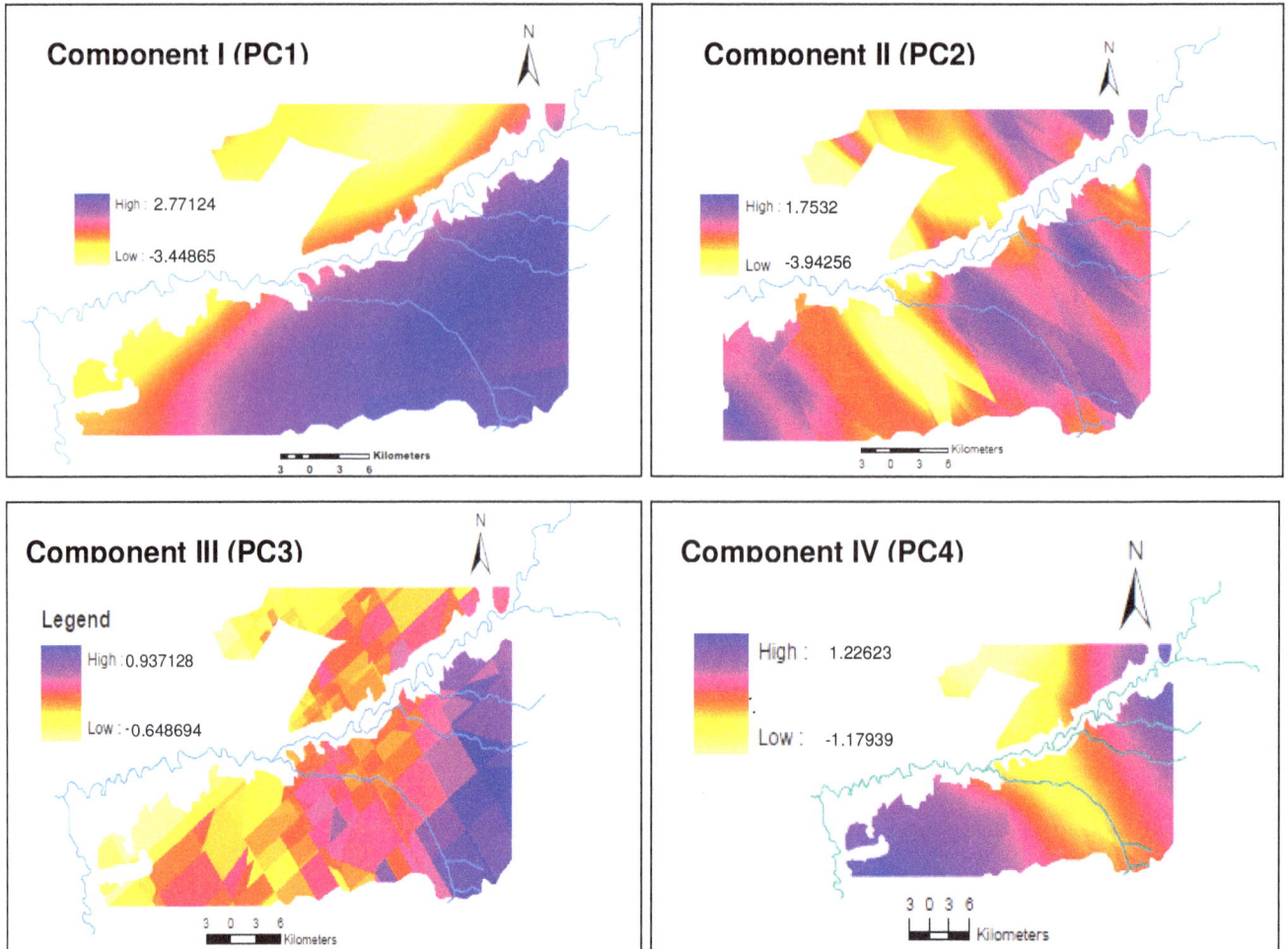

Figure 7. Geographical representation of PC1, PC2, PC3 and PC4 by the ordinary Kriging method.

Given the known geographic localisation of each sample, it is possible to use mapping techniques to efficiently analyse the relationships between geographic position and sample characteristics.

This was performed by using kriging mapping technique (Davis, 1986) representing each individual PC, whose information amount has been described on a thermographic scale. The results of such visualizations are reported in (Figure 7). The wells are represented on a color scale from light to dark blue (increasing positive score) and from light to dark red light (increasing negative score).

The first map describes the behaviour of the scores of PC1 with respect to the geographical co-ordinates: the map shows the clear presence of a hot spot (yellow to red color zone), identifying samples with large negative scores on PC1; these samples are characterised by large values of conductibility, chlorides, hardness (including calcium and magnesium), hydrocarbonates, permanganate index(larger amount of some metal ions as Na+, Ca^{2+}, Mg^{2+}, K^+ NH^{4+}, and dry residue, located in

the area of Beni Amir especially around the city of "Fkih Ben Saleh" which are characterised by many factories like oil industry and the dairies industry and minor entity of western Beni Moussa. The other areas present samples with positive or slightly negative scores on PC1: this is particularly true for the middle and eastern zone of Beni Moussa. The composition of waters by hydrocarbonates, TAC, conductivity and Na^+ is explained by the leaching of geological red detritic facies in the north of the plain and the calcaire levels of the haut atlas in the south of the plain (limestone and dolomites).

The map of PC2 shows the same hot spot zones corresponding to large negative scores. These samples are rich in nitrite, bacteriological contamination (CF, CT, SF) and Potassium around the Souk Sebt city. Similar characteristics, even if of minor entity, can be detected in the samples collected nearby the city of Ouled Ayad, in the Beni Moussa area and Fkih Ben Saleh (yellow to red colour zones). There is also a large zone clearly detectable at the map, characterised by samples with large positive scores on PC2 and are rich in nitrate.

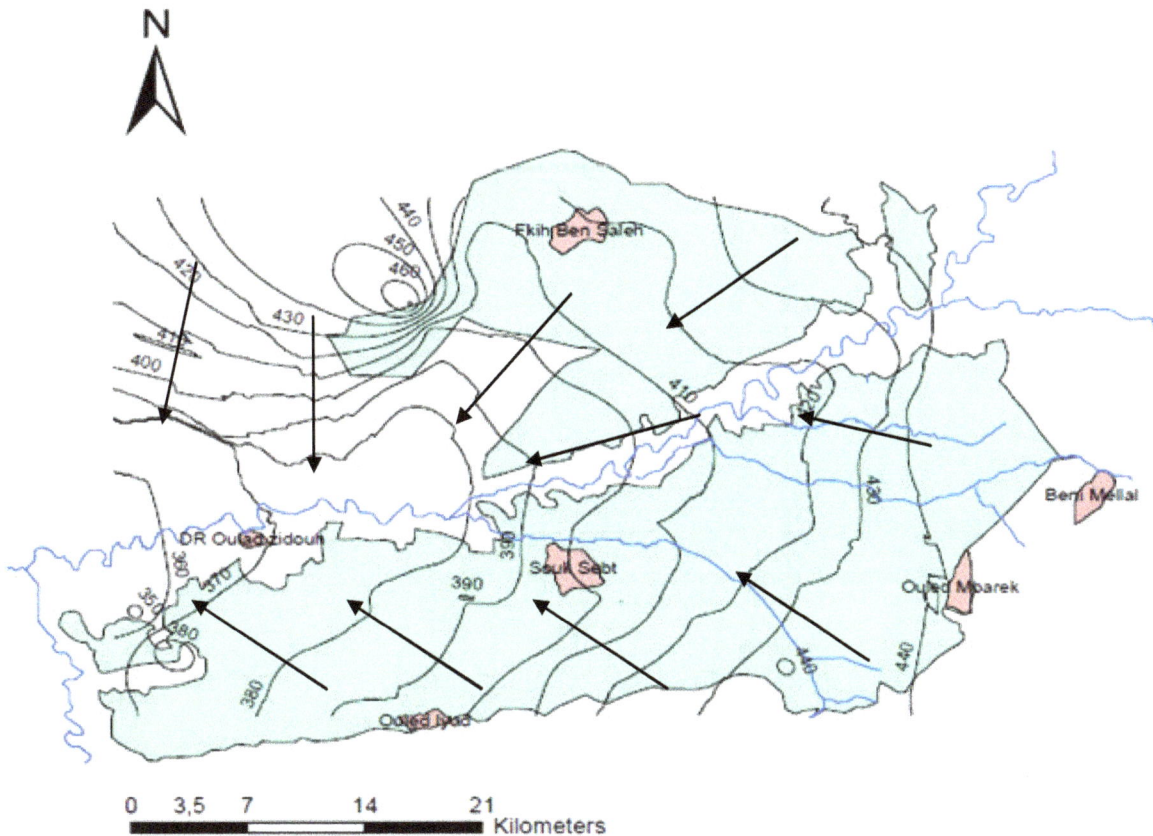

Figure 8. Piezometric map (m), Tadla plain, Morocco (2007).

These samples are collected in the area of Beni Amir and Beni Moussa which are characterised by intensive agriculture releasing a large amount of nitrate exceeding 50 mg/L (Berdai et al., 2004). Comparing the nitrate contamination map with the piezometric map (Figure 8) suggests also that the nitrate contamination occurs mainly in areas of shallow water depths (that is, piezometric heights 5 to 10 m below surface level) or in areas where the soil is permeable.

The map of the third PC shows the presence of a zone with large negative scores (red colour zone) near factories in Beni Moussa area and Beni Amir which are characterised by an intensive agriculture, characterised by waters with large amount of sodium, hydrocarbonates, TAC and total coliforms. Similar characteristics are detected for the Beni Moussa areas. The blue areas present large positive scores on PC3, characterised especially by the pH. These zones correspond to the eastern area of Beni Moussa and the area between Ouled Ayad and Souk Sebt.

In the map of the fourth PC the red colour areas indicate negative scores on PC4 (that is, samples with a small amount of pH). This representation indicates that built-up zones are mainly characterised by large positive values (blue areas), indicating a large amount of total iron (FeT), NH^+ and Mn^{2+}.

Cluster analysis

The obtained diversity-based hierarchical dendrograms for variables and wells are depicted in Figure 8. The analysis of the diversities among variables could be done extracting two main groups of variables, applying a horizontal cut at 80% of diversity (Figure 9a). Considering the loading plots of the first four PCs, group A identified in the dendrogram mainly gathers the variables with negative loadings on PC1: conductibility is therefore related to chlorides, Na^+, dry residue (RS), Mg^{2+}, TH, Ca^{2+}, IP, total iron (FeT), NH^+, K^+, and Mn^{2+} ions are also present in the same group. The analysis of the dendrogram built on variables confirms the conclusions already drawn by PCA, pointing out a group of variables connected with conductibility that show a large negative weight on the first PC.

The dendrogram built on the wells (Figure 9b) shows two main groups with a horizontal cut at 80% of diversity. Group A contains the samples showing particularly large values of some of the variables identified by group A in the corresponding dendrogram: this group is characterised by large negative scores on PC1, showing therefore a large amount of chlorides, Na^+, conductibility, and dry residue. The second group can be further divided in groups B and C, even if this subdivision takes place at

(a)

(b)

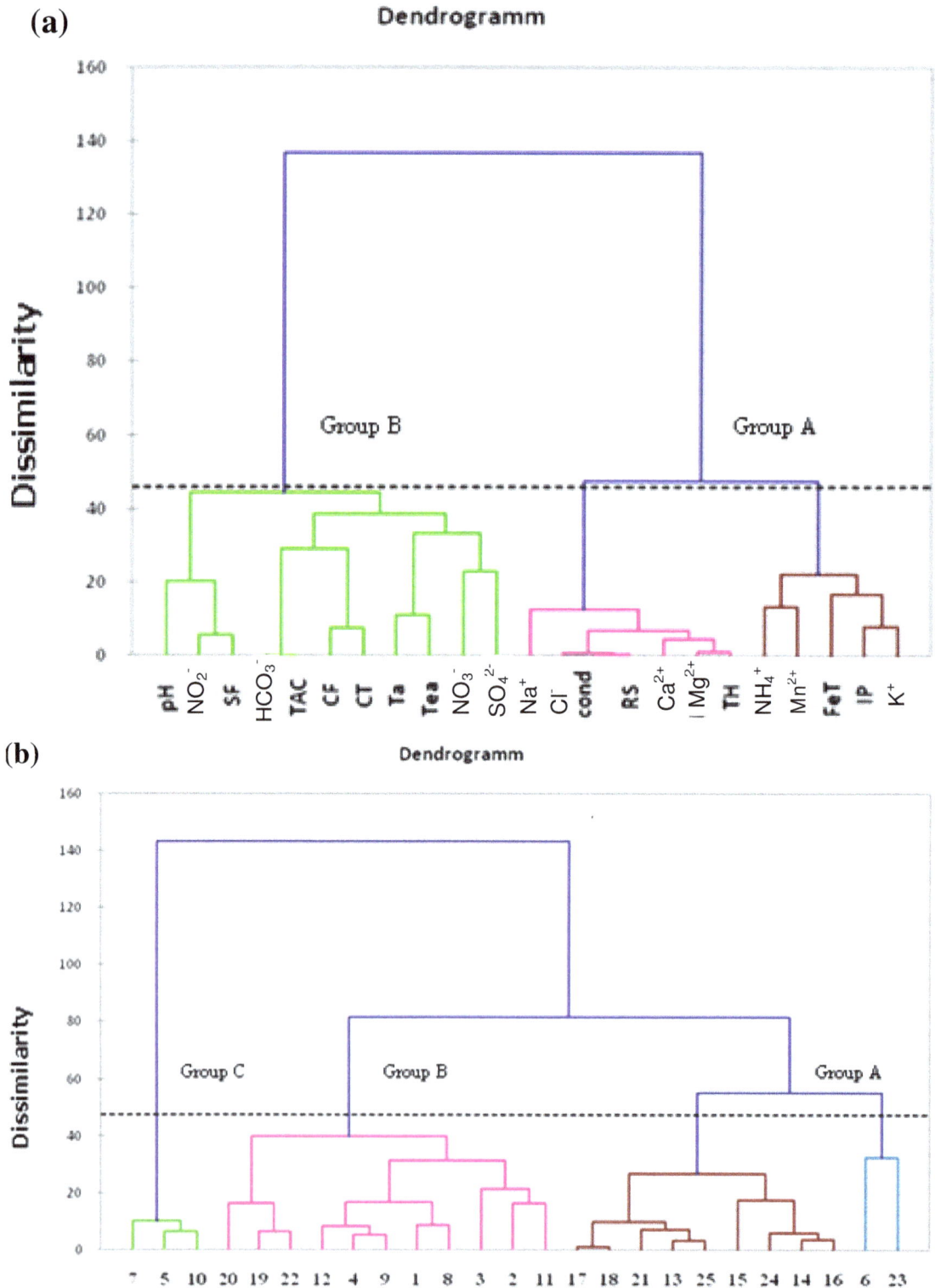

Figure 9. Cluster analysis: (a) grouping original variables (b) grouping wells described by their original variable.

a level of diversity of only 48%. Group B represents samples showing small values for all the variables considered, therefore characterised by no particular behaviour. Group C instead represents a middle course between the anomalous behaviour of samples belonging to group A and the flat behaviour of samples from group B. This analysis therefore confirms the conclusions on wells similarities already pointed out by PCA.

Implications for ground water monitoring

The results of the sampling campaign presented in this paper allow confirming that the groundwater body in the Tadla plain enters in a critical state. The high concentration that has been observed for the nutrient related parameters corroborates with the observed increase of fertilizer use in this area. However, agriculture should not be considered as the single contributor to the observed pollution. Indeed, the water sanitation infrastructure in this region is generally poor. Leakage of nutrients from poorly performing waste water collection systems, the factories installed in the region (sugar, oil and dairy) and waste disposal sites should not be excluded. Similar cautionary remarks were formulated by Laftouhi et al. (2003) for the pollution observed in the Essaouira aquifer in the centre part of Morocco. These authors suspected the presence of different point sources to contribute to the nitrate pollution.

In order to improve the diagnosis of the observed pollution and to quantify the role of agriculture in this pollution, different strategies can be followed. First, the time resolution of the survey could be improved.

A regional survey of the groundwater body was only performed twice campaign per year by the Hydraulic Basin Agency of Oum R'bia in Beni Mellal for all nutrient. and three times per year by the regional office of the agricultural development of Tadla (O.R.M.V.A.T) for salinity and nitrate only.

Comparing the results of both surveys (results not shown) demonstrates that the reaction time of this groundwater body is rather fast. In such cases, more regular surveying is needed to understand the time dynamics of the pollution problem and to alert agriculture and water managers when the groundwater body enters in a critical state. In addition, when time series of pollution would become available, analysis of the time dynamics of the pollution would allow to better assess the origin of the pollution, as agricultural pollution will most likely be determined by seasonal dynamics, in contrast to point pollution which will be more constant in time. A plea is therefore made to repeat the presented survey on a regular time basis and to implement a continuous monitoring of the ground water body. Second, the space resolution of sampling in the survey could be improved.

The observed large nugget effect on the calculated variograms for the components (for example, the PC IV in Figure 6) showed that quite some variability cannot be explained with the adopted sampling density.

Third, other parameters could be analysed during the survey. In the present study, most attention was devoted to basic mineralogical, nutrient and bacteriological related parameters. The correlation analysis between the nutrient and bacteriological parameters on the one hand, and the basic mineralogical parameters on the other hand, did not allow to discriminate between agricultural and non-agricultural pollution sources, as both sources are suspected to give similar signature in the observed pollution of the ground water body.

However other parameters could be added in the future monitoring programmes. It may be suggested to analyse the ground water body at least for residues of plant protection products, and the heavy metals as these parameters would allow to identify more efficiently the origin of the observed pollution. Alternatively, isotope analysis could be proposed to discriminate between the agricultural and non-agricultural pollution.

Fourth, the current knowledge of the agricultural practices in this area is poorly documented. The presented survey, allows to screen the area and to identify sub-regions where the pollution is most problematic. A more detailed analysis of the nutrient balance in those areas by means of advanced agro-ecological models would definitely allow to elucidate clearly the role of agriculture in the observed contamination. This latter strategy is currently followed in an ongoing study.

CONCLUSION AND RECOMMENDATIONS

The results obtained by the application of the different pattern recognition methods proved to be in agreement with each other. From PCA, groups of variables and samples were pointed out, stressing the existence of areas showing particular behaviours with respect to the amount of some of the investigated parameters. The analysis was then refined through the construction of termographic maps of the first relevant PCs.

The maps showed the existence of a yellow to hot and red spots, located in the area of Beni Amir. This area proved to be peculiar from different points of view (it was identified as peculiar from the analysis of all the PCs considered, showing each an independent source of variation): a further geological study analysis showed that this zone is characterised by a thickening north towards the south, of the aquiferous system of Tadla observed on both sides of Oum R' Bia river.

In the Beni Moussa area it was possible to point out even similar contributions due to the presence of cities (with a probable human and industrial polluting effect). The results obtained were further confirmed by cluster analysis.

In the plain, contamination by nutrients may therefore be indicative for bacteriological contamination of groundwater. Wells contaminated by nutrients should therefore receive particular attention in the groundwater monitoring programme. Nutrient contamination may therefore also guide the design and implementation of sanitation operations needed to reduce environmental and health risks.

Different strategies could be developed to reduce the pollution. A first, but elementary, step hereby would consist in diffusing the relevant information about the observed pollution and associated risk to the local actors, that is, the farmers, the local population, the communes and the different administrations dealing with water and agricultural management. Indeed, mitigation strategies will only be effective if local actors are motivated to implement the mitigation strategies. In case agriculture would play a key role, than farmers must correctly be informed about the pollution problem, the role of agricultural sector in this problem and their responsibility.

Hence, extension services of the O.R.M.V.A.T will probably play a key role in this first step. In a second step, mitigation strategies could be implemented. Different fertilizer management options could be presented such as reducing the total fertilizer dose, adapting the fertilizer dose in terms of the actual crop demand and available residue in soil, modifying the timing of the fertilizer dressing, modifying the spatial application in particular for the fruit tree exploitation in the area, and finally using covering crops and catch crops.

ACKNOWLEDGMENTS

The authors sincerely thank the Hydraulic Basin Agency of Oum R'bia in Beni Mellal (Morocco) for their help in providing necessary data.

REFERENCES

Abouzaid H (1980). Qualité des eaux et orientation de la recherche applique´e a` l'office national de l'eau potable pendant le plan quinquennal 1980–1985. CNCPRST. Rabat. 173:72-80.

Barbieri P, Adami G, Favretto A, Lutman A, Avoscan W, Reisenhofer E (2001). Robust cluster analysis for detecting physico-chemical typologies of freshwater from wells of the plain of Friuli (northeastern Italy). Anal. Chem. Acta 440(2):161-170.

Berdaï H, Soudi B, Bellouti A (2004). Contribution a` l'étude de la pollution nitrique des eaux souterraines en zones irriguées: Cas du Tadla, Projet INCO-WADEMED, Actes du Séminaire, Modernisation de l'Agriculture Irriguée, Rabat, du 19 au 23 avril 2004, p. 28.

Box GEP, Hunter WG, Hunter JS, Hunter WG, Hunter JS (1978). Statistics for Experimenters: An Introduction to Design, Data Analysis, and Model Building, Wiley-Interscience, New York, ISBN 0471093157.

Bureau De Recherches Géologiques Et Minières (BRGM) (1993). Etude du système multicouche de la plaine de Tadla. Description des aquifères et analyse du système multicouche. Maroc. Rapport R35610 4S/Eau- 92:129.

Brodnjak-Vonina D, Dobnik D, Novi M, Zupan J (2002). Chemometrics characterisation of the quality of river water. Anal. Chem. Acta. 462(1):87-100.

Caselli M, De Giglio A, Mangone A, Traini A (1998). Characterisation of mineral waters by pattern recognition methods. J. Sci. Food Agric. 76(4):533-536.

Durfor CN, Becker E (1964) «Public water supplies of the 100 largest cities in the US». US-Geological Survey Water Supply Paper, 1812, 364.

Davis JC (1986). Statistics and Data Analysis in Geology, John Wiley & Sons, New York, ISBN 0471080799.

Everitt B (2001). Cluster Analysis (4th edn). Hodder Arnold, London.

Food and Agriculture Organization (FAO) (2002). The State of Food Insecurity in the World. Rome [http://www.fao.org/ DOCREP/003/YO1500E/Y1500E00. HTM], April 2002.

Farnham IM, Johannesson KH, Singh AK, Hodge VF, Stetzenbach KJ (2003). Factor analytical approaches for evaluating groundwater trace element chemistry data. Anal. Chem. Acta 490(1):123-138.

Fetouani S, Sbaa M, Vanclooster M, Bendra B (2008). Assessing ground water quality in the irrigated plain of Triffa (north-east Morocco). Agric. Water Manage. 95(13):3-142.

Harvey RW, Harmas H (2002). Transport on micro organisms in the terrestrial subsurface: in situ and laboratory methods. In: Dans: Manual of Environmental Microbiology, 2ème edition. ASM Press, Washington, DC.

Isaaks EH, Srivastava RM (1990). An Introduction to Applied Geostatistics, Oxford University Press, Oxford ISBN: 0195050134.

Khattabi A (2003). Diagnostic socio-économique: Beni Snassen. Rapp. Inédit, projet MedWetCoast-Maroc, PNUE/Secr. Etat Envir. /Départ. Eaux & Forêts, Maroc, p. 82.

Kholtei S (2002). Caractéristiques physico-chimiques des eaux usées des villes de Settat et de Berrechid et évaluation de leur impact sur la qualité des eaux souterraines. Thèse. Doctoral p. 146.

Krige DG (1978). Lognormal-de Wijsian Geostatistics for Ore Evaluation (Geostatistics), South African Institute of Mining and Metallurgy, Johannesburg, ISBN: 0620030062.

Laftouhi N, Vanclooster M, Jalal M (2003). Groundwater nitrate pollution in the Essaouira Basin (Morocco). CR Geosci. 335:307-317.

Lhadi EK, Mountadar M, Younsi (1993). Contamination des eaux souterraines par les sels et les nitrates de la zone littorale entre Azemour et Bir Jdid. Se´minaire de recherche nationale dans le domaine: Eaux et Environnement, LPEE, Casablanca (Maroc).

Librando V, Drava G, Forina M (1998). 3-way principal component analysis applied to the evaluation of water quality of underground waters in the area of Siracusa. Ann. Chem.-Rome 88(11-12):867-878.

Maier RM, Pepper IL, Gerba CP (2000). Environmental Microbiology. Academic Press, San Diego, CA.

Massart DL, Vandeginste BG, Buydens LM, Lewi PJ, Smeyers-Verbeke J (1997). Handbook of Chemometrics and Qualimetrics: Part A, Elsevier Science Ed., Amsterdam, ISBN: 0444897240.

Massart DL, Vandeginste BGM, Deming SN, Michotte Y, Kaufman L (1988). Chemometrics: A Textbook (Data Handling in Science and Technology), Elsevier Ed., Amsterdam ISBN: 0444426604.

PDAIRE (Plan Directeur d'Aménagement Intégré des ressources en Eau) (2008). Etude d'actualisation du PDAIRE de la zone d'action de l'Agence de l'OER. Mission I, ETAT DES LIEUX ET PERSPECTIVES D'EVOLUTION. Rapport définitif, Royaume du Maroc, p. 4.

Reghunath R, Sreedhara Murthy TR, Raghavan BR (2002). The utility of multivariate statistical techniques in hydrogeochemical studies: An example from Karnataka, India. Water Res. 36(10):2437-2442.

Reisenhofer E, Adami G, Barbieri P (1998). Using chemical and physical parameters to define the quality of karstic freshwaters (Timavo River, North- Eastern Italy): A Chemometric Approach. Water Res. 32(4):1193-1203.

Rodier J (1984). L'analyse de l'eau, eaux naturelles, eaux résiduaires, eau de mer, 7ème édition. Dunod, Paris p. 1353.

Sinan M (2000). Méthodologie d'identification, d'évaluation et de protection des ressources en eau des aquifères régionaux par la combinaison des SIG, de la géophysique et de la géostatistique. Application à l'aquifère du Haouz de Marrakech (Maroc), thèse d'état, université Mohamed V de Rabat, Maroc.

Sinan M, Razack M (2008). An extension to the DRASTIC model to assess groundwater vulnerability to pollution. Application to the Haouz groundwater of Marrakech (Morocco). Revue Environmental Geology. DOI 10.1007/s00254-008-1304-2 (Allemagne).

Sinan M, Razack M (2005). Management of regional aquifers using a combined procedure based on geostatistical and GIS tools (Haouz groundwater of Marrakekech, Morocco). Proceeding de la 6ème Conférence Internationale de l'European Water Resources Association : EWRA 2005 (7 au 10 septembre 2005), Menton, France.

USEPA (2000). CFR title, sous-partie S, parties 141 et 142, RIN 2040-aa97 National Prinking Water Regulation, Ground Water Rule.

Vega M, Pardo R, Barrado E, Debán L (1998). Assessment of seasonal and polluting effects on the quality of river water by exploratory data analysis. Water Res. 32(12):3581–3592.

Vidal M, López A, Santoalla MC, Valles V (2000). Factor analysis for the study of water resources contamination due to the use of livestock slurries as fertilizer. Agric. Water Manage. 45(1):1-15.

World Health Organization (WHO) (2000). Global Water Supply and Sanitation Assessment 2000 report. OMS, UNICEF p. 77.

Wunderlin DA, Díaz MP, Amé MV, Pesce SF, Hued AC, Bistoni MA (2001). Pattern recognition techniques for the evaluation of spatial and temporal variations in water quality. A case study: Suquía River Basin (Córdoba Argentina). Water Res. 35(12):2881-2894.

3

Hydrogeochemical characterization of the coastal Paleocene aquifer of Togo (West Africa)

Gnazou M. D. T.[1*], Bawa L. M[1], Banton O.[2] and Djanéyé-Boundjou G.[1]

[1]Laboratoire de Chimie de l'Eau, Faculté Des Sciences, Université de Lomé, B.P 1515, Lomé, Togo.
[2]Laboratoire d'Hydrogéologie, Université d'Avignon et des Pays du Vaucluse, France.

This paper presents a hydrogeochemical characterization of the Paleocene aquifer in the coastal basin of Togo. During the dry season (March, 2005), piezometric measurements were made on inventoried water wells. Fifty nine (59) water samples were taken for chemical analysis and the results subjected to statistical study. The quantitative variables considered for this purpose were temperature, pH, electrical conductivity, the major and some minor elements. The results showed that the water samples were weakly mineralized. The electrical conductivity did not exceed 1000 µS/cm, but it increased considerably in the sulphate rich samples. Hydrochemistry was dominated by calcium and bicarbonate except where important values of sulphate (967 mg/L) were obtained. The results of chemical analysis allowed the definition of four different hydrochemical facies. The calcic and magnesian bicarbonated facies was the most important. A sodic and potassic bicarbonated facies identified was mainly due to the residence time of water in the reservoir. Principal component analysis showed one pole of mineralization with strong correlations between sulphate, carbonate, magnesium and calcium.

Key words: Hydrochemistry, aquifer, chemical facies, residence time, Paleocene, Togo.

INTRODUCTION

The focus of this study is the aquifer system hosted by the Paleocene limestone formations. This aquifer is intensely exploited for diverse uses (domestic, industrial and agricultural). As a matter of fact, the coastal basin is the most populated part of Togo. One third of the nation's population is accommodated in this encompassing 3450 km² (Bourgeois, 1981), which is about 6% of the total area of the country. It is a region of important economic activities and the home of nascent industry. The concentration of population and industries in such a small area leads to major disequilibria between water supply and demand. Besides the Mono River, the other rivers in the basin are seasonal with extremely low flows. The aquifer is potentially threatened by anthropogenic pollution but it is not well known chemically. The purpose of this study is to determine the hydrogeochemical characteristics of the Palaeocene aquifer. To fulfil the objectives of this study, chemical analyses were conducted on samples collected from piezometers. This characterization allows one to ensure the conservation and the durability of this resource.

STUDY AREA

The study area forms part of the vast Gulf of Benin basin which stretches from East to West along the Atlantic coast from Nigeria to Ghana. In Togo, it is bounded in the North by outcrops of the crystalline basement of Pan-African age, and in the South it opens up to the offshore basin under the Atlantic Ocean. It widens from West to East from the Ghana border to the Mono River which forms the frontier with the Republic of Benin.

Geomorphologically, the coastal basin appears as a collection of inclined plateaus separated by river valleys

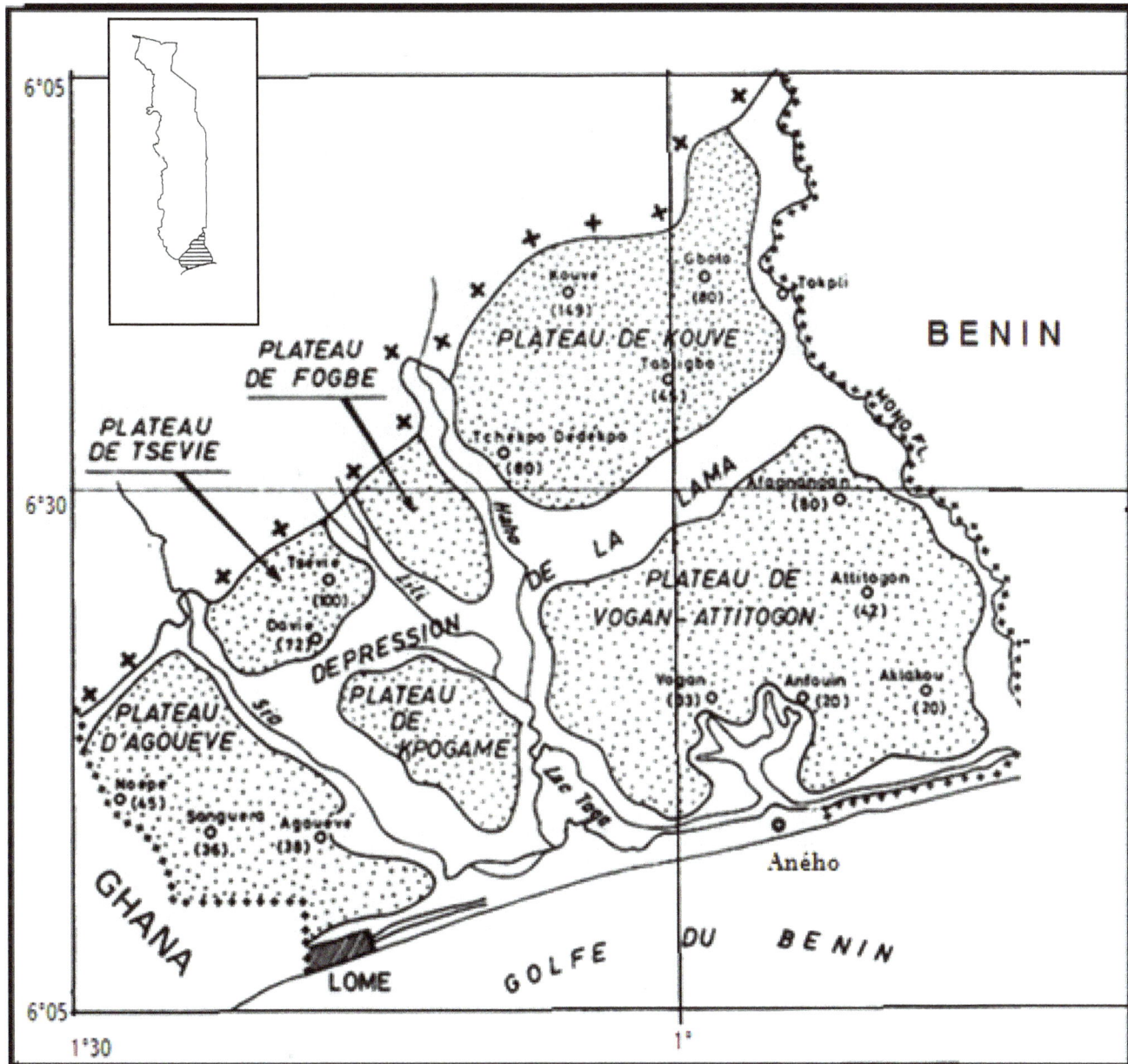

Figure 1. Study area location: hydrological and geomorphological settings (PNUD, 1982).

and located on both sides of a median depression oriented NNE-SSW, known as the "Lama depression". A system of lagoons stretches along the coastal plain. It consists of small lagoons; the most important one is the Lake Togo (Figure 1).

The region, with a subequatorial climate, is characterized by two distinct rainy seasons related to the movement of the "intertropical front". Rainfall in the basin is not uniformly distributed and decreases significantly from NE (1445 mm in Tabligbo) to SW (864 mm in Lomé). The average monthly temperature varies between

25 and 29 °C in Lomé (Gnazou, 2008).

The geology of this part of Togo is known thanks to the numerous hydraulic and petroleum wells drilled in the area, during geophysical and mining surveys (Sylvain et al., 1986). The Post-Paleozoic sedimentary series thus constituted is known from the Maastrichtian to the Quaternary (Johnson, 1987; Da Costa, 2005) (Figure 2).

The Maastrichtian deposits are dominated by quartzo-detrial clayey facies rich in organic matter. The sands found in the North of the basin, may or may not be clayey. In the South, the detrial character diminishes and

LEGEND

QUATERNARY

- Marine offshore bar sands
- Lagoonal-marine sands and clays
- Undifferentiated alluvial deposits
- Fluvio-lacustrine and fluvio-lagoonal deposits
- Low-plateau clayey sands (anté-Holocène)

CENOZOIC

- Continental Terminal

Oligocene-Miocene (?)

- More or less sandy clays and Nummulitid bearing limestones

Middle Eocene (Bartonian)

- Marls, clays and dolomitic organogenous limestones

Terminal Lower to middle Eocene

- Attapulgite clays, phosphatic clayey limestones, sometimes argillo-marly

Basal lower Eocene

- Quartzose detrital marly argillaceous deposits with scattered lenticular dolomitic, glauconitic and gypseous inclusions

Paleocene

- Clayey to clayey sand deposits, marly to sandy marl deposits, and bioclastic limestones

MESOZOIC

Maastrichtian

- Quartz-bearing-detrital and clayey deposits rich in organic matter, sands and clayey sands

◆ Phosphate deposit

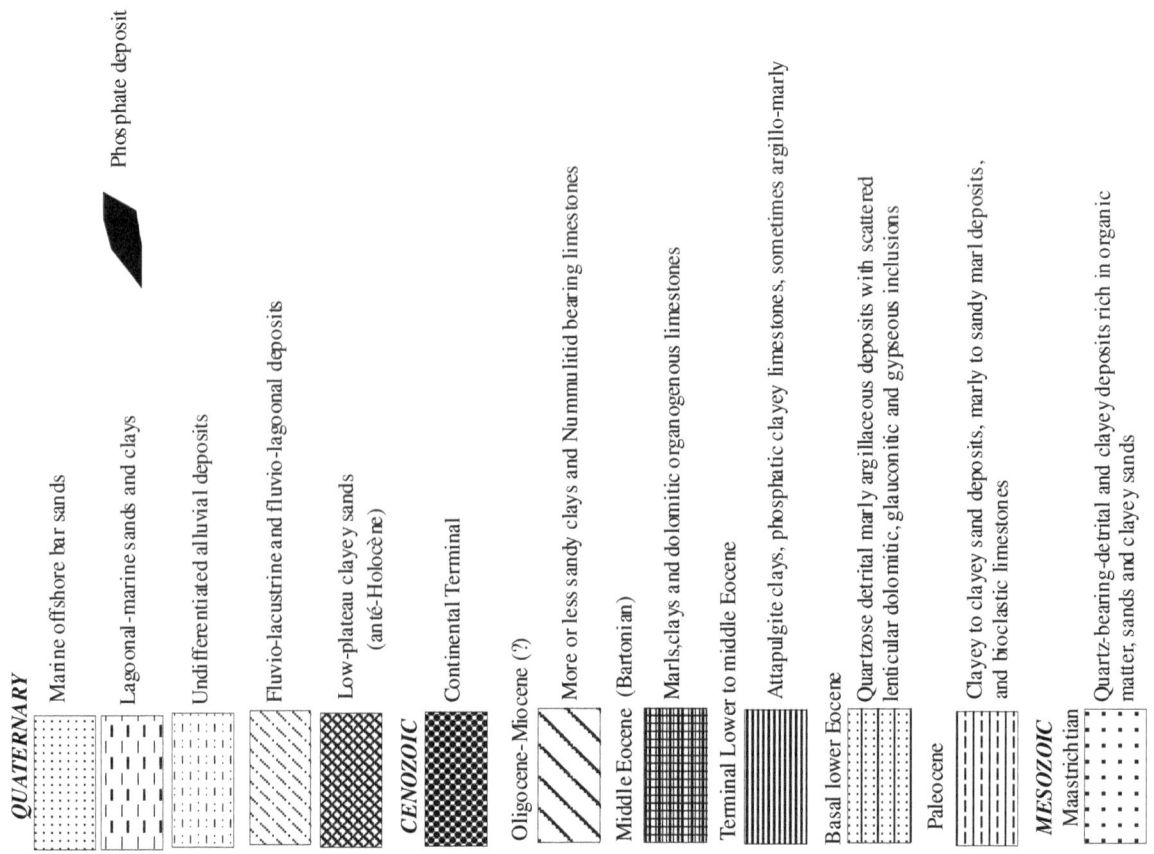

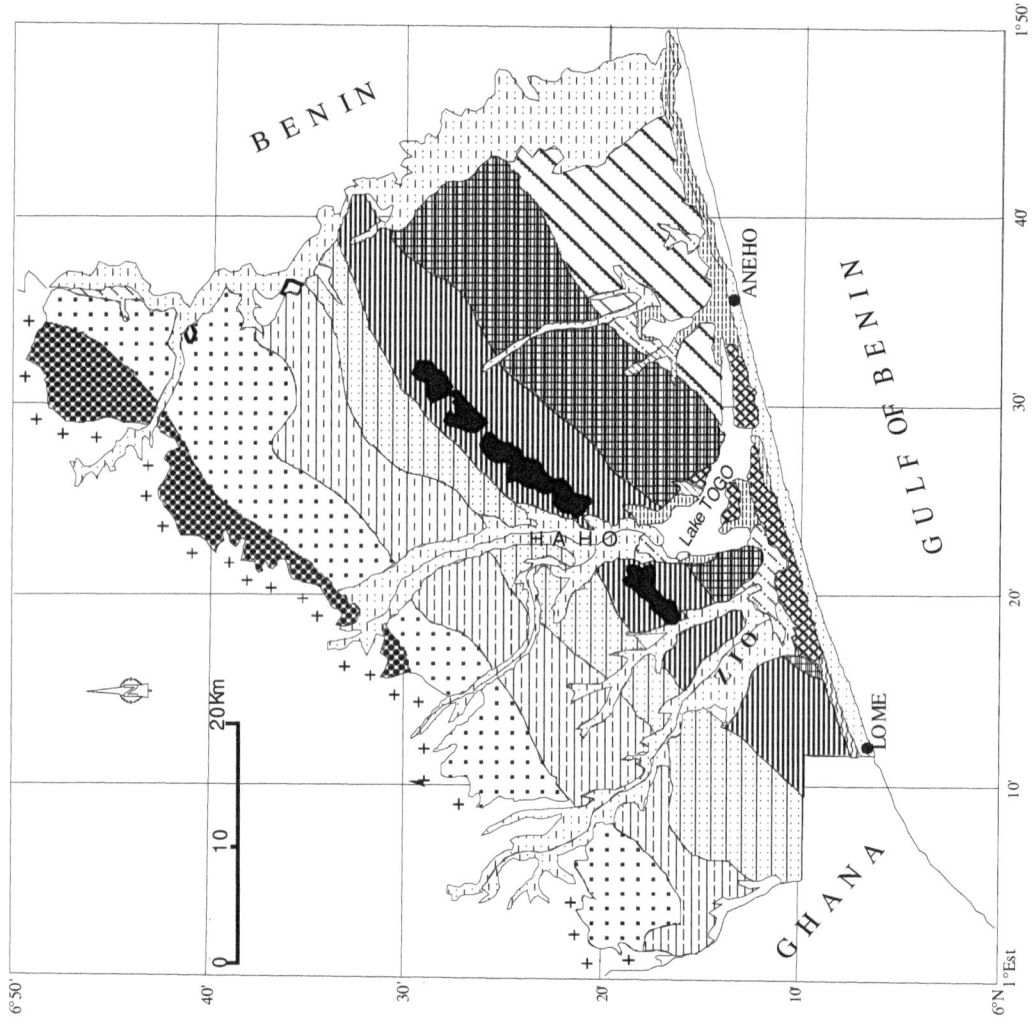

Figure 2. Geological map of the Togo coastal sedimentary basin (Sylvain et al., 1986).

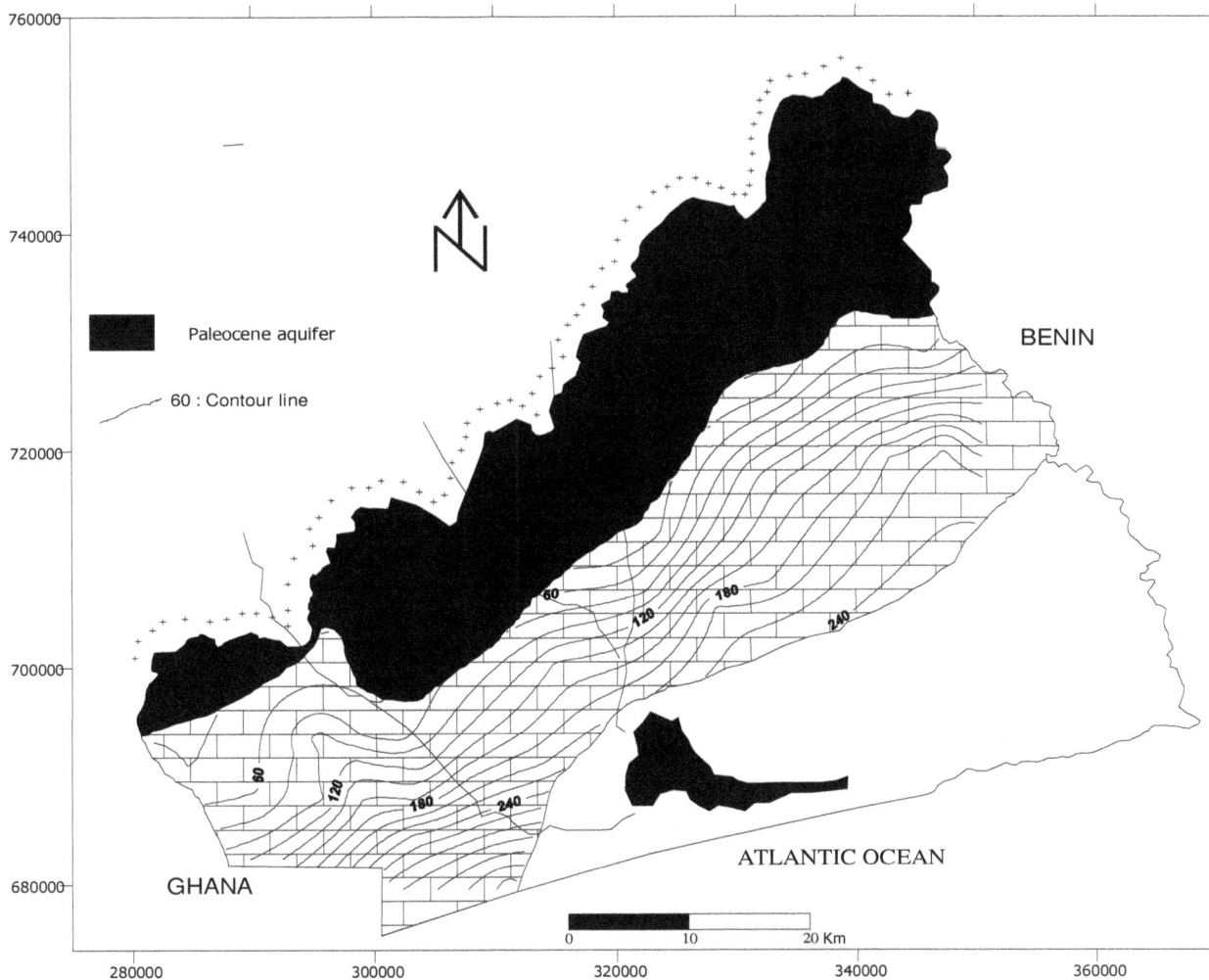

Figure 3. Extension of Paleocene aquifer and substratum geometry (Gnazou, 2008).

is progressively replaced by marls, clayey limestones and dolomites. Glauconite, gypsum and phosphate grains occur in small quantities. The thickness of the Maastrichtian formations increases from NW to SE (Johnson, 1987).

Paleocene sedimentation is characterized by a dominance of biochemical deposits. In the North, it is reduced to slightly organogenous clayey and clayey sand deposits. In the South, it may begin with marly and sandy marl facies. The Togocyamus seefriedi bearing limestone overlies this basal facies. A series of often laminated clayey marls rich in organic matter succeeds the limestones.

The Eocene consists of clayey-marl deposits with glauconite, gypsum and phosphates. The Oligocene formations include nummulitic limestone and sandy clay facies. The Continental terminal consists of sands and clayey silts.

Finally the Quaternary formations are made of beach sands and fluvio-lacustrine deposits.

In the coastal basin formations, four aquiferous horizons are exploited. These productive levels are separated by thick aquicludes. The aquifers are the upper Cretaceous, Paleocene, Continental terminal and the Quaternary coastal sands.

The Paleocene aquifer is sandwiched between clayey and marly Eocene and Maastrichtian formations. From West to East, it forms a 10 to 20 km wide SW to NE trending band along the entire basin. Its northern boundary corresponds to the end of the limestones. In the south, this boundary, situated in Lomé on the Atlantic, rises gradually up to the Mono River in the NE (Figure 3). The top of the limestone is very heterogeneous and consists of Eocene clays and marl. Its depth in the SW varies between 30 and more than 200 m. The reservoir, with a thickness between 5 and 35 m is made mainly of limestone. In some cases, sand may be found at the base of the limestone. Like the top, the substratum, made mainly of Maastrichtian clays, is inclined from North to South.

The Paleocene aquifer is recharged mainly in the North of the basin where it is in contact with the shallow aquifer of the Continental terminal (BCEOM/BRGM, 1983). Water flows in the Paleocene are from the North toward the South and are considered to be specific to the aquifer, because of the morphology of the substratum which plunges in the direction of the Atlantic Ocean (Gnazou, 2009).

METHODS

Fifty nine (59) groundwater samples were collected from the Palaeocene aquifer in March 2005 to investigate the chemical quality of the water. The data on the location of water wells that tap the Paleocene were collected from the data bank of the Hydraulics Department of Togo. In order to render each sample representative, we took a number of precautions. Thus a sufficient pumping time preceded sampling (Rodier et al., 2005). Water samples were pre-treated in the field before storage. All samples taken were passed through a 45 µm diameter filter. Filtration allowed the removal of contained particle and bacteria that could modify contents (Atteia, 2005). To the samples destined for cation analysis, we added the nitric to maintain the metals ions in solution (Rodier et al., 2005). The geographic locations of the sampling sites with a global positioning system (GPS) are shown in Figure 4.
Field parameters such as temperature, pH, turbidity and electrical conductivity were measured at each site using WTW (Wissenschaftliche Technische Werkstätten) type instruments. Water samples were stored in a refrigerator at a temperature of 4°C and analyzed within 72 h of sampling. The employed analytical methods, summarized in Table 1, were based mainly on the French association standards (AFNOR, 1986).

The data were analysed using appropriate computer programs and software: Xlstat (Fahmy, 2006), Excel, Golden Software Surfer. Piper and binary diagrams and statistical methods were utilized to characterize the Paleocene aquifer hydrogeochemically. Piper's diagram allowed a global visualization of the hydrochemical facies by identifying the chemical evolution of the different parameters used. The statistical study by principal component analysis allowed to determine the relations between the different parameters, while the binary diagrams allowed, by simple correlation between two elements, to investigate the processes responsible for the mineralization.

RESULTS AND DISCUSSION

Physicochemical characterization of Paleocene waters

The detailed results of the analysis are shown in Table 2. To facilitate interpretation the results are regrouped in statistical form in Table 3. Minimum, maximum, mean, and standard error values of the physical and chemical

parameters identified for the purpose of this study are summarized in Table 2. These results showed that the water temperature varied between 29.0 and 34.0°C. These values, higher than air temperature (27.4°C), reflect the high confining nature of this aquifer. This is explained by the deeper occurrence of the aquifer (50 to more than 200 m). The highest temperatures (>33°C) were recorded in the SW part of the basin where the aquifer depth exceeds 250 m.

The pH values varied between 6.6 and 8.5 with a mean near to the neutral value. The tendency toward alkalinity in some of the samples may be explained by the dissolution of carbonates composing the reservoir rock. Electrical conductivity values oscillated between 244 and 3213 µS/cm with a mean of about 1000 µS/cm. These values are representative of waters experiencing slight to excessive mineralization. Furthermore, the high value of the electrical conductivity standard error illustrates the important variations in the aquifer mineralization.

Areas with conductivities exceeding 1000 µS/cm are located mainly in the Eastern part of the basin and are associated with high sulphate content (Figure 5). Considering the hydrogeological context, the potential sources of salinity are the evaporitic deposits present in the surrounding geological formations (gypsum in the Eocene). This is possible because of the leakage observed in the basin aquifers (PNUD, 1975).

Nitrate concentrations were observed to be low throughout the sampled area; the maximum value was 20 mg/L.

These observations may be due to the confined and deep nature of the aquifer which limits the downward migration of surface nitrogen compounds to the aquifer. On the other hand, the confinement of the aquifer system favours anoxic conditions that enhance denitrification with a reduction in nitrate concentration. This is responsible for the relatively higher nitrite and ammonium ion contents observed in the deepest wells. From the point of view of toxicity, nitrite, like nitrate can cause a methemoglobinizing effect (Rodier et al., 2005).

Sulphate concentrations varied between 5.2 and 962 mg/L. In the western part of the basin, these concentrations varied between 130 and 225 mg/L. The highest concentrations (greater 250 mg/L) were observed in the Eastern part of the basin (Figure 6), the main source of sulphate being the dissolution of evaporites such as gypsum ($CaSO_4$, $2H_2O$) contained in Eocene formations, or in the laminated Paleocene argillites that make up the top of the aquifer.

Chloride contents fluctuated between 20 and 450 mg/L (Figure 7). Since the investigated wells are located far from the ocean, this high chloride content in the Paleocene aquifer can be explained by paleosalinity (Akouvi, 2000).

Fluoride and cadmium analyses were indispensable to defining the impact of the Eocene phosphate deposit

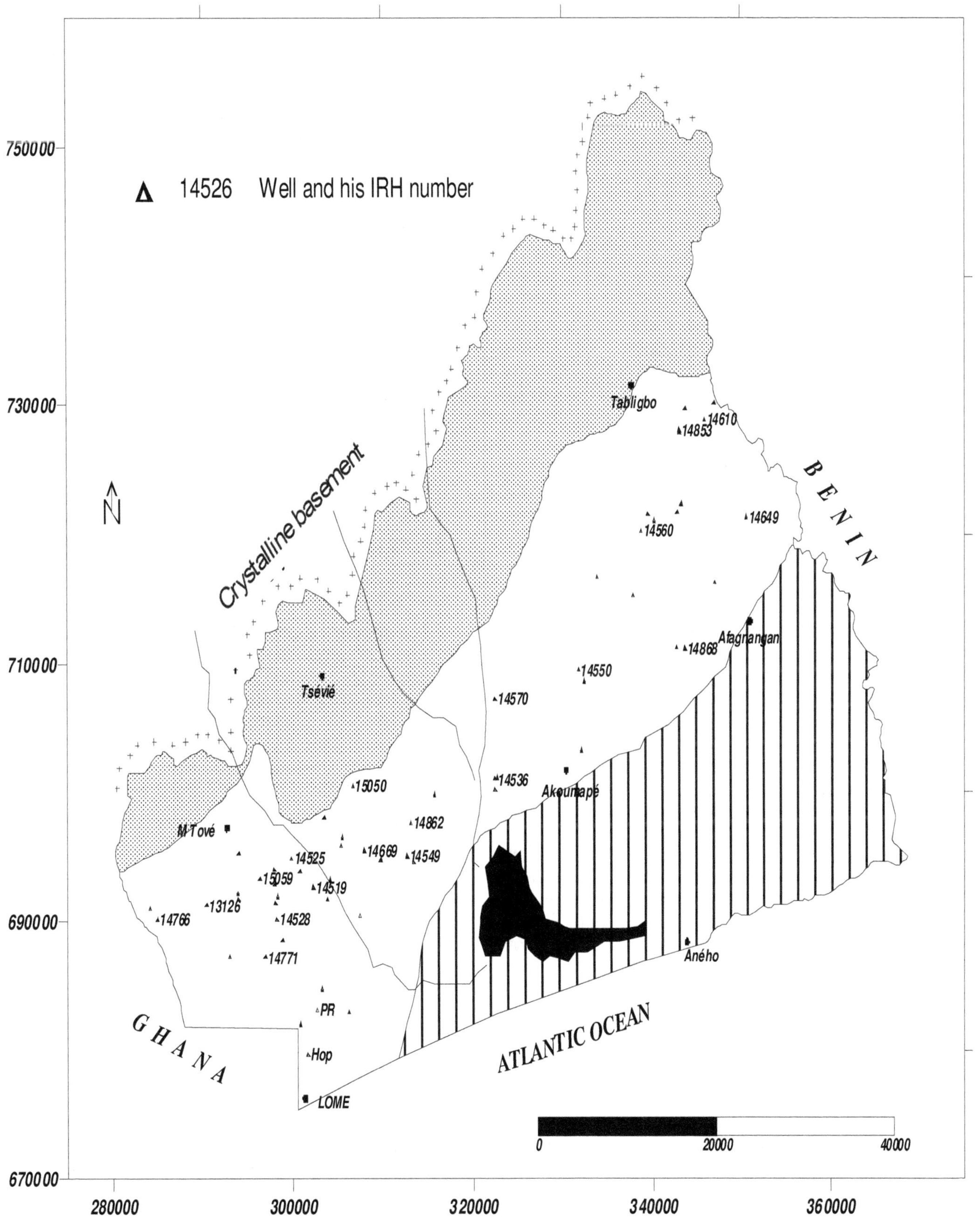

Figure 4. Geographic location of sampling sites.

Table 1. Analytical materials and methods.

Parameters	Methods	AFNOR Norms	Method accuracy	Material
Turbidity	Nephelometry	NFT90-33	±0.01NTU	DRT100B, model 20012 turbidimeter
Temperature	Electrometry	NFT90-100	±0.1 °C	WTW Cond 330i Conductimeter
pH	Electrometry	NFT90-008	±0.01	pH 330i pH-meter WTW
Electrical conductivity	Conductimetry	NFT90-031	±0.5%	WTW Cond 330i Conductimeter
Cl^-	Argentimetry	NFT90-014	±0.5 mg/L	-
HCO_3^-	Acidimetry	NFT90-036	±0.25 mg/L	-
SO_4^{2-}	Nephelometry	NFT90-009	1 to 2%	GENESYS 10 Spectrophotometer
NO_3^-	Molecular Spectrophotometry	NFT90-012	1 to 2%	GENESYS 10 Spectrophotometer
NO_2	Molecular Spectrophotometry	NFT90-013	1 to 2%	GENESYS 10 Spectrophotometer
NH_4^+	Molecular Spectrophotometry	NFT90-015	1 to 2%	GENESYS 10 spectrophotometer
Ca^{2+}	Complexometry EDTA	NFT90-016	-±0.5 mg/L	-
Mg^{2+}	Complexometry EDTA	NFT90-016	±0.24mg/L	-
Na^+	Atomic absorption spectrophotometry	NFT90-020	± 0.04mg/L	Perkin-Elmer, model 2380 Spectrophotometer.
K^+	Atomic absorption spectrophotometry	NFT90-020	±0.02 mg/L	Perkin-Elmer, model 2380 Spectrophotometer
SiO_2	Molecular Spectrophotometry	NFT90-007	1 to 2%	GENESYS 10 Spectrophotometer
F^-	Potentiometry	NFT90-004	± 0.2mV	JENWAY Ion meter 3345 Potentiometer
Fe	Molecular Spectrophotometry	NFT90-017	1 to 2%	GENESYS 10 Spectrophotometer
Cd^{2+}	Atomic absorption spectrophotometry	T90-112	0.02mg/L	Perkin-Elmer, model 2380 Spectrophotometer.

($Ca_5(PO_4)_3F$) on the underlying Paleocene waters. Phosphate ore is composed mainly of fluorapatite with fluoride and cadmium contents of 0.15% and 49 ppm respectively (Gnandi, 1998; Tchangbédji et al., 2003; Bawa et al., 2006). Fluoride and cadmium, if present in the Paleocene would constitute indicators of groundwater pollution by the overlying Eocene phosphate and exchange between aquifers (Continental terminal, Eocene and Paleocene). Fluoride concentrations in the groundwater varied from 0.1 to 0.7 mg/L. These concentrations are below the recommended World Health Organization drinkable water standard of 1.5 mg/L (WHO, 2006). This is an indication that phosphate has no impact on the Paleocene aquifer water quality. Furthermore cadmium concentrations were below the limit of detection (<0.02 mg/L).

Regarding the alkaline-earth elements, groundwater calcium content varied between 28 and 380 mg/L while that for magnesium was between 4.8 and 164 mg/L. One notes a similarity in the spatial distribution of these elements with high localized concentrations in the East of the basin (Figures 8 and 9).

The presence of those elements in high concentration increases water hardness. Using Langelier's empirical method, thirty two (32) samples were identified to exhibit a hard and five (5) a corrosive character. The remaining three samples reveal a character close to calco-carbonic equilibrium. For domestic consumers hard water means increased soap use for cleaning.

Silica concentrations in the water samples lay between 3.1 and 42.0 mg/L. These low concentrations are due to the carbonated nature of the aquifer and the very low solubility of silica in natural water (Banton and Bangoy, 1997). The silica present in the samples results from the dissolution of the clays (silicates) that form the top and the bottom of the aquifer, or that are contained in the carbonated matrix. One other possible source of silica in the Paleocene waters is its recharge by the sandy of the Continental terminal.

Ion contents varied between 0.0 and 4.4 mg/L, the iron presence is often associated with high water turbidity. High concentrations were often obtained in abandoned wells. This situation favours casing corrosion.

Groundwater chemical facies

Groundwater chemical facies were determined from the Piper diagram using the computer software "Diagrammes" (Simler, 2005). The advantage of using this diagram is that it allows many analyses to be represented on a single graph and facilitates their grouping by family. From these analyses the following observations emerged. The ternary diagrams of anions and cations indicated that the majority of samples were oriented towards the bicarbonate and calcic poles respectively. Projection of these points in the lozenge allowed the identification of the following pattern (Figure 10):

Table 2. Physico-chemical parameters of the Paleocene groundwaters.

Localities	N°IRH	Coordinates X (m)	Coordinates Y (m)	T°C	EC (µs/cm)	Turb (NTU)	pH	HCO₃ (mg/L)	Cl (mg/L)	SO₄²⁻ (mg/L)	Ca²⁺ (mg/L)	F (mg/L)	NO₃ (mg/L)	NO₂ (mg/L)	NH₄⁺ (mg/L)	Mg²⁺ (mg/L)	Na⁺ (mg/L)	K⁺ (mg/L)	Cd²⁺ (mg/L)	Fe total (mg/L)	SiO₂ (mg/L)
Atiémé II	14513	293975	691673	29.5	437.7	0.65	7.59	244	32.04	18.88	66	0.14	0.25	0.00	0.09	17	19.00	3.60	<0.02	1.81	17.05
Madjikpéto	14511	293915	692120	31	459.8	0.84	7.73	244	40.04	14.49	84	0.17	0.22	0.00	0.10	4.8	23.08	4.03	<0.02	1.11	16.36
Apessito	14753	294015	695220	30	526.3	2.59	7.78	207.4	45.05	18.63	61	0.3	1.70	0.00	0.21	4.8	22.68	3.39	<0.02	2.45	13.20
Yohonou	15059	296374	693224	30.5	498.6	0.02	7.95	305	35.04	12.29	86	0.27	0.88	0.22	0.02	12	17.80	4.74	<0.02	0.84	13.52
Sogbos EPP	14506	298033	692758	30	1351.8	0.3	7.82	378.2	110.12	14.88	91	0.34	3.89	0.01	0.35	38.4	73.90	10.61	<0.02	0.95	19.45
Sogboss	14767	298109	691311	33	714.7	0.08	7.77	329.4	80.09	38.39	72	0.31	3.53	0.09	0.31	55.2	38.94	6.15	<0.02	0.74	14.78
Nyamassi	14524	298355	691865	34	687.0	0.1	7.94	323.3	85.09	18.15	56	0.31	2.83	0.19	0.03	33.6	47.10	9.27	<0.02	0.83	13.77
GuénouKopé	14514	304121	693136	32	797.8	0.11	7.93	381.25	82.09	10.59	48	0.28	1.76	0.00	0.04	38.4	54.38	5.09	<0.02	0.92	21.98
Fidokpoe	14516	303870	691601	32.5	675.9	0.32	7.73	353.8	62.07	18.15	40	0.31	3.45	0.06	0.23	33.6	37.72	5.86	<0.02	1.52	13.45
Zongo	14519	302287	692560	31	692.5	0.4	7.86	359.9	65.07	15.71	64	0.31	2.31	0.04	0.56	28.8	39.34	6.74	<0.02	1.49	13.01
Apedokoe	14504	293030	687158	32	1263.1	0.41	7.73	402.6	175.19	15.95	104	0.29	2.59	0.18	0.43	57.6	79.18	6.56	<0.02	0.41	23.11
Logopé	14772	298922	688493	32	886.4	0.8	7.74	305	107.12	23.76	84	0.15	2.87	1.91	0.11	36	55.16	13.33	<0.02	0.22	15.93
Vakpo	14771	296968	687168	32	864.2	0.6	7.77	366	88.10	29.37	72	0.21	1.16	0.03	0.30	28.8	39.96	7.56	<0.02	0.70	19.97
Akoin	14773	300840	693818	32	642.6	4.15	7.68	366	55.06	13.00	124	0.32	4.63	1.53	0.02	9.6	34.43	10.42	<0.02	0.90	14.92
Akepe	14219	284120	690923	31	626.0	0.5	7.91	366	58.06	14.24	72	0.29	1.44	0.41	0.06	14.4	26.13	7.84	<0.02	0.26	17.95
Alinka	14525	299853	694805	30	542.9	0.2	8.06	298.9	60.07	13.27	72	0.30	1.95	0.45	0.05	9.6	23.14	8.72	<0.02	0.18	16.69
Alinka	13934	297975	693986	34	831.0	8.89	7.67	340	63.00	20.83	76	0.24	1.32	0.17	0.01	13	28.00	4.32	<0.02	0.21	19.00
Rama	14671	305480	696472	33	609.4	0.5	7.72	366	39.04	29.12	88	0.19	3.22	0.01	0.12	21.6	24.29	3.55	<0.02	0.48	21.49
Kladjémé	14862	313057	697571	34	675.9	0.53	7.77	390.4	42.05	23.27	80	0.16	1.32	0.00	0.10	12	31.66	3.06	<0.02	0.00	15.18
Adodo	15050	306635	700439	33	637.1	0.5	7.06	240.2	20.02	11.80	76	0.14	0.22	0.00	0.02	9.4	12.31	1.57	<0.02	0.24	25.03
Tonoukouti	14756	305331	695811	35	880.9	0.35	7.79	335.5	45.05	20.59	80	0.15	0.89	0.00	0.01	19.2	26.33	3.56	<0.02	0.00	13.91
Sewouvi	14669	307901	695445	31	903.0	0.5	7.7	402.6	80.09	24.98	96	0.26	0.85	0.00	0.01	31.2	44.63	5.68	<0.02	2.46	21.49
Kpotavé	14363	303491	697951	34	321.3	7.5	7.03	146.4	22.02	12.05	60	0.33	1.00	0.00	0.04	24	12.77	2.23	<0.02	1.18	21.74
Adidome	14766	284950	690100	30	243.8	0.56	6.64	134.2	20.02	7.17	56	0.22	2.41	0.00	0.04	9.6	11.70	1.21	<0.02	0.19	27.55
ESTAO	Etao	300859	681972	34	819.9	0.54	7.92	305	100.51	25.46	56	0.35	4.11	0.61	0.02	28.8	63.73	12.32	<0.02	0.00	19.22
Hopital	Hop	301672	679611	35	919.6	0.62	7.93	305	125.14	21.56	52	0.21	6.51	0.61	0.14	16.8	81.62	15.08	<0.02	0.13	31.09
SALT	Salt	306197	682875	34.5	952.9	0.53	7.98	334	140.35	9.85	48	0.41	8.45	0.00	1.70	26.4	89.73	22.12	<0.02	0.06	14.92
Présidence	PR	302676	683054	33.5	842.1	0.4	8.47	346.48	109.12	11.56	29	0.16	4.53	0.00	1.50	33	85.00	13.79	<0.02	0.09	23.26
Libye	Lib	303258	684689	34	842.1	0.68	8.06	329.4	116.13	16.20	88	0.64	4.78	0.00	1.20	43.2	71.05	14.84	<0.02	0.03	19.72
Klobatémé	14762	307407	690376	32	736.8	0.32	7.97	329.4	75.08	35.71	48	0.60	4.75	0.57	0.35	19.2	48.94	17.67	<0.02	0.29	18.96
Lowé	13778	309742	694747	31	847.6	0.61	7.88	377.4	80.39	46.93	81	0.09	1.30	0.03	0.10	29	48.00	6.62	<0.02	0.66	30.08
Abao	14761	313382	694538	31	775.6	0.46	8.04	366	79.09	37.17	100	0.26	1.62	0.04	0.11	4.8	45.03	8.21	<0.02	0.39	25.78
AkodessEPP	14538	322655	701048	32	964.0	0.07	7.67	362	105.12	36.44	104	0.4	3.48	0.31	0.02	16.8	71.00	17.00	<0.02	0.10	19.89
Zéglé	14549	312642	695016	30	819.9	0.47	7.79	411	76.08	30.59	68	0.26	1.27	0.00	0.11	38.4	50.71	11.00	<0.02	0.09	16.10
Akadjamé	13332	315678	699745	38	714.7	1.97	8.21	274.5	88.10	5.71	72	0.19	2.97	0.00	0.38	7.2	54.70	11.08	<0.02	1.89	16.67

Table 2. Cont.

Agbleta	14553	322445	700135	32	797.8	8.03	311	76.58	64	0.17	3.86	0.11	0.04	33.6	15.00	<0.02	0.07	21.09
Akodéssewa	14536	322445	700951	35	858.7	7.89	358	109.62	104	0.54	3.16	0.06	0.94	7.2	17.00	<0.02	0.03	13.39
Vodze	14570	322446	707146	29	775.6	7.67	417	33.54	92	0.21	1.29	0.06	0.07	24	11.00	<0.02	0.40	13.20
Kpakpladzev	14554	332329	708523	32	759.0	7.58	658.8	34.04	120	0.09	0.78	0.11	0.06	40	7.60	<0.02	0.60	13.01
Tokamé	14550	331685	709414	31	709.1	7.55	414.8	36.04	96	0.11	0.50	0.06	0.02	9.6	4.60	<0.02	0.46	13.96
Kpotémé	14602	343240	722318	31	897.5	7.74	451.4	58.06	96	0.16	1.52	0.00	0.01	28.8	10.54	<0.02	1.00	15.92
Dévémé	14594	346991	716201	30.5	609.4	7.16	183	73.08	84	0.37	2.45	0.02		9.6	6.22	<0.02	0.19	15.67
Gbodjomé	14649	350571	721287	31.3	886.4	7.57	402.6	68.88	76	0.27	1.72	0.02	0.36	60	11.00	<0.02	0.43	10.24
Akladjénou	14633	346927	730171	30.7	797.8	7.51	500.2	37.90	80	0.24	0.33	0.00	0.03	7.2	6.35	<0.02	0.83	8.85
Ahlemeyi	14624	343657	729741	31.0	952.9	7.56	475.8	24.98	116	0.23	5.39	0.00	0.01	52.8	7.84	<0.02	0.00	14.91
Sonou Djog	14853	343038	728033	30.2	865.0	7.28	414.8	56.00	92	0.31	6.20	0.50	0.00	43.2	9.00	<0.02	0.39	3.10
Cci		343036	727833	33.7	839.0	7.17	317	94.00	28	0.19	3.70	0.00	0.00	31	16.00	<0.02	0.40	18.05
Klikamé	13126	290471	691184	31.0	421.0	6.79	207.4	25.82	57.22	0.16	3.10	0.00	0.00	6.27	3.37	<0.02	2.29	14.24
Anonkui	14528	298282	690105	32.1	1309.0	7.2	390.4	197.23	131.9	0.45	0.80	0.00	0.00	58.59	9.59	<0.02	4.38	22.16
Sokomé	13927	331986	703202	31.4	3213.2	7.02	530.2	737.56	380	0.13	0.82	1.48	0.17	144	10.54	<0.02	0.44	13.45
Dagbati	13170	333776	716621	30.8	2105.2	7.89	658.8	777.56	212	0.32	0.22	0.10	0.03	132	38.24	<0.02	0.00	18.57
Logové	14868	343624	711112	31.4	1529.0	7.35	878.4	293.66	144	0.23	0.98	0.42	0.00	67.2	35.60	<0.02	0.00	31.51
Hélépémé	14568	340183	721026	30.5	2603.8	7.51	445.3	722.93	268	0.21	0.22	0.05	0.02	156	13.00	<0.02	3.18	16.68
Nyékonakpoe	14567	339468	721521	30.6	1551.2	7.66	610	16.20	176	0.50	0.22	0.02	0.01	52.8	8.11	<0.02	1.67	16.19
Ouatchidomé	14560	338700	720211	31.1	2991.6	7.6	488	961.95	316	0.14	0.22	0.00	0.05	164.64	39.19	<0.02	0.51	15.54
DjokotoF1	14571	337811	715148	31.7	2204.9	7.71	447.74	476.59	216	0.36	16.28	0.00	0.01	115.2	24.73	<0.02	3.14	18.95
KpessouM493	13932	342769	721621	30.1	1795.0	8.35	463.6	211.97	162	0.57	7.58	0.47	0.02	83	7.01	<0.02	1.30	26.65
Mome Hagou	14610	345826	728846	31.5	2170.0	7.9	833	87.00	219	0.30	19.80	0.00	0.01	79.5	8.00	<0.02	0.19	42.00
Klo-Gagnon	14592	342712	711178	30.4	1561.0	7.85	746	61.00	197	0.30	20.00	0.00	0.11	59	8.00	<0.02	0.51	36.00

Table 3. Minimum, maximum, mean, and standard error values of the physico-chemical parameters measured.

Parameters	T°C	EC (µS/cm)	pH	Ca²⁺ mg/L	Mg²⁺ mg/L	Na⁺ mg/L	K⁺ mg/L	HCO₃ mg/L	SiO₂ mg/L	SO₄²⁻ mg/L	Cl⁻ mg/L	NO₃ mg/L	F mg/L
Minimum	29.0	243.8	6.6	28.0	4.8	11.7	1.2	134.2	3.1	5.7	20	0.20	0.1
Maximum	34.0	3213.2	8.5	380.0	164.6	289.0	39.2	878.4	42.0	962	450.5	20.0	0.7
Mean	31.4	1003.5	7.7	103.1	38.8	64.6	10.5	388.6	18.6	97.1	100.51	3.2	0.3
Standard error	1.3	610.4	0.4	67.0	37.1	59.0	8	143.6	6.5	205.4	79.6	4.1	0.2

Parameters	NO₂⁻ mg/L	Turbidity NTU	Fe mg/L	Cd²⁺ mg/L	NH₄⁺ mg/L
Minimum	0.0	0.03	0.0	<0.02	0.0
Maximum	1.9	65.6	4.4	<0.02	1.7
Moyenne	0.2	2	0.8	<0.02	0.2
Standard error	0.4	8.5	0.9	<0.02	0.4

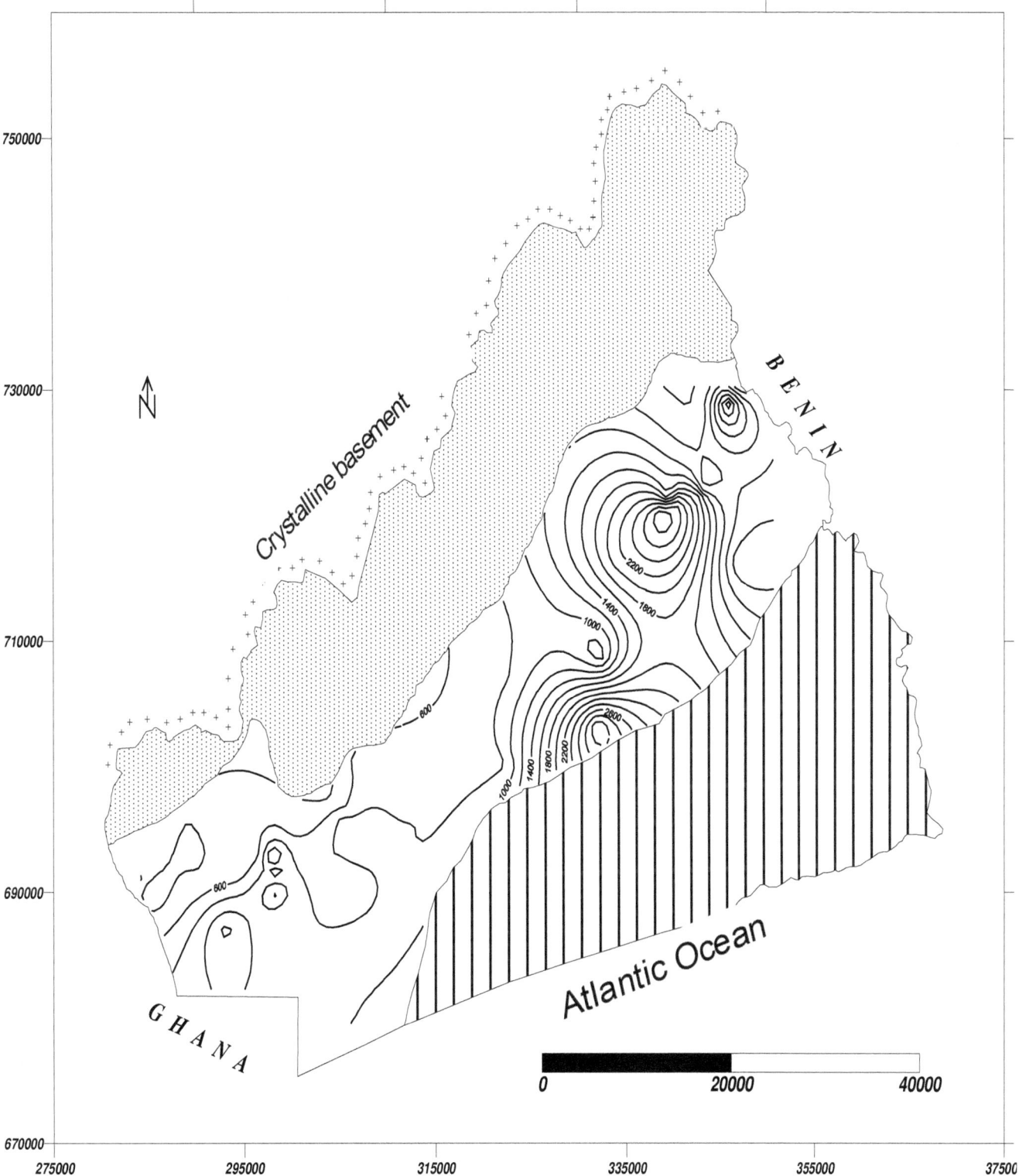

Figure 5. Spatial distribution of conductivity.

(i) The calcic and magnesian bicarbonated facies that evolved toward the sodic and potassic bicarbonated facies in the South-West.

(ii) The calcic and magnesian sulphated facies. This

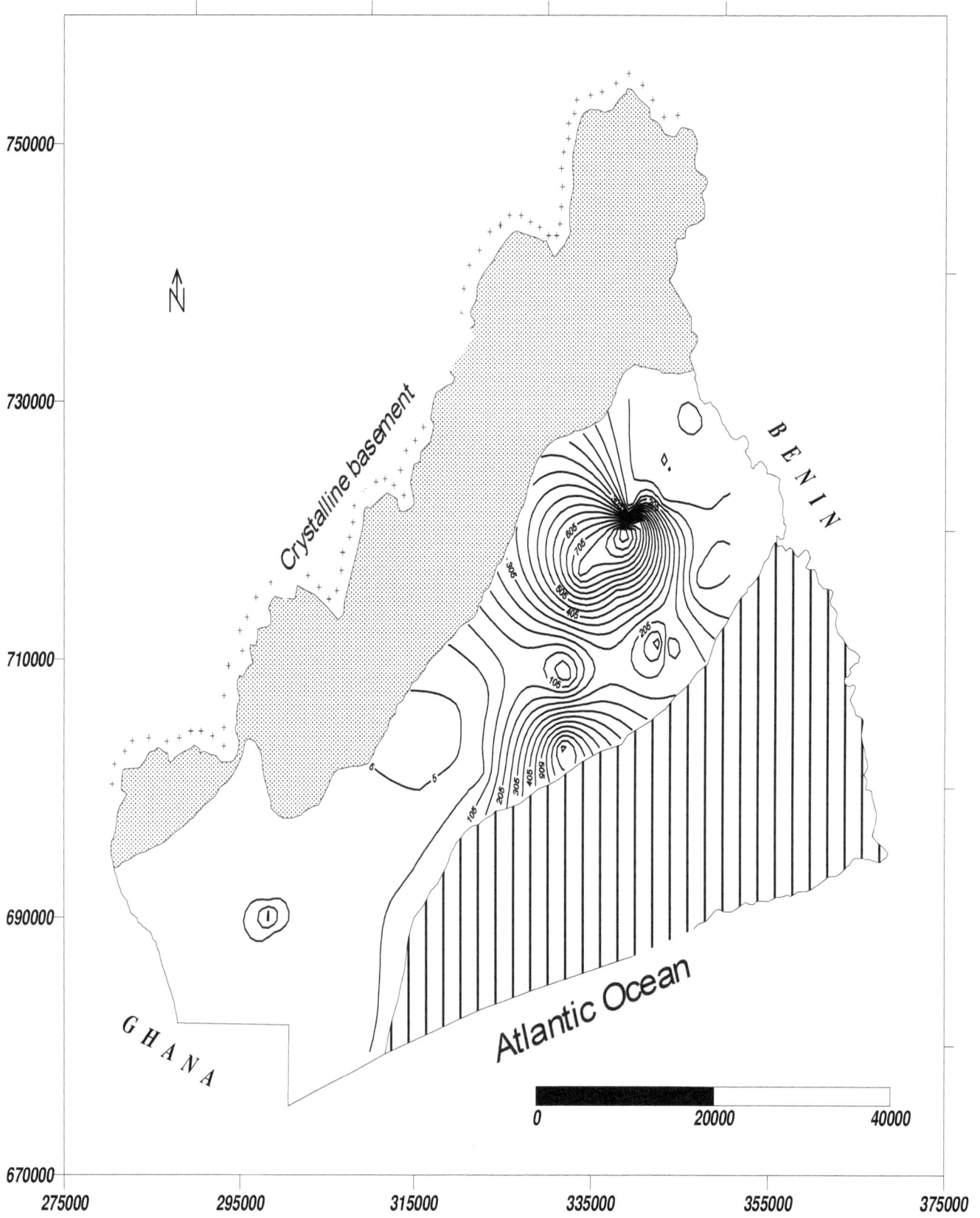

Figure 6. Spatial distribution of sulphate.

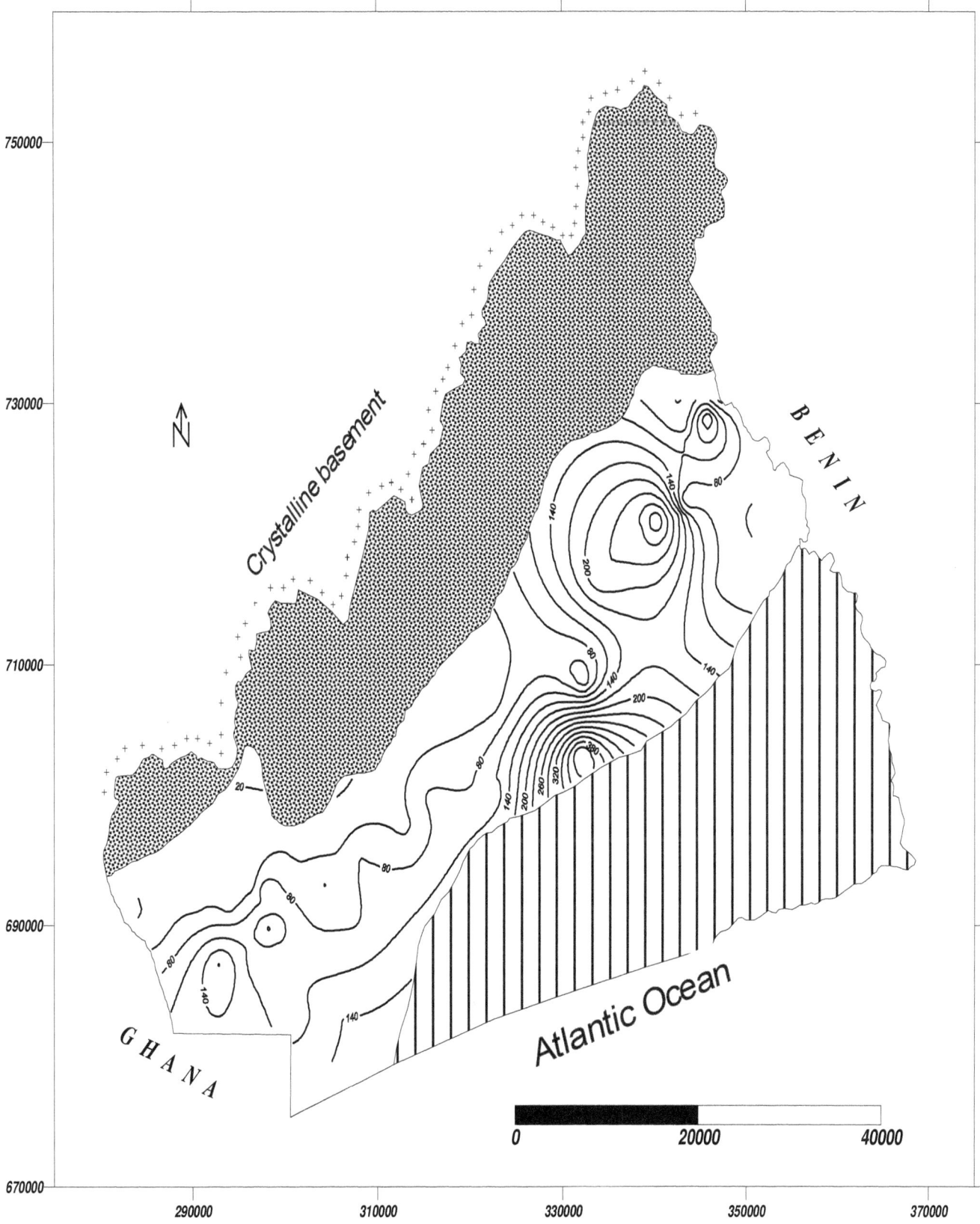

Figure 7. Spatial distribution of chloride.

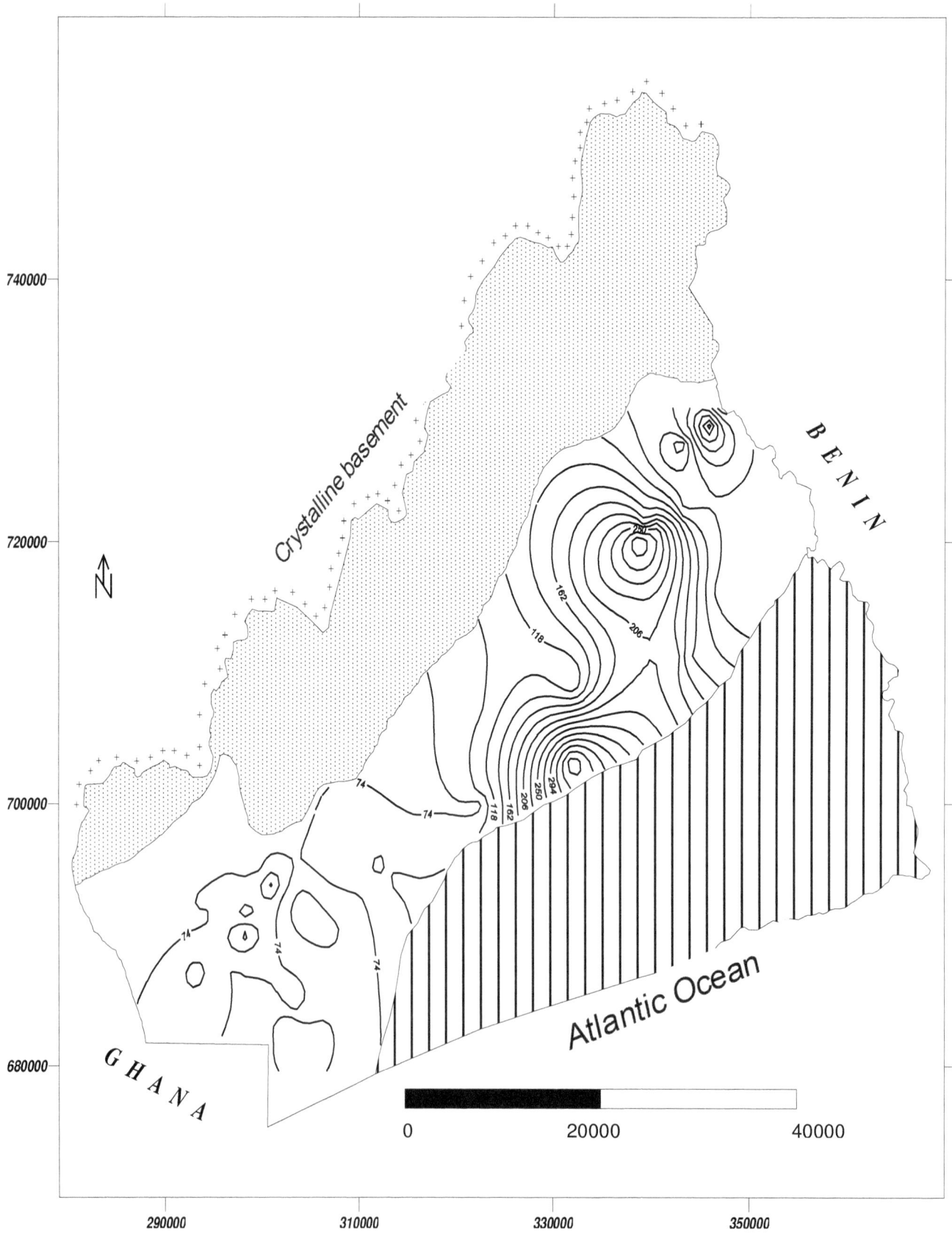

Figure 8. Spatial distribution of calcium.

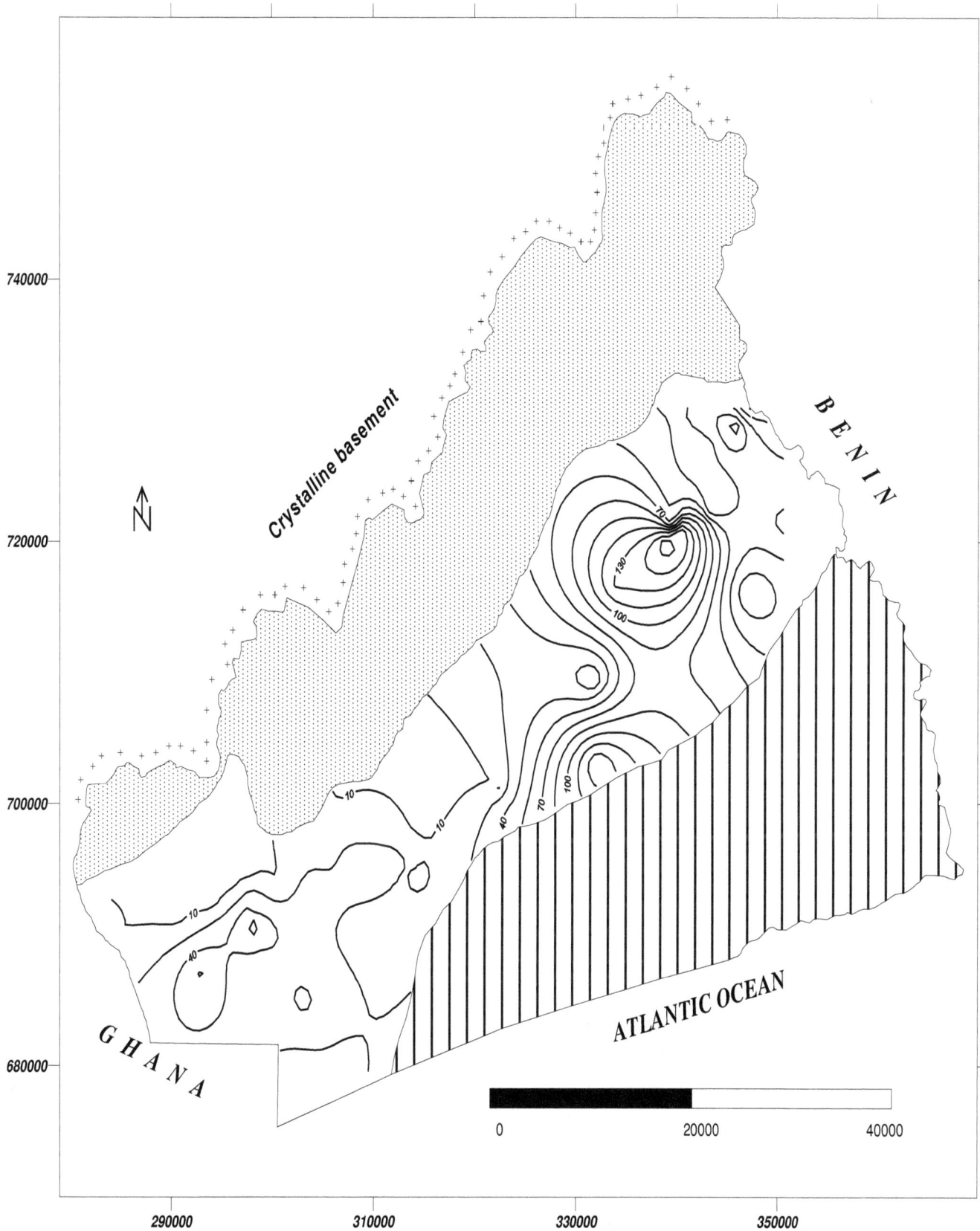

Figure 9. Spatial distribution of magnesium.

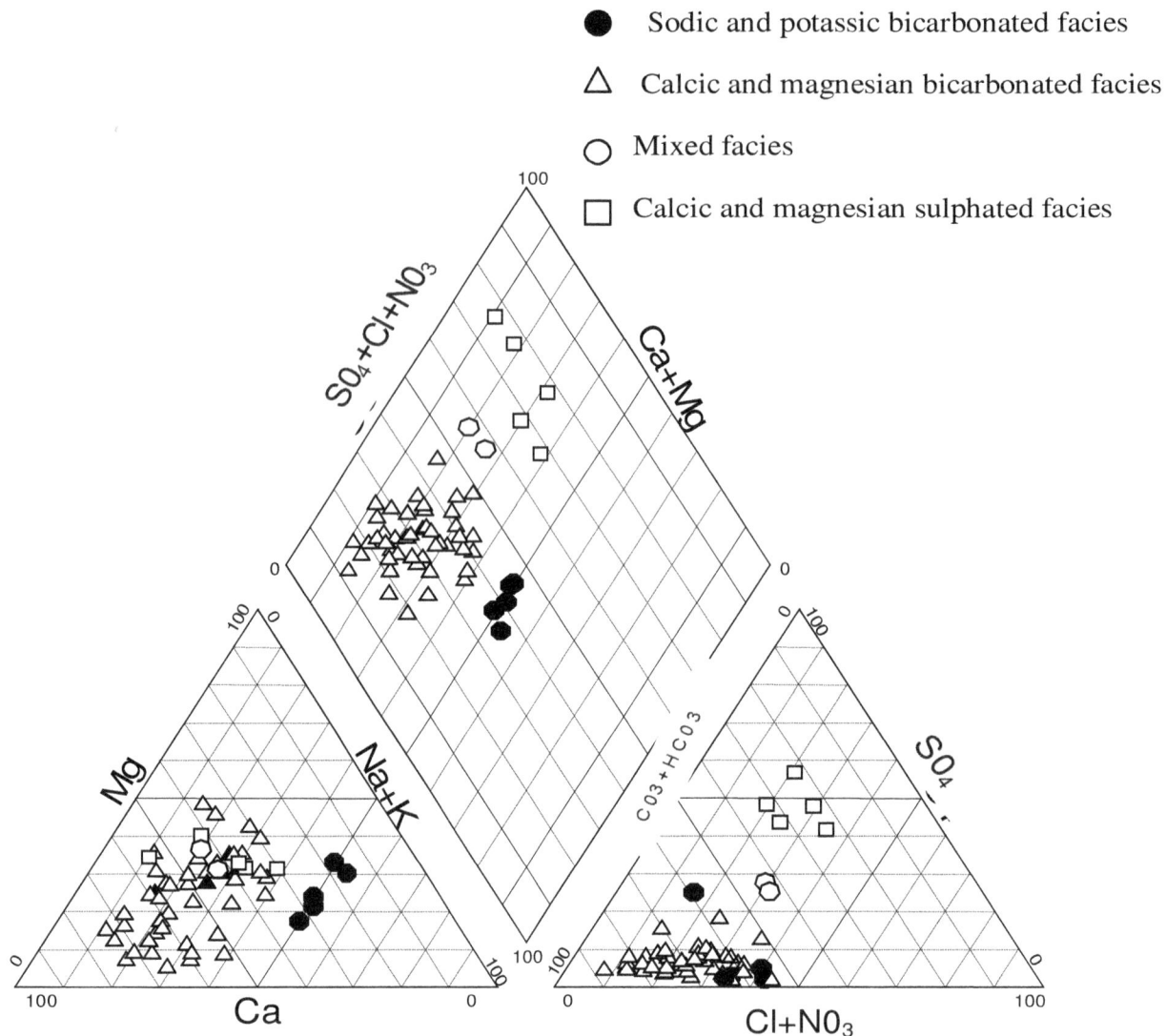

Figure 10. Piper's diagram: representation of hydrochemical pattern of Paleocene groundwater aquifer.

facies is found exclusively in the East of the study area (Figure 11).

(iii) The mixed anionic and cationic facies with the exception of a single point oriented toward the calcic pole. Projection of these points in the lozenge shows that these are calcic and magnesian and chloride-sulphated waters.

To explain the presence of sodic and potassic bicarbonated facies in the South-West of the study area, correlations between the major cations such as Ca2+, Mg2+, Na+, and K+ and well depth were analyzed. These indicated that calcium concentrations decreased with aquifer depth. Similarly magnesium concentrations decreased with aquifer depth but to a lesser extent (Figure 12).

Conversely, sodium and potassium concentrations tended to increase with depth (Figure 12). Long periods of residence and the associated water-rock interactions in the Paleocene reservoir could contribute to cationic substitution reactions. This is a typical exchange phenomenon that takes place at the contact with clay minerals (illite, smectite) present in the aquifer (Drever, 1997). Decreasing Ca2+ concentrations associated with increasing Na+ and K+ is observed by Edmunds et al. (1987) and Edmunds and Smedley (2000) in some British aquifers. For the Paleocene aquifer, this phenomenon can be considered a qualitative indicator of the age of thewaters. These results are in agreement with those obtained by Akiti (1980), and AKouvi (2000) which show, using isotopic techniques, the aging from North to South of the Paleocene water.

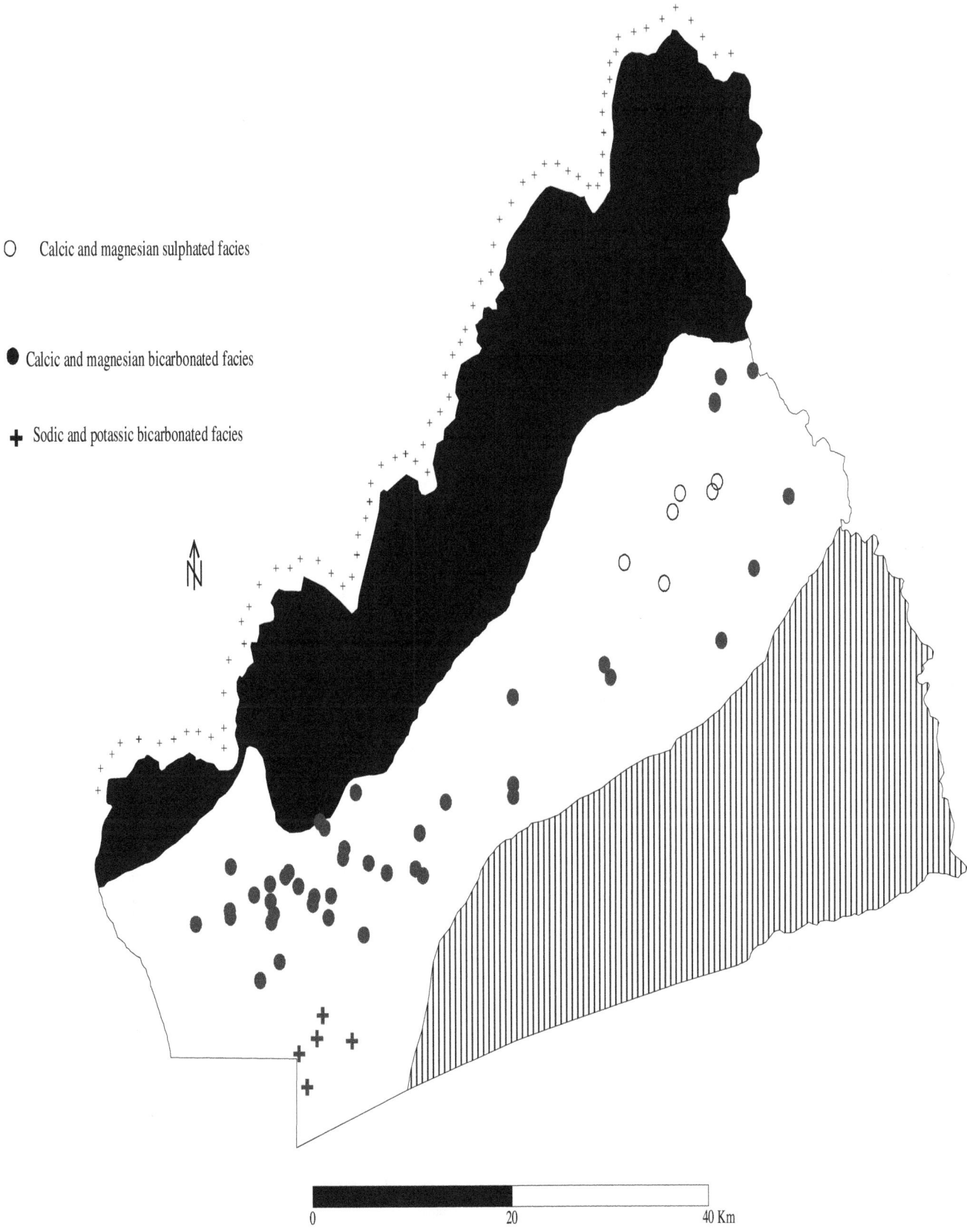

Figure 11. Spatial distribution of hydrochemical facies.

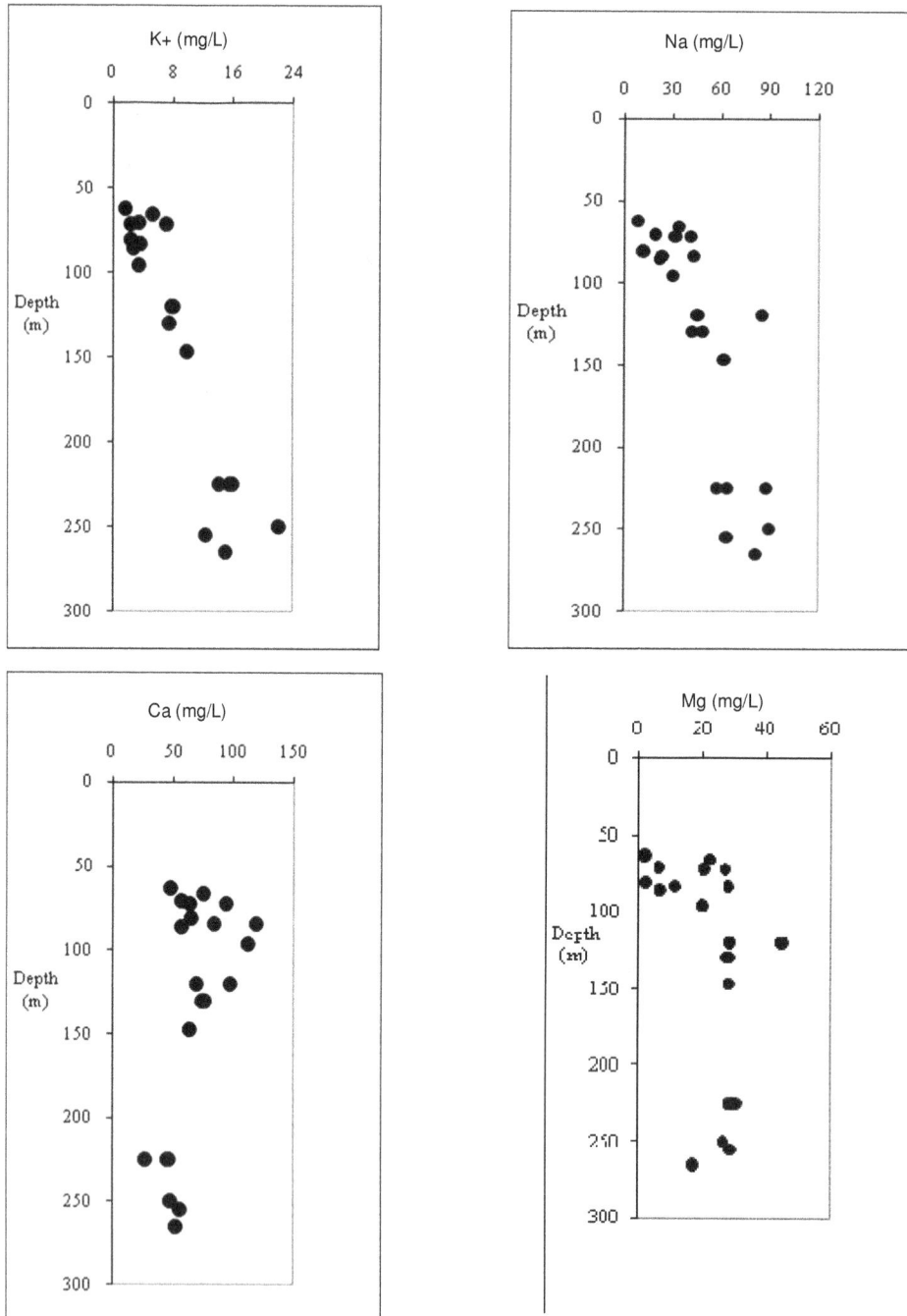

Figure 12. Distribution of cation (K^+, Na^+, Ca^{2+}, and Mg^{2+}) concentrations with depth in the Western part of the aquifer.

It should also be added that for Akiti (1980), the presence of high contents of chloride, sodium, calcium and magnesium in the water suggests a mixing of Paleocene water and that contained in the gneissic substratum. This is not the case in the Togolese part of the aquifer where the substratum is clayey.

Principal component data analysis

Principal Component Analysis (PCA) is a mathematical technique that allows the reduction of a complex system of correlations to a smaller number of dimensions. This study employed this multidimensional analysis because

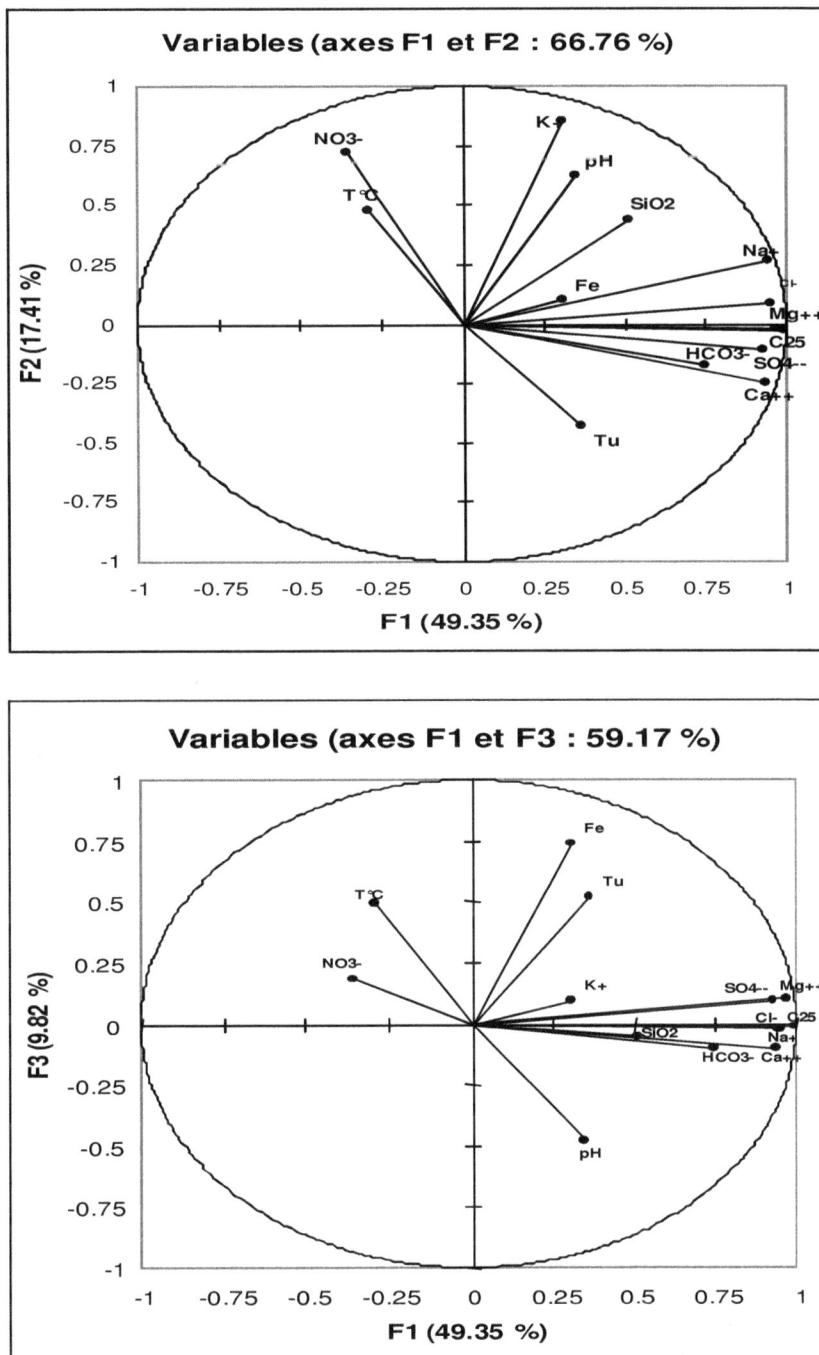

Figure 13. Diagram of variables: plans I-II and I-III (14 parameters).

of the great number of samples and the many chemical parameters considered. PCA considers all the variables and their relationships simultaneously, revealing relationships that are not always evident in the usual graphical representations or simple correlations (Lefebvre, 1980). The program used for data analyses was Xlstat (Fahmy, 2006). The variables considered in this study are

temperature, pH, electrical conductivity, turbidity, major and minor ions; that is a total of fourteen parameters. Variability percentage (67%) for the first two factors (F1 and F2) was not high enough. To avoid a misinterpretation of the graphs we posted the F1 and F3 axes (Figure 13).

Axis F1 (49% of the variance) is characterized by

Table 4. Correlation coefficients of chemical parameters (in bold type the significant correlation coefficients).

Variables	T°C	pH	C25°	HCO_3^-	Cl^-	SO_4^{--}	Ca^{++}	Mg^{++}	Na^+	K^+	SiO_2	NO_3^-
T°C	1											
pH	-0.01	1										
C25°	-0.28	0.32	1									
HCO_3^-	-0.17	0.13	0.732	1								
Cl^-	-0.19	0.35	0.94	0.73	1							
SO_4^{--}	-0.31	0.24	0.94	0.56	0.81	1						
Ca^{++}	-0.46	0.22	0.95	0.73	0.87	0.92	1					
Mg^{++}	-0.19	0.27	0.97	0.71	0.89	0.94	0.89	1				
Na^+	-0.15	0.47	0.93	0.61	0.95	0.85	0.83	0.88	1			
K^+	0.28	0.56	0.29	-0.04	0.34	0.25	0.06	0.31	0.54	1		
SiO_2	-0.08	0.36	0.43	0.38	0.56	0.30	0.31	0.41	0.55	0.42	1	
NO_3^-	0.33	0.12	0.35	-0.45	0.30	0.33	0.48	0.35	0.13	0.52	0.13	1

conductivity, chloride, sulphate, calcium, sodium, bicarbonate. This axis showed a mineralization that is mainly lithological in origin. The elements correlated with this axis are indicators of carbonated and evaporitic environments. This axis therefore indicates the water time residence in the aquifer.

Axis F2 (17% of the variance) is characterized by nitrates, potassium, pH, turbidity and to a lesser extent silica. This axis is defined by pollution elements.

Finally, axis F3 (10% of the variance) is defined by temperature, iron, turbidity and to a lesser extent nitrate. Axis F3 showed that sample turbidity was due to iron concentration. As with axis F2, this axis is characteristic of pollution elements.

Table 4 shows the significant correlation coefficients for 59 samples. One notes that water mineralization represented by electrical conductivity, is determined mainly by major ions ($r > 0.7$). The contribution of nitrates and potassium is low ($r < 0.3$).

Calcium and magnesium were well correlated with bicarbonates. This indicated the calcareo-dolomitic nature of the matrix. A spot test performed by Monciardini et al. (1986) on organogenic limestones that constitute the matrix shows that they are composed of 81% calcite and 9% dolomite. Silica was also correlated with potassium and sodium ($r = 0.55$ and 0.41 respectively) indicating the presence of clay minerals in the carbonate matrix.

Conclusion

The results of this study have allowed bringing out the main chemical characteristics of the Paleocene aquifer. Chemical parameters and Principal Component Analysis indicate that groundwater composition of the Paleocene aquifer is related to the geological nature of the formations traversed.

The confinement of the deep groundwater in the Paleocene aquifer, sometimes in contact with distinctive rocks (notably gypseous) confers on them in many places specific hydrochemical characteristics

Cation distribution with depth showed calcium and magnesium depletion of the waters to the detriment of sodium and potassium ions as depth increases. This evolution can be considered an indicator of residencetime of water in the reservoir.

REFERENCES

AFNOR (1986). Water, testing methods. Afnor Ed. 624 p.

Akiti TT (1980). Geochemical and isotopic study of some aquifers in Ghana: Accra plain gneiss, Limestone in the South-East of the Volta, the upper region granite. Doctoral thesis, University of Paris (ORSAY), 232 pp.

Akouvi A (2000). Geochemical and hydrogeological study of the groundwaters of tropical coastal basin. Impact on the management, protection and preservation of water resources in Togo (West Africa). Doctoral Thesis, Paris VI University, 163 p.

Atteia O (2005). Chemistry and pollution of underground waters. Tec and Do Ed., Lavoisier. 398 p.

Banton O, Bangoy LM (1997). Hydrogeology, environmental multiscience of underground waters. Québec University/AUPEL Press, 460 p.

Bawa ML, Djanéyé-Boundjou G, Boukari Y (2006). Characterization of two industrial effluents in Togo: environmental impact study. Afr. Sci., 2(1): 57-68.

BCEOM/BRGM (1983). Lome mater Supply. Groundwater resource. Synthesis of hydrogeological data. Report 83. AGE 040 DGH/BRGM. 37 pp.

Bourgeois M (1981). Water resources planning maps of Ivory Coast, Ghana, Togo, Benin and Cameroun. BRGM bulletin 2nd Series, Section III, n° 4, 1980- 81. pp. 396-379.

da Costa YDP (2005). Biostratigraphy and Paleogeography of the Togo sedimentary basin. Doctorat Thesis, University of Lome. 2 volume 476 pp.

Drever JI (1997). The geochemistry of natural waters: surface and groundwater environments. Third edition. Prentice Hall. Upper Saddle River, NJ 07458. 436 p.

Edmunds WM, Smedley PL (2000). Residence time indicators in groundwater: the East Midland Triassic sandstone aquifer. Appl. Geochm., 15: 737-752.

Edmunds WM, Cook JM, Darling WG, Kinniburg DG, Miles DL, Bath AH, Morgan-Jones M, Andrew JN (1987). Baseline geochemical conditions in the chalk aquifer, Bershire, UK: a basis for groundwater quality management. Appl. Geochm., 2: 251-274.

Fahmy T (2006).www.xlstat.com. Addinsoft (1993–2006).

Gnandi K (1998). Cadmium and other inorganic pollutants in the soils and sediments of the coastal region: a geochemical study. Doctoral Thesis, Friedrich-Alexanded'Erlangen-Nuremberg University, RFA.

Gnazou MDT (2008). Hydrodynamic, hydrogeochemcal, isotopic study and modeling of the aquifer of the coastal sedimentary basin of Togo. Doctoral Thesis, Univ. Of Lomé, 204p + annexes.

Gnazou MDT, Bawa LM, Banton O, Djaneye-Boundjou G (2009). Hydrodynamic characterization of the Paleocene aquifer in the coastal sedimentary basin of Togo. Afr. J. Environ. Sci. Technol., 3(5): 141-148.

Johnson AKC (1987). The phosphatic coastal basin of Togo (Maastrichtien-Eocene). Doctoral thesis, University of Bourgogne (France) and University of Benin (Togo). 360 pp.

Lefebvre JC (1980). Introduction to multidimensionnal statistical analysis. Masson, Paris, 2^e Edition, 259p.

Monciardini C, Tchota K, Slansky M, Podevin G, Marteau P, Le Nidre Y, Farjanel G, Chateauneuf JJ, Castaing C, Carbonel G, Blondeau A, Andreiff P (1986). Geological synthesis of the cretaceous-tertiary basin. Prospection for peat, lignite, coal and other industrial materials. Unpublished BRGM report 86 TGO 056 GEO

UNDP (1975). Prospection for groundwater in the coastal zone (TOGO): conclusions and recommendations. DP/UN/TOG-70-511/1. United Nations, New York, 83 pp.

UNDP (1982). Water development Strategy, resources and water demand planning. Central Hydraulic Laboratory of France. 11 notes and 11 plates.

Rodler J, Bazin C, Broutin J-P, Chambon P, Champsaur H, Rodi L (2005). Water analysis : natural waters, residual waters, sea waters. 8th edition Dunod, Paris, 1383 p.

Simler R (2005). Logiciel d'hydrochimie/Diagrammes. http//www.lha.univ-avignon.fr/logiciel.htm

Sylvain JP, Aregba A, Assih-Edeou P, Castaing C, Chevremont J, Collart J, Monciardini C, Marteau P, Ouasane I, Tchota K (1986). Explanatory note of the 1/200000 geological map. Lome sheet. 1st edition, DGMG/BNRM report N° 5. 64 pp.

Tchangbédji G, Djételi G, Kili K, Tchassanti OA (2003). Extraction of some metallic elements in natural phosphates by granulometry and demagnetization: case of the Hahotoe phosphates (Togo). J. Rech. Sc. Univ. Bénin (Togo) pp. 111-120.

4

Geoelectric evaluation of groundwater prospect within Zion Estate, Akure, Southwest, Nigeria

Akintorinwa, O. J.* and Olowolafe T. S.

Department of Applied Geophysics, Federal University of Technology, Akure, Nigeria

Electrical Resistivity method involving Vertical Electrical Sounding (VES) have been used to evaluate the groundwater potential and aquifer protective capacity of the overburden units within Zion Estate area. Sixty-eight Vertical Electrical Soundings were carried out using Schlumberger configuration with AB/2 varying from 1 to 65 m. The generated geoelectric section from the interpretation of the sounding curves revealed four layers - the top soil, the weathered layer, the partially weathered/fractured basement and the fresh basement. The weathered layers constitute the major aquifer unit in the area but are generally thin. The groundwater potential map revealed that about 85% of the study area falls within the low groundwater potential rating while about 10% constitutes the medium groundwater potential rating and the remaining 5% constitutes high groundwater potential rating. Hence the groundwater potential of the area is generally rated to be low. The overburden protective capacity map of the study area shows that about 75% of the area falls within the poor overburden protective capacity, while the remaining 25% constitutes the moderate/weak protective capacity rating. This suggests that the area is characterized by low longitudinal conductance which informed weak protective capacity rating of the area. Therefore the study area can be vulnerable to pollution from contaminant sources such as industrial waste, septic tanks, underground petroleum storage tanks and landfills when located close to the study area.

Key words: Groundwater, geoelectric, sounding, Schlumberger.

INTRODUCTION

In many developed and developing countries there is not only a heavy reliance on ground water as a primary source for the supply of drinking water but also as a source of water for both agricultural and industrial uses. The reliance on groundwater is such that it is necessary to ensure that there are significant quantities of water. The continuous increase in population growth and of large industrial and agricultural complexes has led to large demand for water within Akure and its environs. There is inadequate supply of water at Zion estate as water from hand-dug wells, the only existing borehole and a stream can no longer serve the general needs as a result of their seasonal variations and hence, there is need for proper geophysical investigation of the area. The present study aims to evaluate the groundwater potential of the study area for proper groundwater

development. Groundwater resources account for about 98% of the world's fresh water needs and is fairly evenly distributed throughout the world. Groundwater provides a reasonable constant supply which is not completely susceptible to drying up under natural conditions unlike surface water (Korzun and Komitel, 1978).

In the basement terrain, groundwater development may be primarily restricted to the aquifer in the weathered overburden or complemented by fractured crystalline rocks which are mainly of Precambrian age (Olayinka and Olorunfemi, 1992; Wright, 1992). The concealed basement rocks may contain faulted areas, incipient joints and fractured systems derived from earlier tectonic events. The detection of and delineation of these hydrogeologic structures may facilitate the location of groundwater prospect zones in typical basement settings (Omosuyi et al., 2003). Fractured crystalline bedrock remains good sources for drinkable water but siting of highly productive wells in these rock units remains a challenging and expensive task because fractured

*Corresponding author. E-mail: orllyola@yahoo.com.

developments on the regional scale are both heterogeneous and anisotropic. However, fractured viable aquifers wholly within the fractured bedrock are of rare occurrence because of the typically low storativity of fracture systems (Clark, 1985).

This study presents the use of an electrical resistivity method as a geophysical tool for the delineation of basement aquifers which are potential reservoirs for groundwater. Electrical resistivity survey is widely employed in the delineation of basement regolith and location of fissured media and associated zones of deep weathering in crystalline terrains (Benson and Jones, 1988; Hazell et al., 1988; Olayinka, 1990; Olayinka et al., 2004). Drilling programmes for groundwater development in areas of basement terrain are generally preceded by detailed geophysical investigations.

Description of the study area

Figure 1 shows the location of the study area along Ilesha-Akure expressway, Akure, southwestern, Nigeria. It lies within latitude 7° 18' 17" N and 7° 18' 59" N and longitude 5° 08' 28" E and 5° 08' 54" E. The study area is about 32 km^2. The terrain is gently undulating with topographic elevation ranging between 330 and 376 m above sea level. The area is traversed by Ala stream which flows approximately in the east-west direction.

The area is underlain by the Precambrian basement complex of southwestern Nigeria (Rahaman, 1989). The lithologic units identified in the area include migmatite-gneiss and charnokites. The charnokites are the predominant rock units in the study area, covering more than half of the area and underlies the southern and north-eastern parts of the area. The other parts of the area are underlain by the magmatite-gneiss (Figure 2). The study area is situated within the tropical rain forest region, with a climate characterized by wet and dry seasons. The wet season usually occurs from March to October and is dominated by heavy rainfall. The dry season occurs from November to March when the area is under the influence of north-easterly winds. The annual rainfall ranges between 1000 and 1500 mm. The annual temperature is from 18 to 34°C with relatively high humidity during the wet season and low humidity during the dry season (Iloeje, 1981).

MATERIALS AND METHODS

The Vertical Electrical Sounding (VES) using Schlumberger configuration was adopted for the study. A total of 68 stations were occupied across the area (Figure 1). The electrode spacing (AB/2) was varied from 1 to 65 m. The R50 DC resistivity meter was used for the data acquisition and the position of the occupied sounding stations in Universal Traverse Mercator (UTM) was recorded using the GARMIN 12 channel personnel navigation Geographic Positioning System (GPS) unit. The field curves were generated by plotting of the apparent resistivity values against the electrode spacing (AB/2). The curves were interpreted using the partial curve

matching technique. The geoelectric parameters obtained from manual interpretation of VES data were refined using the software algorithm RESIST version 1.0 (Van der Velpen, 1988). Second order geoelectric parameters called Dar Zarrouk parameters were determined from the iterated geoelectric parameters (Mallet, 1947). The second order parameter of interest in this study is the longitudinal conductance (S_l). This second order parameters are derived using equations developed by Zohdy et al. (1974) such as:

For n layers, the total longitudinal unit conductance is given by:

$$S = \sum_{i=1}^{n} \frac{h_i}{\rho_i} = \frac{h_1}{\rho_1} + \frac{h_2}{\rho_2} + \frac{h_3}{\rho_3} + \dots \dots \frac{h_n}{\rho_n} \qquad (1)$$

where: h_i is the layer thickness, ρ_i is layer resistivity and the number of layers from the surface to the top of aquifer varies from i = 1 to n.

RESULTS AND DISCUSSION

The results of this work are presented as sounding curves, geoelectric sections, maps, charts and tables. Curve types identified ranges A, H, K, KH, QH, and HK (Table 1). The predominant curve type is the A curve type having percentage frequency of 41.2%, H curve type has 40% and KH curve type has 8.8% of the total occurrence while HKH, QH, and K have 7.4, 1.5 and 1.5% respectively (Figure 3). Typical curve types are as shown in Figure 4.

Geoelectric sequences

Figures 5a to 5d show four geoelectric sections generated in the N-S, W-E, SW-NE and NW-SE directions respectively. The geoelectric sections show the variations of resistivity and thickness values of layers within the depth penetrated in the study area at the indicated VES stations. The geoelectric sections revealed four subsurface geologic/geoelectric layers consisting of topsoil, weathered layer, partly weathered/fractured basement and fresh basement. The topsoil is relatively thin and the thickness ranges between 0.3 and 7.6 m while the resistivity values range from 27 to 1829 Ωm. This indicates that the predominant composition of the topsoil is clay, sandy clay, clayey sand and bedrock which were outcropping in some places and leads to high values of topsoil resistivity in some places. The weathered layer thickness ranges from 0.6 to 12.4 m and the resistivity values range from 37 to 848 Ωm, which indicate that the weathered layer material composition is of clay, sandy clay, clayey sand and laterite. The partially weathered/fractured basement resistivity values range between 51 and 78 Ωm which either indicates high degree of fracturing or water saturation. It is of infinite thickness where it is the last observable layer and where it is underlain by fresh basement, the thickness ranges from 6.6 to 32.9 m. The fresh basement is the last

Figure 1. Geophysical data acquisition map of the study area.

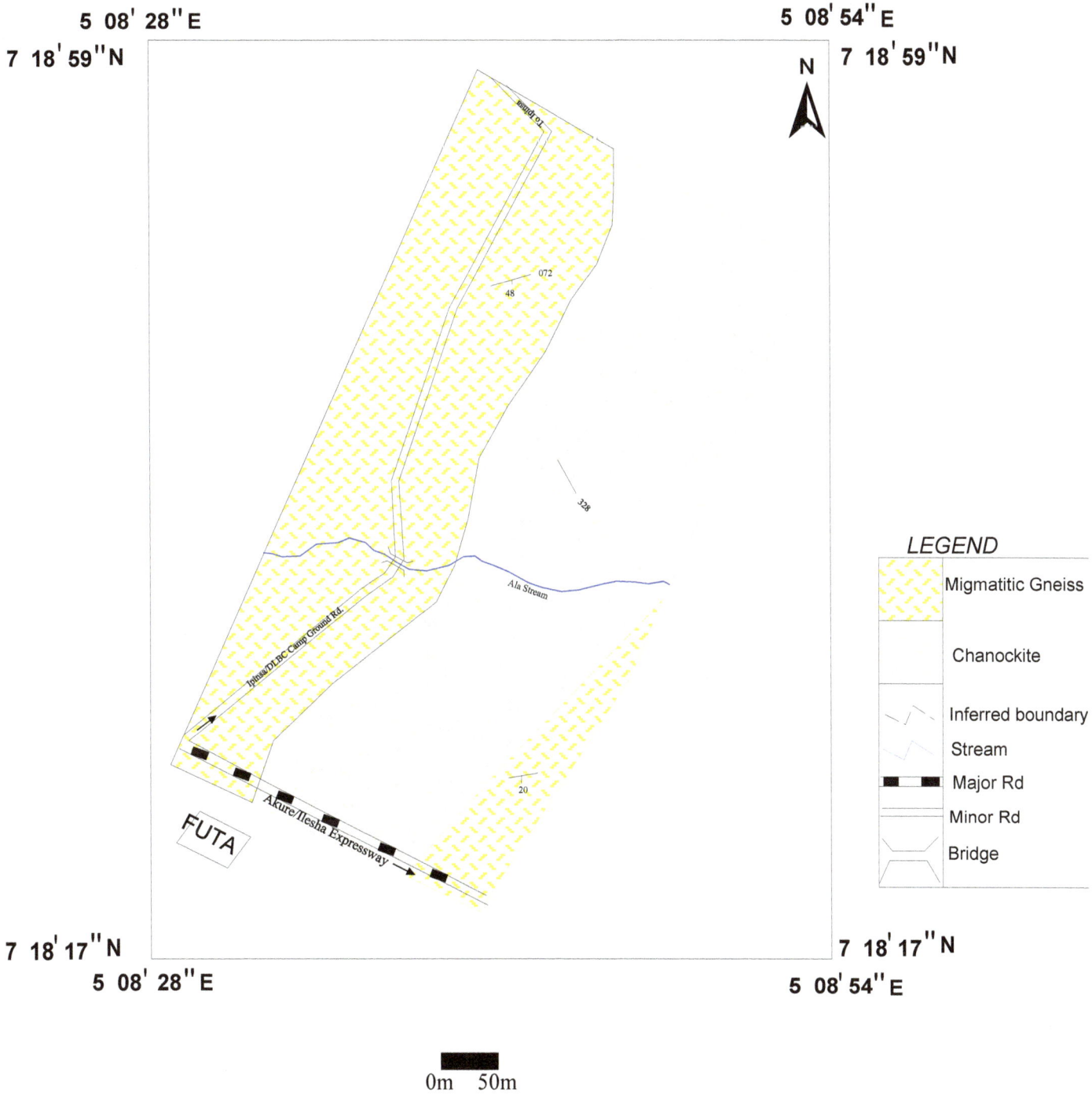

Figure 2. Geological map of the study area.

observable layer with resistivity values ranging between 950 Ωm and infinity. At most VES station, it is infinitely resistive because of its crystalline nature.

Isopach and isoresistivity maps of the topsoil

The thickness and the resistivity maps of the topsoil are presented in Figures 6 and 7 respectively. The isopach map of the topsoil reveals that the thickness distribution of the topsoil within the study area ranges from 0.3 to 5.1 m but generally less than 3.5 m. This indicates that the topsoils are generally thin with no hydrogeological significance. The isoresistivity map of the topsoil is shown in Figure 7. It shows that the resistivity values range from 17 to 758 Ωm (mostly < 500 Ωm). This reveals the high

Table 1. Summary of VES interpretation results.

VES	Resistivity (Ωm)					Thickness (m)				Curve type	Longitudinal conductance (S) (mhos)
	ρ_1	ρ_2	ρ_3	ρ_4	ρ_5	h_1	h_2	h_3	h_4		
1	202	91	2687			1.0	8.3			H	0.10
2	539	145	1049	10	∞	0.5	0.8	2.0	0.9	HKH	0.10
3	550	234	992			3.1	38.6			H	0.17
4	658	483	27	∞		4.3	0.9	5.3		QH	0.20
5	236	65	505	62	∞	0.5	0.8	2.1	3.2	HKH	0.07
6	517	211	1130			0.8	7.2			H	0.04
7	231	69	∞			0.8	1.6			H	0.03
8	151	1291	2313			1.8	0.7			A	0.01
9	225	64	3048			0.7	1.7			H	0.03
10	93	362	∞			3.5	6.3			A	0.06
11	162	72	821			0.9	3.2			H	0.05
12	290	412	101	950		0.9	1.7	16.8		KH	0.17
13	347	219	414			1.3	19.8			H	0..09
14	520	91	208			1.0	1.6			H	0.02
15	256	69	4452			2.0	10.6			H	0.16
16	303	36	5434			3.4	4.2			H	0.13
17	165	578	1470			1.8	5.8			A	0.02
18	239	78	1463			0.5	2.3			H	0.03
19	145	508	1800			2.1	6.3			A	0.03
20	207	1403	2134			2.1	1.5			A	0.01
21	588	277	5685			0.6	1.8			H	0.01
22	242	1718	4209			1.6	0.5			A	0.01
23	346	471	∞			0.6	8.2			A	0.02
24	380	475	299	1850		1.6	3.6	9.0		KH	0.04
25	109	36	2559			0.9	6.5			H	0.19
26	284	58	749			2.0	5.1			H	0.09
27	17	1250	6984			0.3	0.1			A	0.02
28	350	128	4751			0.8	1.2			H	0.01
29	340	221	1855			0.8	1.4			H	0.01
30	286	411	1270			1.6	1.0			A	0.01
31	758	201	3065	360	∞	0.6	1.1	5.7	13.3	HKH	0.05
32	88	274	35	∞		0.3	2.3	5.4		KH	0.17
33	226	52	190	36	594	0.5	0.8	2.5	5.5	HKH	0.18
34	41	105	244			2.0	1.3			A	0.06
35	145	508	840			1.9	1.5			A	0.02
36	174	845	439			0.8	11.8			K	0.02
37	188	1829	165	625		0.5	1.1	3.1		KH	0.02
38	123	1711	∞			1.8	1.2			A	0.02
39	150	96	3513			2.6	2.1			H	0.04
40	300	163	7133			0.7	3.8			H	0.03
41	269	47	1753			0.9	1.4			H	0.04
42	199	390	4042			3.8	12.4			A	0.05
43	121	299	78	637		0.7	6.9	13.6		KH	0.20
44	208	848	78	2762		0.9	2.9	32.9		KH	0.43
45	298	37	543	51	3291	0.4	0.4	1.3	6.6	HKH	0.14
46	99	457	∞			5.1	2.5			A	0.06
47	133	292	2305			1.8	2.4			A	0.02
48	127	254	6732			4.2	4.6			A	0.05
49	99	597	∞			2.9	1.9			A	0.03
50	251	505	∞			1.7	3.0			A	0.01

Table 1. Contd.

51	219	663	∞	1.7	3.7	A	0.01
52	36	360	∞	1.9	0.4	A	0.05
53	282	39	∞	0.5	1.2	H	0.03
54	27	307	∞	0.3	7.3	A	0.04
55	165	184	2000	2.5	3.0	A	0.03
56	103	124	2153	1.7	3.2	A	0.04
57	440	132	3400	0.9	3.7	H	0.03
58	133	588	∞	1.1	6.0	A	0.02
59	215	437	5683	2.9	2.0	A	0.02
60	223	1102	8504	1.0	6.9	A	0.01
61	611	2406	∞	0.8	11.5	A	0.01
62	373	69	1680	0.8	2.9	H	0.04
63	133	293	1335	2.8	2.8	A	0.03
64	309	53	1478	0.7	3.5	H	0.07
65	210	73	7148	1.6	5.5	H	0.08
66	730	213	1165	0.6	4.9	H	0.02
67	292	64	1777	0.6	2.2	H	0.04
68	71	23	5693	1.3	1.7	H	0.09

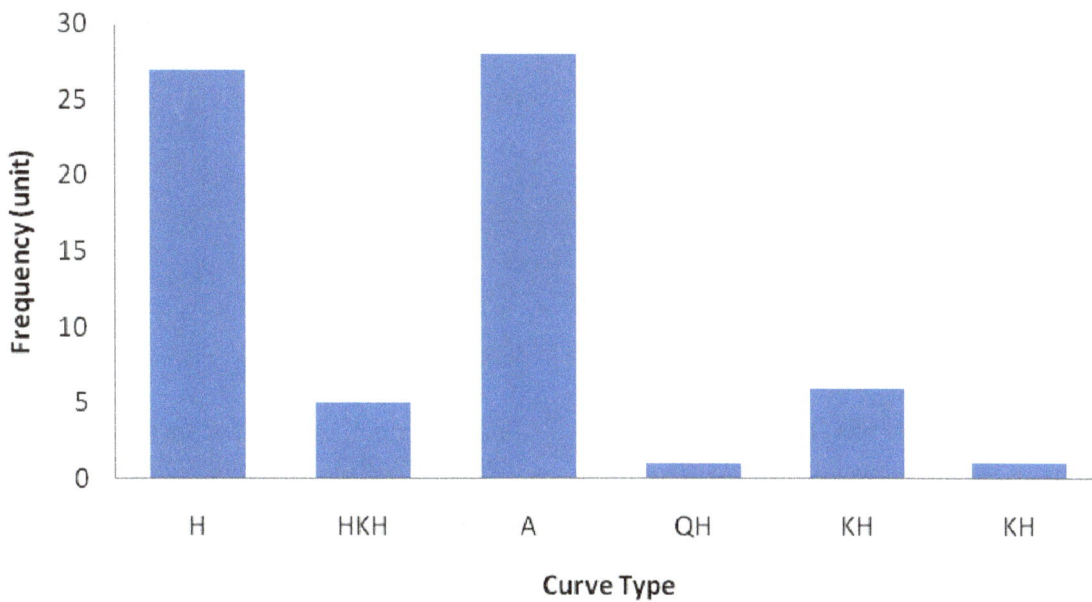

Figure 3. Frequency distribution of the observed curve types.

heterogeneous variation in the composition of the topsoil from clay, sandy clay, clayey sand and laterite. This lithology has limited hydrogeological significance due to its small thickness.

Isopach and isoresistivity maps of the weathered layer

Figure 8 isopach map of the weathered layer within the study area. It shows that the thickness of the weathered layer over the study area varies from 0.4 to 29 m but generally less than 10 m, indicating thin weathered layer across the area. The highest thickness (localized) occurs at the portion marked X_1 (Figure 8). The resistivity values in the study area vary from 100 to 600 Ωm (Figure 9). This indicates low clay content and shows a fairly saturated formation which is able to accumulate water and is also able to transmit it. The weathered layer constitutes the main aquifer unit in the area. However,

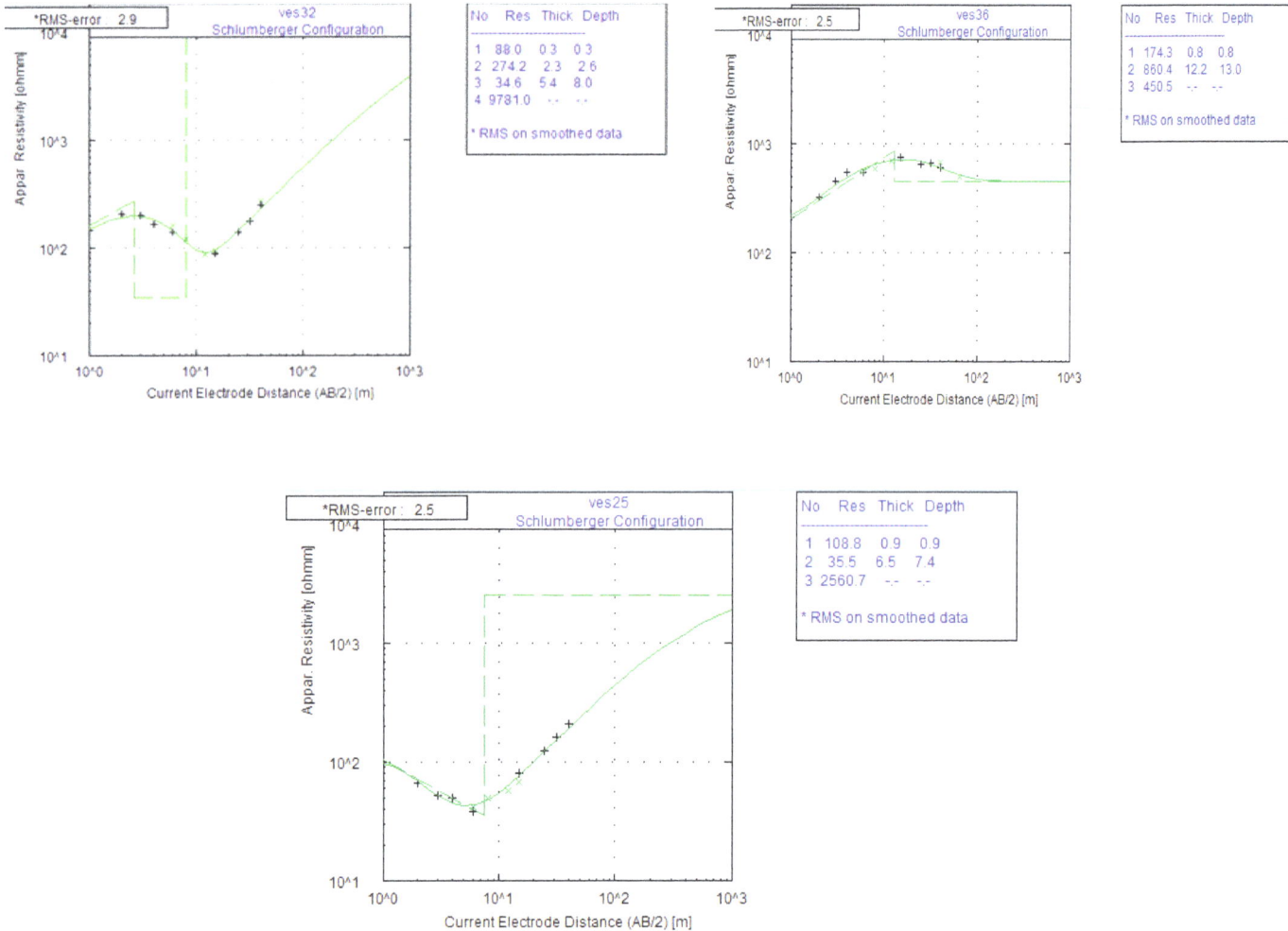

Figure 4. Typical VES curves from the study area.

because of its thinness it has little hydrogeological significance.

Isopach map of the overburden

Figure 10 shows the depth to the top of the fresh bedrock beneath the sounding stations. The overburden is assumed to include the topsoil and the weathered layer. The depth to the bedrock varies from 4 to 35 m. From the area marked X₂, it can be said to have a relatively thick overburden (>20 m).

Groundwater evaluation

Figure 11 shows the groundwater potential map of the study area. The evaluation of the groundwater potential of the study area is based on the thickness of the overburden and the resistivity of the weathered layer, since the nature of the weathered layer and its thickness

are important parameters in the groundwater potential evaluation of a basement complex terrain (Clark, 1985; Bala and Ike, 2001). The horizon is also regarded as a significant water bearing layer especially if significantly thick and the resistivity parameters suggest saturated conditions (Shemang, 1993; Bala and Ike, 2001). The groundwater potential of the study area was zoned into high, medium and low potentials.

In this study, zones where the thickness of the overburden (which constitutes the major aquifer unit) is greater than 25 m and of low clay content (average resistivity values between 200 and 300 Ωm) are considered zones of high groundwater potentials. Areas where the aquifer thickness ranges from 10 to 25 m with less clay contents are considered to have medium groundwater potential and the areas where the aquifer thickness is less than 10 m are considered to have low groundwater potential.

From Figure 11, about 85% of the study area falls within the low groundwater potential rating, while about 10% of the study area constitutes the medium groundwater

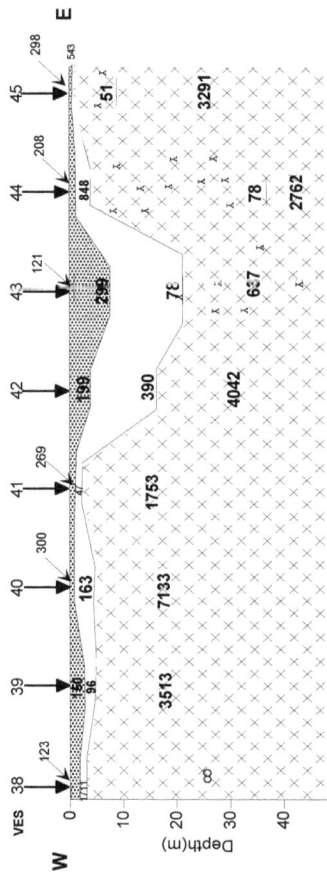

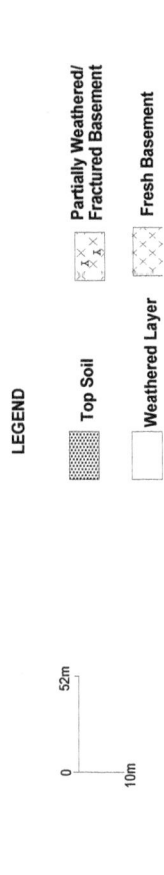

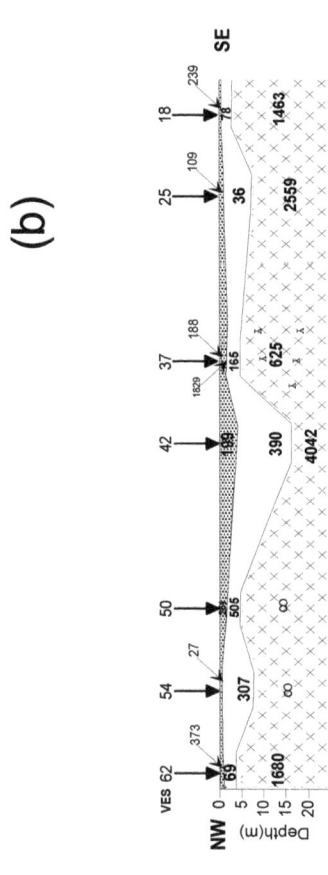

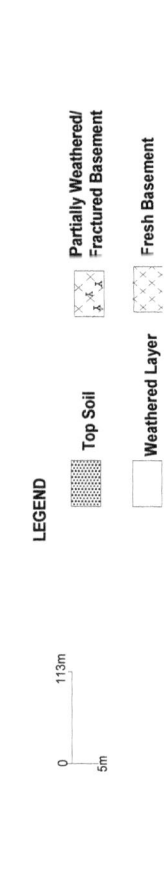

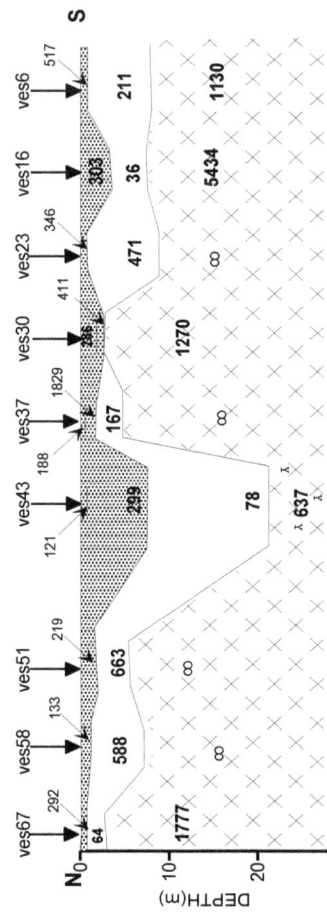

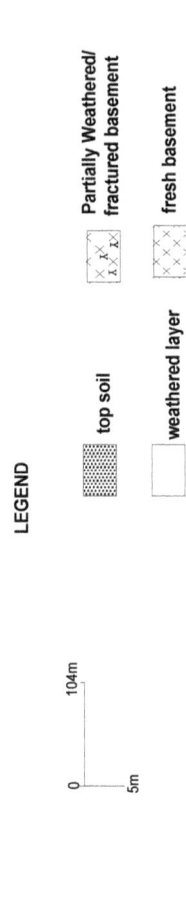

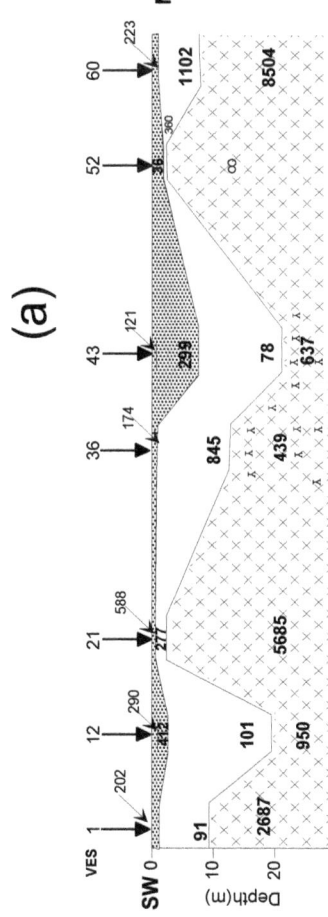

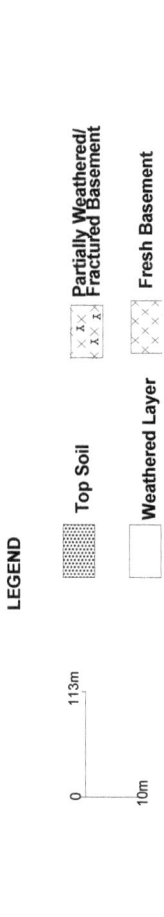

Figure 5. Geoelectric sections along (a) North South (N-S) (b) West-East (W-E) (c) Southwest-Northeast (SW-NE) and (d) northwest southeast (NW-SE) direction.

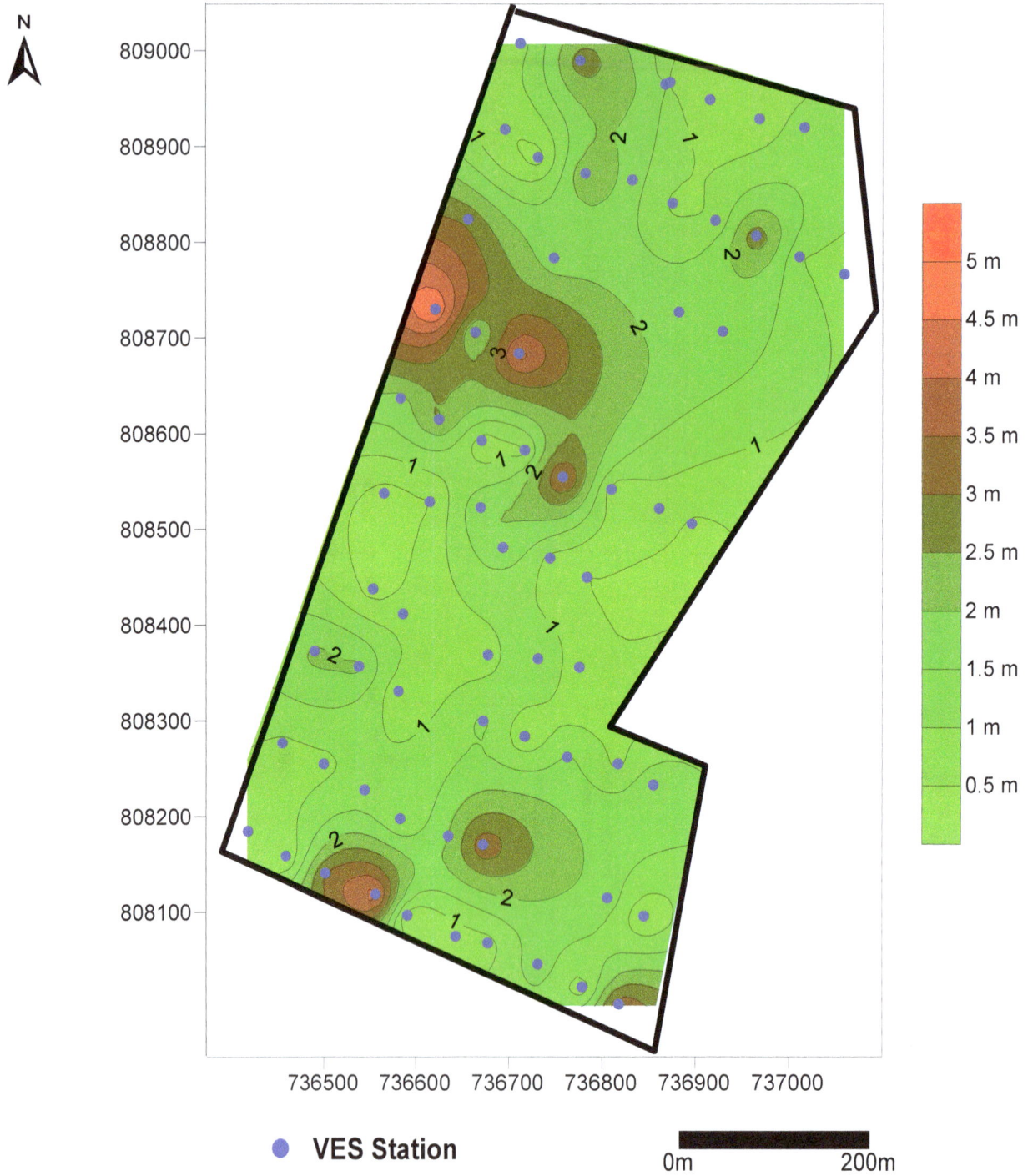

Figure 6. Isopach map of the top soil.

potential rating and the remaining 5% has high groundwater potential rating. The fractured basement is unconfined, thin to moderately thick and localized (delineated beneath VES 36, 37, 43 and 44). The resistivity values of the fractured layer are relatively high (439 - 637) indicating possibly low fracture density which is also unsaturated.

Aquifer protective capacity evaluation

The longitudinal conductance map generated in the study area is as shown in Figure 12. This map was used for the overburden protective capacity rating of the study area. The longitudinal conductance values obtained from the area range from 0.01 to 0.4 mhos (Figure 12).

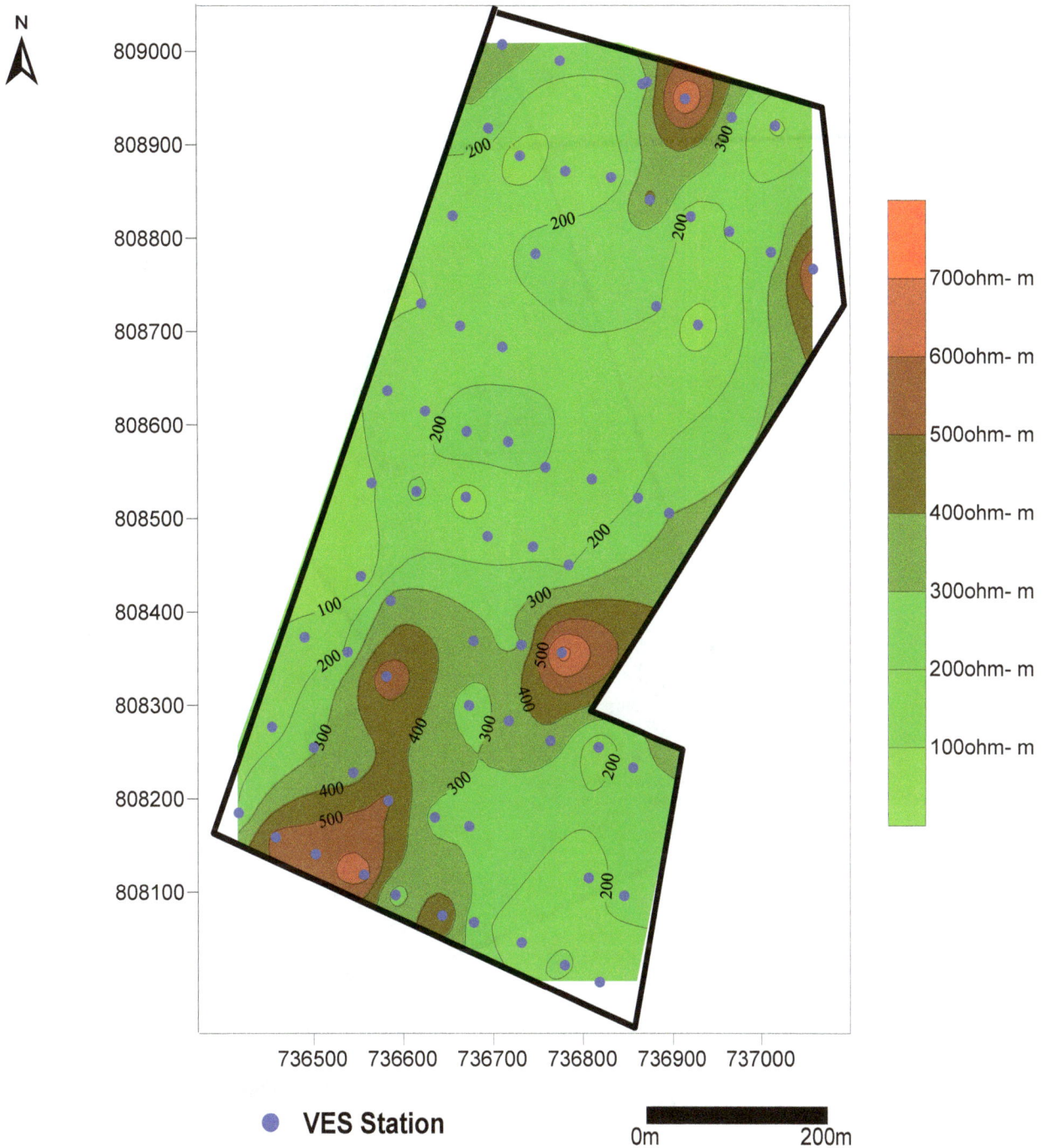

Figure 7. Isoresitivity map of the top soil.

The protective capacity of an overburden could be considered proportional to the longitudinal conductance (Oladapo and Akintorinwa, 2007). Clayey overburden, which is characterized by relatively high longitudinal conductance, offers protection to the underlying aquifer. According to Oladapo and Akintorinwa (2007), the protective capacity of the overburden could be zoned into good, moderate and weak protective capacity. Zones where the conductance is greater than 0.7 mhos are considered zones of good protective capacities (Figure 13 and Table 2). The portions having conductance values ranging from 0.2 to 0.69 mhos were classified as zones of moderate protective capacity (Figure 13 and Table 2), the zones whose conductance ranged from 0.1 to 0.19 mhos were classified as of weak protective capacity while the zones where the conductance is less than 0.1 mhos

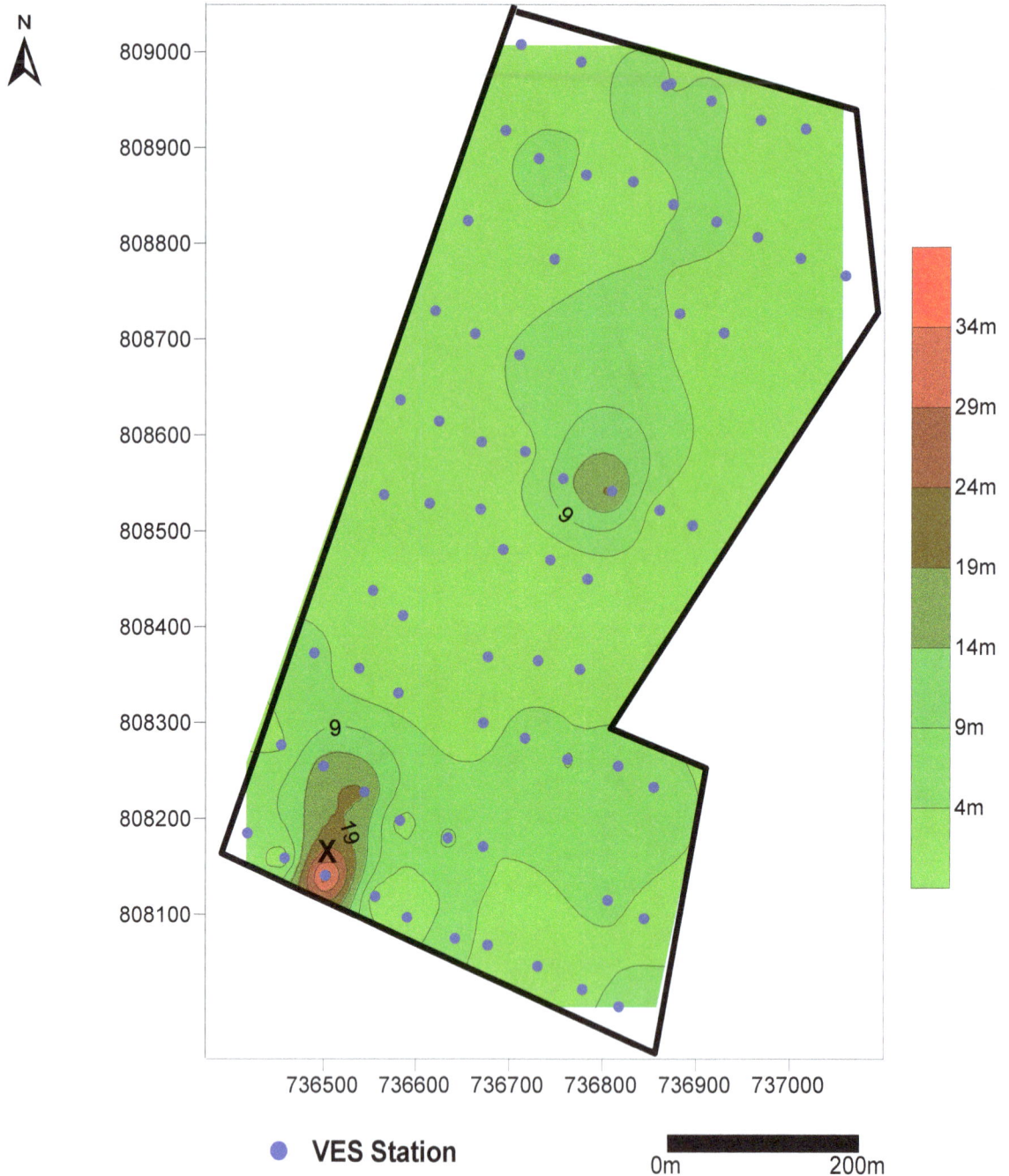

Figure 8. Isopach map of the weathered layer.

were considered to have poor protective capacity (Figure 13 and Table 2). The map of the study area shows that about 75% of the area falls within the poor protective capacity, while the remaining 25% constitutes the moderate/weak protective capacity rating (Figure 13). This suggests that the area is characterized by low longitudinal conductance which informed weak protective capacity rating of the area. Therefore the study area is vulnerable to pollution from contaminant sources such as industrial waste, septic tanks, underground petroleum storage tanks and landfills if located close to the study area.

Conclusions

Electrical Resistivity Method involving Vertical Electrical Sounding (VES) has proved useful in the evaluation of groundwater potential and overburden protective capacity of Zion estate. The Vertical Electrical Sounding data were

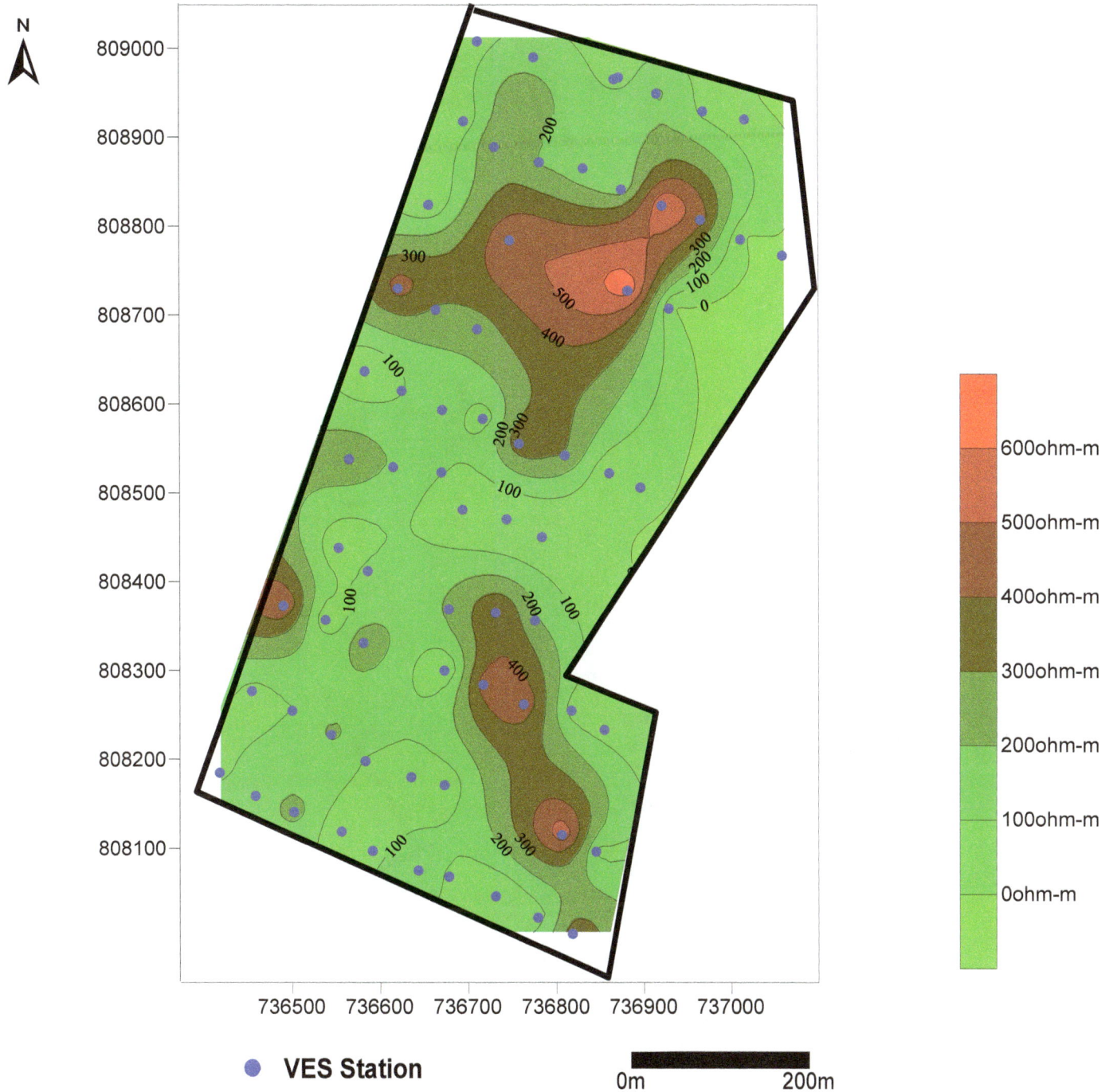

Figure 9. Isoresistivity of the weathered layer.

quantitatively interpreted using partial curve matching and the results were refined using WinResist software. The first order geoelectric parameters obtained from the interpretation of the vertical electrical sounding data and the second order Dar Zarrouk parameter (longitudinal conductance S_i) were used to generate various maps and sections which were analyzed with respect to the groundwater potential and overburden protective capacity of the study area. The results show that the weathered layer (which generally has a small thickness <10 m) constitutes the major aquifer unit in the area, has little hydrogeologic significance. However, a localized part of the area marked X has significant hydrogeologic significance with weathered layer thicknesses up to 34 m.

Generally the area was zoned into high, medium and low groundwater potential and about 85% of the area falls within the low groundwater potential rating, while about 10% constitutes the medium groundwater potential rating

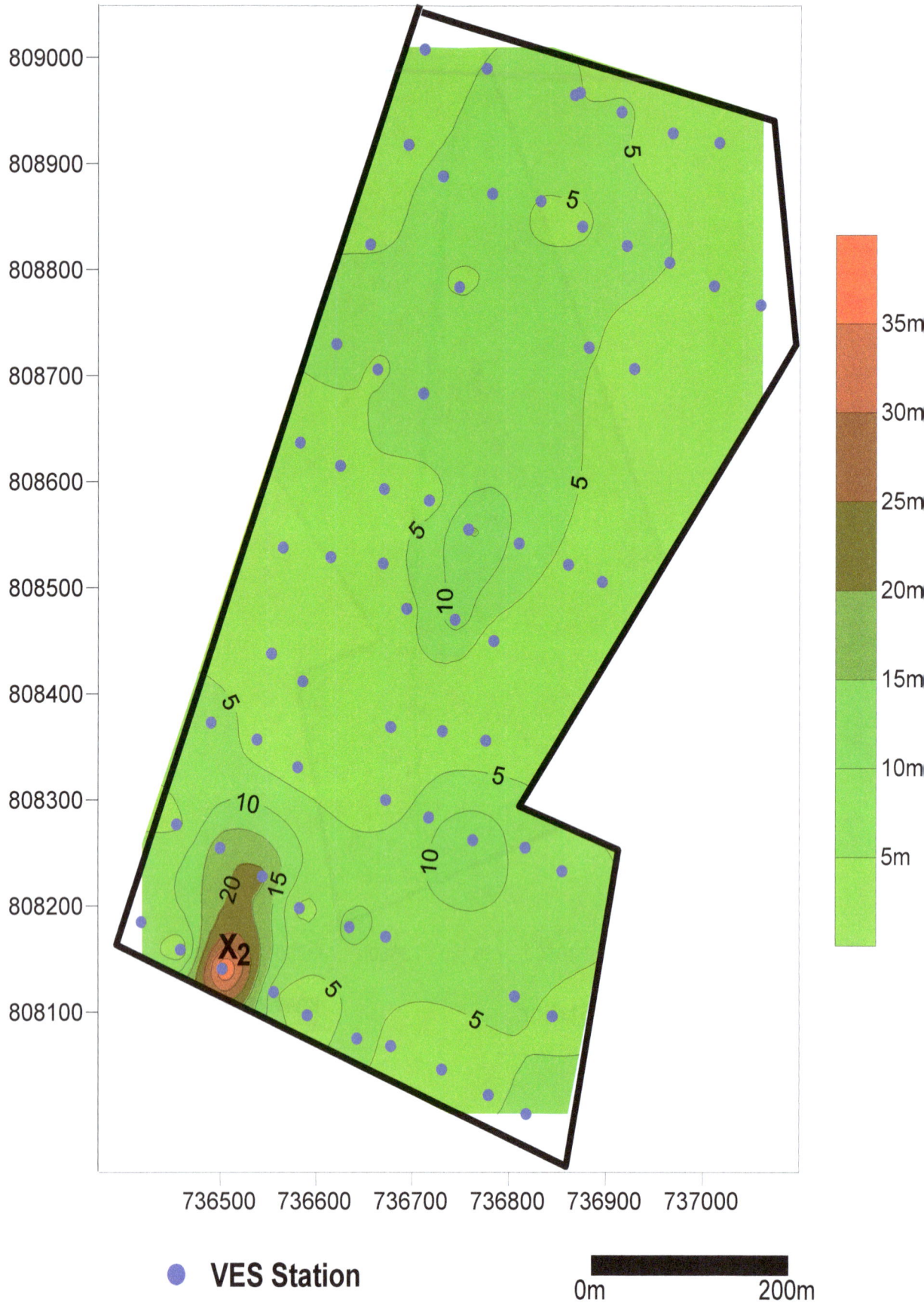

Figure 10. Isopach map of the overburden.

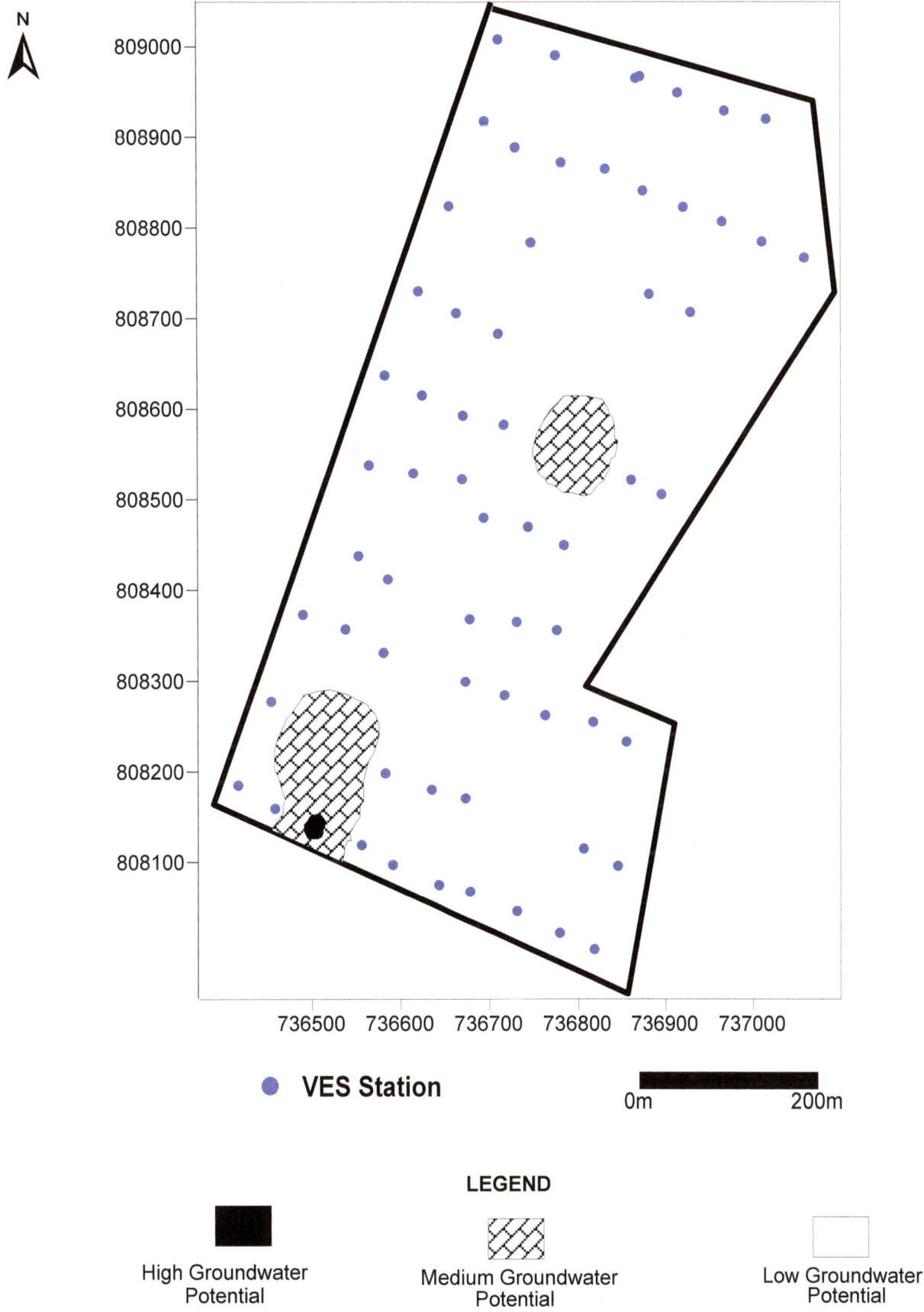

Figure 11. Groundwater potential map of the study area.

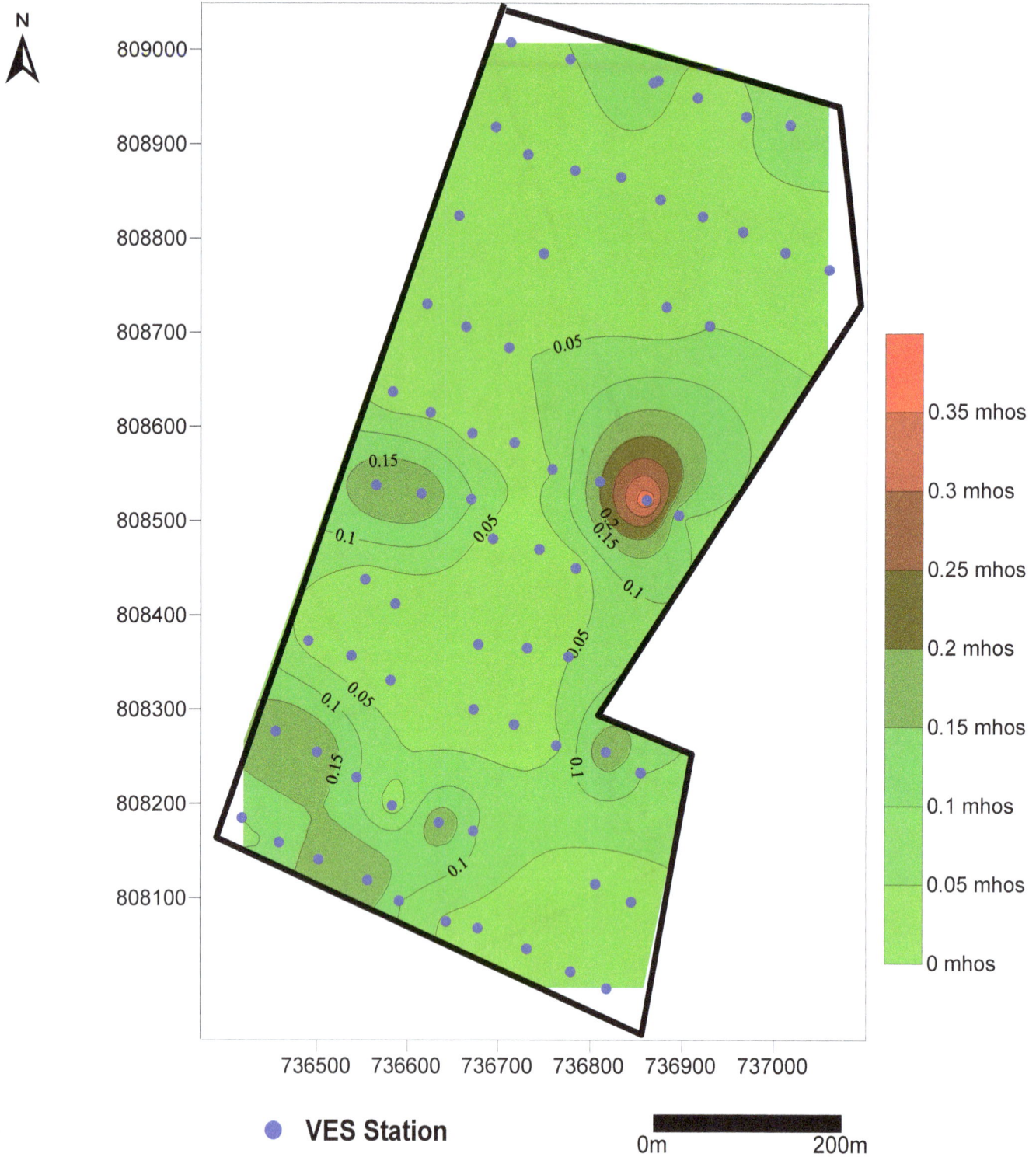

Figure 12. Longitudinal conductance map of the study area.

and the remaining 5% constitutes high groundwater potential rating. Hence the groundwater potential of the area was generally rated to be low. The VES stations underlain by high and medium groundwater potential zones are envisaged to be viable for groundwater development within the area. The protective capacity of the overburden in the study area was zoned into moderate, weak and poor protective capacity. About 75%

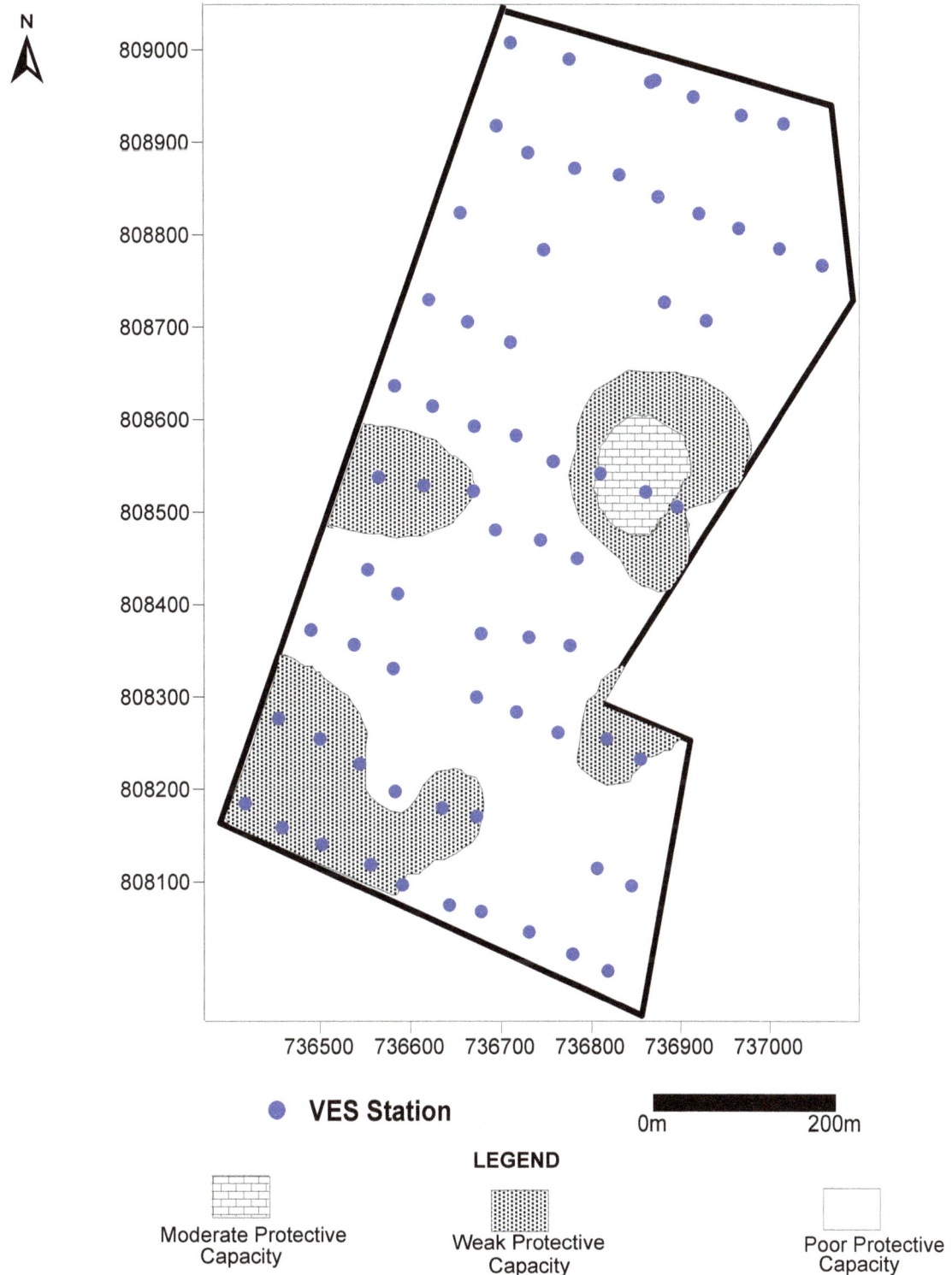

Figure 13. Overburden protective map of the study area.

of the area falls within the poor protective capacity, while the remaining 25% constitutes the moderate/weak protective capacity rating. This suggests that the area is characterized by weak protective capacity rating. Therefore, the study area is vulnerable to pollution from contaminant sources such as industrial waste, septic tanks, underground petroleum storage tanks and landfills if located close to the study area.

The results of this research have provided reliable information on the groundwater potential and overburden

Table 2. Modified longitudinal conductance/protective capacity rating (Oladapo and Akintorinwa, 2007).

Longitudinal conductance (mhos)	Protective capacity rating
>10	Excellent
5 - 10	Very good
0.7 - 4.9	Good
0.2 - 0.69	Moderate
0.1 - 0.19	Weak
<0.1	Poor

protective capacity necessary for proper planning and development of groundwater within the study area. It is therefore recommended that groundwater exploration and development be restricted to the area zoned into medium/high groundwater potential. Also the location of septic tanks, petroleum storage tanks, shallow subsurface piping utilities and other contaminant facilities should be confined to zones of moderate protective capacity.

REFERENCES

Bala AE, Ike EC (2001). The aquifer of the crystalline basement rocks in Gusau area, Northwestern Nigeria. J. Min. Geol. 37(2):177-184.

Benson S, Jones CR (1988). "The combined EMT/VES Geophysical method for siting borehole" pp. 54-63.

Clark L (1985). Groundwater Abstraction from Basement Complex Areas of Africa. J. Eng. Geol. London 18:25-34.

Hazell JRT, Cratchley CR, Preston AM (1988). "The Location of Aquifers In Crystalline Rocks and Alluvium in Northern Nigeria Using Combined Electromagnetic and Resistivity Methods". Quart. J. Eng. Geol. 21:159-175.

Iloeje NP (1981). A new geography of Nigeria (New revised edition). Longman Nig. Ltd., Lagos, p. 201.

Mallet R (1947). The Fundamental Equations of Electrical prospecting. Geophys. Prospect. 12:529-556.

Mundel JA, Lother L, Oliver EM, Allen-Long S (2003). Aquifer vulnerability analysis for Water Resources Protection. Indiana Department of Environmental Management (IDEM), 'Source Water Assessment Plan' p. 25.

Oladapo MI, Akintorinwa OJ (2007). Hydrogeophysical Study of Ogbese Southwestern, Nigeria. Global J. Pure Appl. Sci. 13(1):55-61.

Olayinka AI (1990). "Electromagnetic Profiling for Groundwater in Precambrian Basement Complex Areas of Nigeria". Nordic Hydrol. pp. 205-216.

Olayinka AI, Olorunfemi MO (1992). "Determination of geoelectrical Characteristic in Okene Area and implication for boreholes setting". J Min. Geol. 28:403-412.

Olayinka AI, Amidu SA, Oladunjoye MA (2004). "Use of Electromagnetic Profiling and Resistivity Sounding for Groundwater Exploration in the Crystalline Basement Area of Igbeti, Southwestern Nigeria". Global J Geol. Sci. 2(2):243-253.

Omosuyi GO, Ojo JS, Enikanselu PA (2003). "Geophysical Investigation for Groundwater around Obanla – Obakekere in Akure Area within the Basement Complex of South-Western Nigeria". J. Min. Geol. 39(2):109-116.

Rahaman MA (1989). Review of the basement geology of southwestern Nigeria: In Geology of Nigeria (Kogbe CA, ed.). Elizabeth Publishing. Co. Nigeria pp. 41-58.

Shemang EN (1993). Groundwater potentials of Kubami River Basin, Zaria, Nigeria, from DC resistivity study. Water Res. (1 and 2):36-41.

Korzun VI, Komitel M (1978). World water balance and water resources of the earth. UNECO p. 663

Van der-Velpen BPA (1988). "RESIST Version 1.0." MSc Research Project, ITC: Delft, Netherlands.

Wright EP (1992). The hydrogeology of crystalline basement aquifers in Africa. Geological Society, London, Special Publications 66:1-27.

Zohdy AAR (1974). "The Auxiliary Point Method of Electrical Sounding Interpretation and its Relationship to Dar Zorrouk Parameters". Geophysics 30:644-650.

5

Application of Geographic Information System (GIS) in sustainable groundwater development, Imo River Basin Nigeria

Michael A. Nwachukwu[1], **Aslan Aslan**[2] and **Maureen I. Nwachukwu**[3]

[1]Department of Environmental Technology, Federal University of Technology Owerri Nigeria.
[2]Earth and Environmental Studies, Montclair State University New Jersey, U.S.A.
[3]Department of Geosciences, Federal University of Technology, Owerri Nigeria.

There is proliferation of shallow substandard private water wells, poor distribution of public water wells, poor planning, and poor management of public wells in the Imo River basin Nigeria. Over 60% of water wells developed in the basin is either abortive or not functional. To investigate this, 110 vertical electric soundings (VES), 50 down-hole logs, and 44 pumping test data have been integrated into geographic information system (GIS) in this study. Map of 44 county areas of the basin was produced. Thematic maps showing mean values of water table, aquifer resistivity, aquifer thickness, and a groundwater prospect map of the basin in five potential areas were also produced. Correlating the GIS map with regional geologic map defined three groundwater prospect zones in the basin. Sustainable practices recommended are *government-private partnership* for public wells, and *private-private partnership* for private wells. Surface water development is recommended in zone 3, against groundwater development. Geophysics/GIS groundwater prospect model shows significant correlation with geology, confirming the effects of geology to groundwater development in the basin.

Key words: Private/Public water wells, geophysics data, geographic information system (GIS) models, abortive wells, geology, sustainability.

INTRODUCTION

Unavailability of groundwater prospect map of the basin area for public consumption is thought to have contributed greatly to the increasing failure of water well projects in certain parts of the Imo River basin Nigeria (Figure 1a and 1b). For example, groundwater prospect map is a vital tool to government in water well allocation decision making process. By this, individuals, organizations, communities, and governments could visualize areas mapped as difficult for groundwater development and then be committed to an alternative water sources in such areas. This will reduce incidents of failure, abortive, or abandoned public wells and non-functionality of domestic water wells estimated at over

850 in the basin. It is therefore very necessary to do everything possible that can reduce incidents of abortive wells, substandard domestic wells, and non-functional wells. With good groundwater prospect map, government would award contracts of water well projects to only viable areas, excluding areas where difficult subsurface geology does not allow successful groundwater development. Groundwater prospect map will allow fair distribution of public water wells to communities. Groundwater prospect map will reduce economic wastes, and environmental degradation, as well as disappointment and hardship to the affected communities; based on a model of distribution with no consideration for areas with

Figure 1. Map of Nigeria showing Imo River basin.

groundwater development difficulty.

The result is economic waste, and environmental degradation, as well as greater disappointment to the affected communities. Citizens are therefore constrained to the proliferation of shallow substandard wells often recharged by surface water through fractures, and harvested rain water stored in ground tanks of all manners. This may be accountable to the persistence of water related diseases in the basin.

Proliferation of shallow private/commercial wells of poor standards by individuals also implies financial incapability for standard wells. This is due to lack of functional public water supply, and the excessive desire of citizens to be self dependent. The most contaminated wells are usually the shallow hand-dug rather than drilled, and having poor casing material (Comely, 1987). Incidentally most of the shallow private and commercial wells in the basin are hand-driven, and some constructed with inferior casings. According to Nwachukwu et al. (2010b), the greatest problem of manual drilling is the impunity at which the operators declare the drilling terminated, soon as the crew penetrates the water table, or run out of energy. Ibe et al. (2007), Nwachukwu et al. (2010a), and Nwachukwu et al. (2010c), have confirmed that environmental pollution in the Imo River basin increases from the shaly north to the sandy south. They hold that human activities also follow a similar trend, representing the primary source of water pollution in the Imo River basin. The final analysis implicates shallow wells in the sandy south to be

more vulnerable to pollution than the equivalent in the north.

Geology

The basin is a 140 km N-S trending sedimentary syncline located at the mid south-eastern part of Nigeria (Figure 1). It is rich in oil and gas and has many of Nigerian oil fields. The basin has rich deposit of clay minerals, sand, gravel, and lignite. Based on near surface lithology, the Imo River Basin was divided into two zones (Nwachukwu et al., 2010a). The northern zone consisting of group of shale formations have greater drainage density, than the southern zone belonging to the sandy Benin Formation (Figure 10). Age of the shaly zone range from *paleocene* to *Albian,* and the sandy zone is *Olgocene* to *Recent.* Prospective horizon for groundwater development in the basin varies from location to location. Age and characteristics of the geologic units are discussed by Reyment (1965), Onyeagocha (1980), and Ekweozor and Unomah (1990).

Recent investigation confirmed the presence of 40 to 70 counts of coliform per 100 ml of water in six out of ten sampled shallow wells around Owerri (Nwachukwu et al., 2010b). This may have significant health implications. The average cost of treating severe typhoid fever and diarrhea (alias *stooling* and *vomiting*) to families which may result in sudden deaths is enormous. It is a major source of poverty, draining both the household and the state economy. The overall effects, including the loss of man hour in both public and private sectors is alarming, and difficult to be estimated due to lack of records. A recent survey has confirmed the presence of one or two abortive wells in the vicinity of three to five functional well in the area. Most of the private and public wells became abortive immediately on completion. Others became non-functional 2 to 5 years after completion, or abandoned during construction. If this situation continues unchecked, the basin may encounter a more serious disaster with quality of water resources, well interference and loss of space or reservation for future groundwater development.

Singh and Prakash (2004) integrated thematic maps through GIS for identification of groundwater potential zones. Kamaleshwar et al. (2000) integrated some physical parameters and terrain characteristics with GIS and remote sensing to extract groundwater prospect zone in Dala-Renukoot area of Uttar Pradesh, India. They arrived at five groundwater prospect categories. Kumar et al. (2010) has delineated groundwater potential zones using remote sensing and GIS techniques in Kurmapalli Vagu basin, India. They emphasized that systematic planning of groundwater development using modern techniques is essential for the proper utilization and management of this precious but shrinking natural resource. Similarly, they obtained five categories of groundwater potential zones ranging from very good to poor in the basin.

In this study, surface and borehole electrical resistivity and SP, and pumping data are integrated in GIS to map and explain the groundwater prospect zones of the basin. Low success rate of drilling productive wells is a common challenge in hard rock environment. The use of remote sensing with ground or well information is becoming effective method in improving success rate. Gezahegn (2012), integrated thematic layers generated from satellite images, existing maps and well data to delineate the groundwater potential zones of Upper Tumet catchment, Western Ethiopia. The resulting groundwater prospect map revealed that large part of the study area has poor groundwater potential which is in good agreement with field measurements.

Shahid and Kumar Nath (2008) observed that geophysical data help in locating the ground water potential in any hydrogeological setup. The property and thickness of various litho-units obtained from geophysical survey at different locations if integrated can yield a ground water potential model of higher reliability. Omosuyi (2010) conducted a geoelectric assessment of groundwater prospect and vulnerability of overburden aquifers at idanre, southwestern Nigeria and confirmed the suitability of the near surface aquifer for groundwater development.

Water table depth is an important map layer, parameter, and input variable for a wide variety of environmental models and groundwater development

decision support tool. Water table maps are used to estimate ground water flow direction and velocity; vulnerability assessments or pollution transport. It is commonly used in environmental decision making such as locating landfills and wastewater disposal sites. The creation of water table maps is a well accepted practice in ground water investigations. Jasrotia and Singh (2005) used water table thematic layers and other parameters integrated through the DRASTIC model within a GIS environment to demarcate vulnerable zones of a watershed.

Vulnerability of water table aquifer south of the basin area

Figures 2 and 3 largely show contaminant transport to the southeast after (Nwachukwu et al. (2010c). The southward Otamiri River flow, and groundwater flow facilitates this transport (Figure 2). South-southeast tracking and capture zone of contaminants (Figure 3) suggest greater groundwater pollution. Groundwater flow, and contaminant transport is reduced in the south-end portion of the area with horizontal and vertical flows at the swamp (Figure 2b). There is indiscriminate dumping of solid waste, and disposal of urban sewage on farm land as fertilizer farm input. This habit increases bioavailability of trace metals and other contaminants in the southern portion of the basin. Shallow water table in the south west flank of the basin enhances contaminant transport and pollution of the water table \aquifer in the basin area. These conditions greatly endanger shallow water wells in the Imo River basin.

Using environmental impact models, Ibe et al. (2007) also reported that pollutants are transported southwards by surface water flow. According to them, the pollutants could easily migrate to the water table and immediately contaminate groundwater. Results of this study emphasize moderate to high vulnerability of the water table aquifer in the southern flank of the basin from where the shallow wells are recharged (Figure 4).

Khalek and Omran (2008) integrated remote sensing, geophysics and GIS to evaluate groundwater potentiality in Sohag region of Egypt. Three groundwater potentiality zones in the Quaternary aquifer of Sohag region were demarcated. This study has evidently demonstrated the capabilities of the integrated approach in delineation of the different ground- water potential zones. The approach can be used elsewhere with appropriate modifications for identifying candidate well locations and in creating a GIS-based hydrogeologic model of a selected area. Other scholars have integrated different other parameters to map groundwater potential zones successfully. Thus Mayilvagama et al. (2011) integrated thematic layers of lithology, geomorphology, drainage, soil, lineament, land use and surface water body to delineate groundwater potential zones in Thurinjapuram watershed. The result

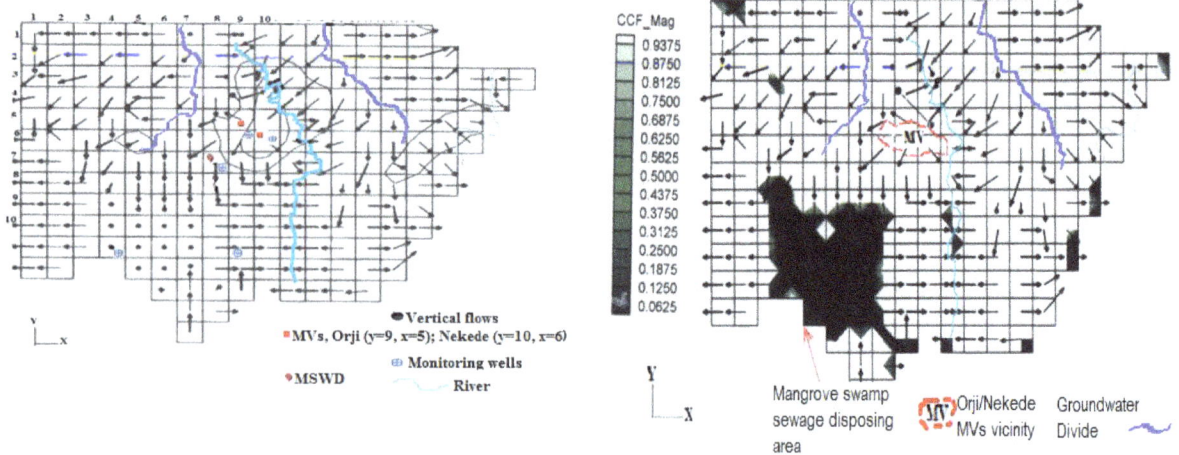

Figure 2. Groundwater flow model by MODFLOW model code, show direction of groundwater flow south of Imo River basin

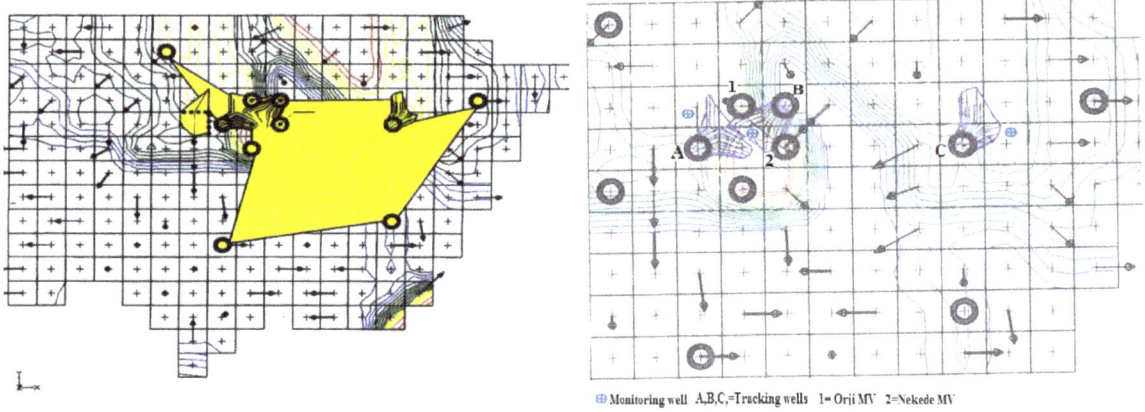

Figure 3. Faith and transport of contaminants obtained by MODPATH model code show (**a**) Particle capture zone (**b**) Particle tracking wells (ABC) south of Imo River basin.

showed three different groundwater potential zones namely 'good', 'moderate' and 'poor'.

MATERIALS AND METHODS

A total of 110 vertical electrical soundings (VES) data at two per county area, 50 geo-electrical down hole litho-logs, and 44 pumping test data were integrated in this study. VES and well-log data were provided by GEOPROBE Int'l Consultants Ltd Owerri, being data collected between 2000 and 2008. Status of the wells, yield and other important information were obtained from Imo State Water Development Agency (IWADA), Anambra-Imo River Basin and Rural Development Authority, Abia State Water Corporation, UNICEF, and some private individuals. The 110 VES were conducted using Allied Omega resistivity meter and Abem Terrameter (SAS 1000). The well log data was collected with Kerk Borehole Resistivity meter, Abem Terrameter and Scintrex instruments. These down hole loggers were used to measure

apparent resistivity and the natural self-potential (SP). VES is the determination of vertical variation of resistivity through earth material, and is very reliable for groundwater exploration particularly in moderate terrain with less subsurface complexity as in the Imo River basin. VES field technique was based on the Schlumberger array conducted under good weather conditions. Readers may consult Telford, et al. (1990), Dobrin and Savit (1988), Reynolds (1997), USDOT (2004), and Kearey et al. (2002). These scholars have discussed the applications of electrical resistivity method in groundwater exploration and mapping in favour of vertical electric sounding (VES) by Schlumberger array.

Applicable GIS layers

A total of 44 reference wells, one in each county area were investigated. Location coordinate of each of the reference wells were measured with a geographic positioning system (GPS). The selected wells are those found within or near each county headquarter. Measured coordinates in degrees, minutes and

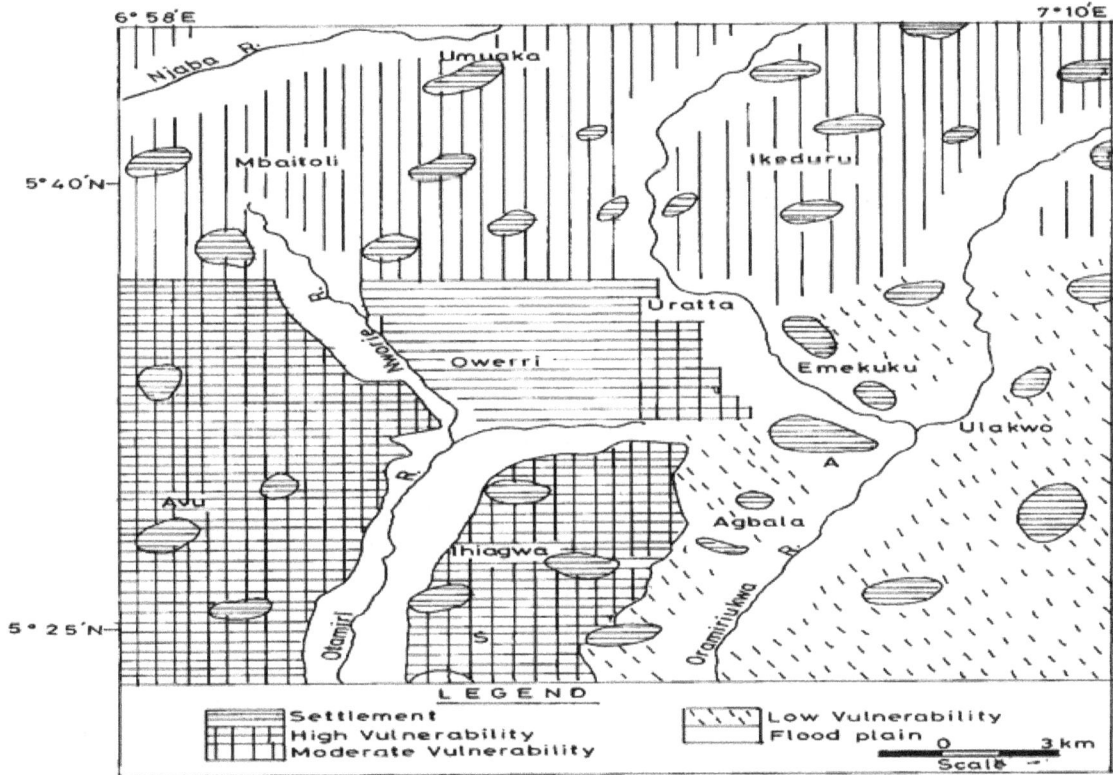

Figure 4. Soil and water vulnerability map of the study area produced by Ibe et al. (2007)

seconds (DMS) were converted to digital degree (DD) in excel sheet by the formula: (degrees) + (min/60) + (s/3600) = digital degrees.

The resultant decimal degrees as shown in Table 1 were applied to ArcMap version 9.3, so each reference well could display its ID, and other attributes. The following GIS layer maps were made: *County areas, Mean Resistivity, Mean Thickness, Water table,* and *Groundwater Prospect.* The resultant groundwater prospect area map was compared and correlated with existing geologic map of the basin, to establish three groundwater prospect zones in the basin.

RESULTS AND DISCUSSION

Data analysis presented in Table 1 is described as the project data sheet. Water table (WT) and well depth (WD) are based on acceptable average values across each county. The stated lithology is for the prospective aquifer thickness (T) and average aquifer resistivity also refer to the prospective aquifer to the well depth (WD). The prospective well depth is found as the economic depth of standard water well for each county. Coordinates are shown in digital degree, as converted from degree, minutes and seconds and followed by the expected yield.

The 44 county area map of the Imo River basin (Figure 5) was obtained by merging Imo and Abia States local government maps over a digitized coordinate grid with geometric correction produced as GIS map 1. Average

aquifer resistivity based on the 110 VES with at least two in each county as shown in Table 1 was contoured to produce the GIS resistivity map (Figure 6). The map indicates areas where the prospective aquifers have equal to similar resistivity, which is expected to relate to hydraulic conductivity of the aquifer. Previous studies by Ekwe et al. (2006), Ungemach et al. (1969) and Kelly (1977) have shown that electrical soundings can be used to estimate aquifer transmissivity and hydraulic conductivity, and that both possess empirical relationship.

Average aquifer thickness derived from the 110 VES and the 50 well-log data constituted the thickness map (Figure 7). The average thickness values were used to construct an isopach contour of the basin, and the corresponding GIS thematic map was made. The map indicates areas where the prospective aquifers have equal to similar thickness, which relates to the expected depth of standard water well in the affected area of the basin. Thematic maps (Figures 6, 7, 8 and 9) are only preliminary guiding tools to water well development stakeholders. They do not in any way eliminate the mandatory site specific groundwater pre-drilling and down-hole geophysical studies for water well development in the basin.

Average of water table depths obtained in each of the reference wells were considered as average depth to water. The average values were used to construct a water

Table 1. Project data analysis.

ID	County	WT	WD	T	Res	Lithology	x	y	Yield
21	Aboh Mbaise	200	350	150	1870	Sandstone	5.451111	7.233333	M-H
14	Ahiazu Mbaise	220	350	130	2020	Sandstone	5.751111	7.384166	M-H
20	Ezinihite	200	360	160	1960	Sandstone	5.466389	7.325111	M-H
7	Ihite Uboma	300	400	100	2200	Sandstone	5.511111	7.166666	M-H
8	Ehime Mbano	240	350	110	2280	Shaly-Sst	5.483056	7.551111	M-H
18	Ideato North	260	400	140	2210	Shaly-Sst	5.773611	7.158055	L-M
10	Ideato South	300	450	150	3270	Shaly-Sst	5.883333	7.133333	L-M
16	Ikeduru	240	350	110	2900	Sandstone	5.566667	7.100001	M-H
15	Isiala Mbano	280	400	120	2150	Sandstone	5.218889	6.917777	L-M
2	Isu	240	360	120	1620	Shaly-sand	5.796389	7.038888	M
17	Mbaitoli	200	340	140	2780	Sandstone	5.583333	7.050001	M-H
13	Njaba	290	400	110	2390	Sandstone	5.703333	6.770555	M-H
36	Ngor-Okpala	130	300	170	1308	Sst/G bed	5.800833	7.058333	P
3	Nkwerre	210	350	140	2640	Sandstone	5.750001	7.116666	L-M
5	Nwangele	300	380	80	1420	Shale top	5.797222	7.038888	L-M
19	Obowo	240	360	120	1790	Fractured	5.100001	7.092501	M-H
33	Oguta	80	240	160	1330	Sandstone	5.711667	6.809444	P
35	Ohaji/Egbema	120	370	150	1210	Sand/G bed	5.800833	7.916666	P
27	Okigwe	260	400	140	2090	Sand/G bed	5.483056	7.550001	L-M
9	Onuimo	230	350	120	3010	Shale/Sst	5.816667	7.200001	L-M
4	Oru East	280	380	100	1872	Sandstone	5.796389	7.038888	M
6	Oru West	260	380	120	3060	Mixed sand	5.711667	6.809444	M-H
11	Orlu	280	420	140	2940	Sand/G bed	5.783333	7.033333	M-H
12	Orsu	300	450	150	2090	Sandstone	5.816667	6.933333	M-H
1	Owerri Urban	100	280	180	980	Mixed sand	5.485001	7.035001	P
22	Owerri North	160	300	140	911	Sand/G bed	5.801389	7.051388	P
34	Owerri West	90	280	190	850	Sand	5.504167	7.021666	P
38	Aba North	120	360	240	1242	Sand	5.333333	7.316666	P
40	Aba South	100	300	200	1109	Sand	5.100001	7.350001	P
32	Arochukwu	240	320	80	3020	Shaly-Sst	5.416667	7.500001	L-D
30	Bende	230	320	90	3500	Shaly-sand	5.566667	7.633333	M
26	Ikwuano	200	320	120	3280	Sandstone	5.433333	7.566666	M
25	Isiala Ngwa N	130	300	170	1210	Sand	5.116667	7.366666	P
37	Isiala Ngwa S	100	300	200	1287	Sand	5.185278	7.601944	P
29	Isiukwuato	250	360	110	2100	Sandstone	5.533333	7.483333	L-M
39	Obi Ngwa	120	320	200	1250	Sand	5.149722	7.330277	P
31	Ohafia	180	280	100	3050	Sandy-Shale	5.616667	7.833333	L-D
41	Osisioma Ngwa	130	340	210	1150	Sand	5.416667	7.500001	H H
42	Ugwunabo	100	300	200	962	Sand	5.185278	7.601944	P
44	Ukwa East	120	320	200	850	Sand	5.116667	7.366666	P
43	Ukwa West	100	320	220	840	Sand	5.104722	7.142501	P
23	Umuahia North	300	400	100	3280	Sandy-shale	5.533333	7.483333	L-M
24	Umuahia South	180	300	120	1720	Sand	5.508056	7.481666	M-H
28	Umu Nneochi	250	400	150	2270	Sandstone	5.104722	7.142501	L-M

*WT = Average water table, WD = Recommended well depth, Res = Mean resistivity value. P = Prolific), M-H = Medium to high yield, L-M = Low to medium yield, L-D = Low yield to dry well. G bed = Gravel bed, Sst = Sandstone and T = Thickness of prospective horizon.

table contour from which a GIS thematic map layer showing six water table depth classifications in the basin area was produced (Figure 8). All the GIS layers were overlaid, with different symbols assigned to the identified features in each layer. The resultant composite coverage was classified into five groundwater prospect zones (Figure 9).

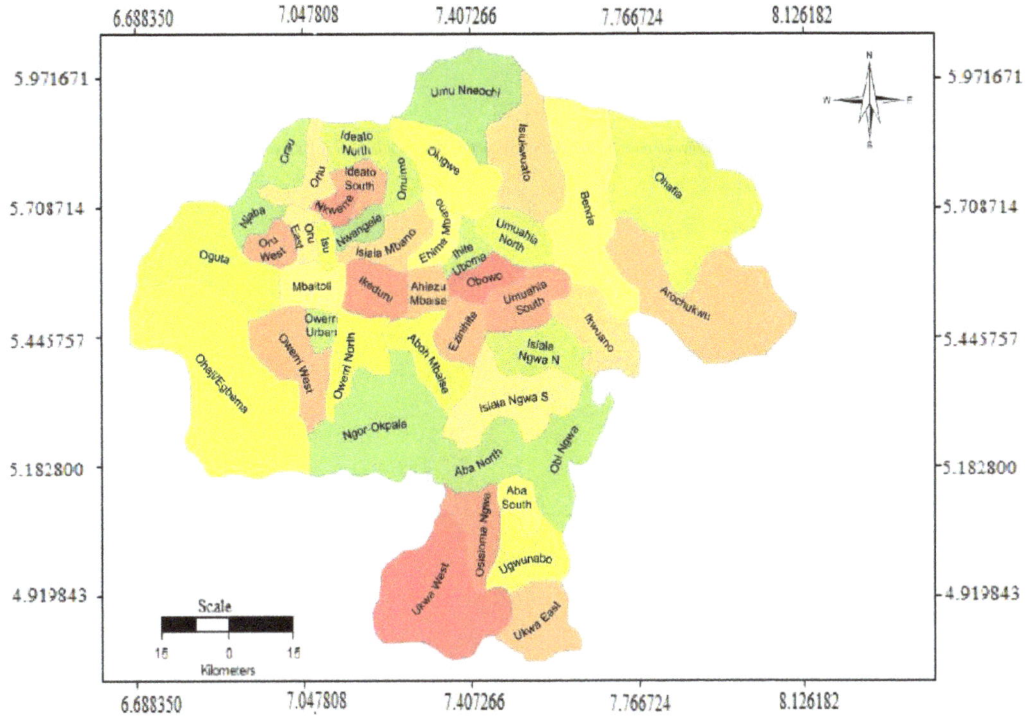

Figure 5. Thematic map of the Imo River basin showing layer of the 44 county areas.

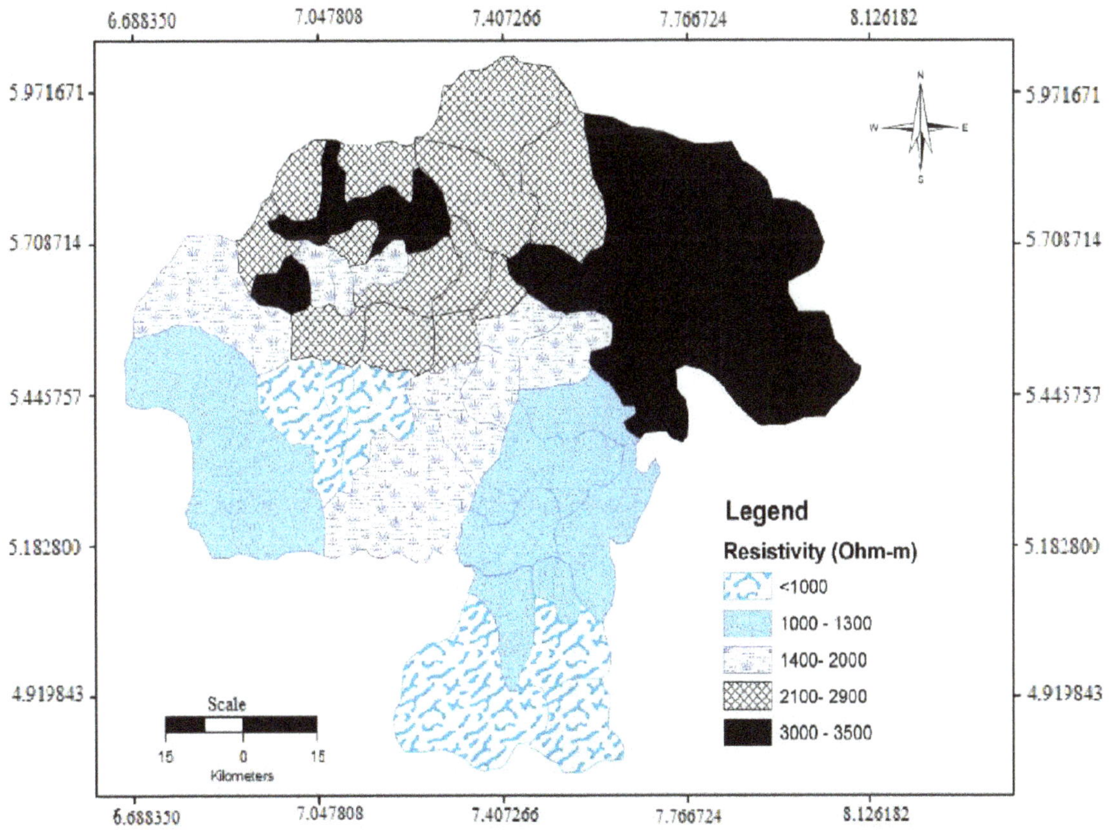

Figure 6. Thematic map of the Imo River basin showing layer of average resistivity

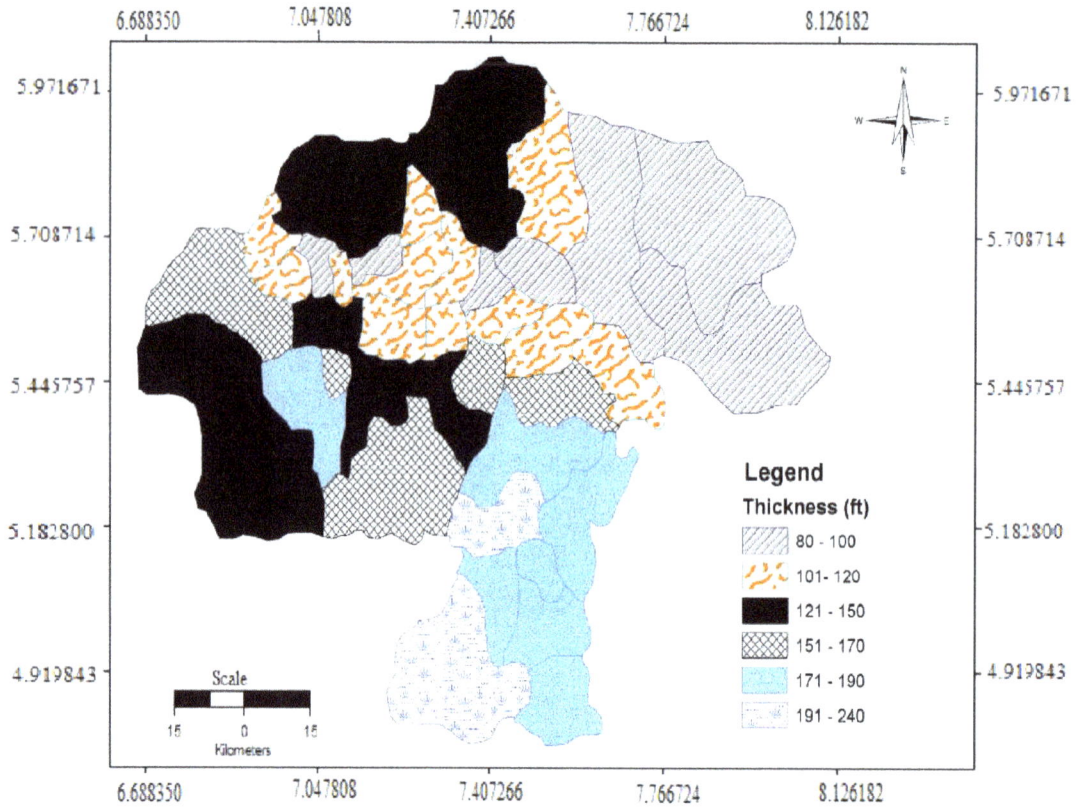

Figure 7. Thematic map of I mo River basin showing layer of average aquifer thickness.

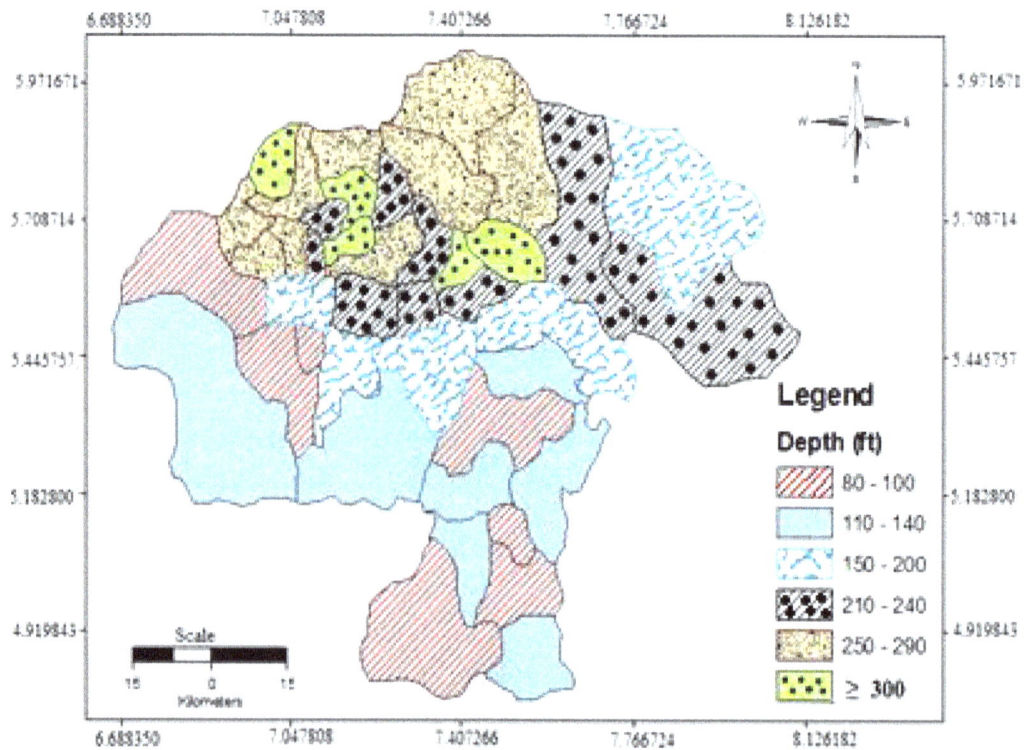

Figure 8. Thematic map of Imo River basin showing layer of average water table.

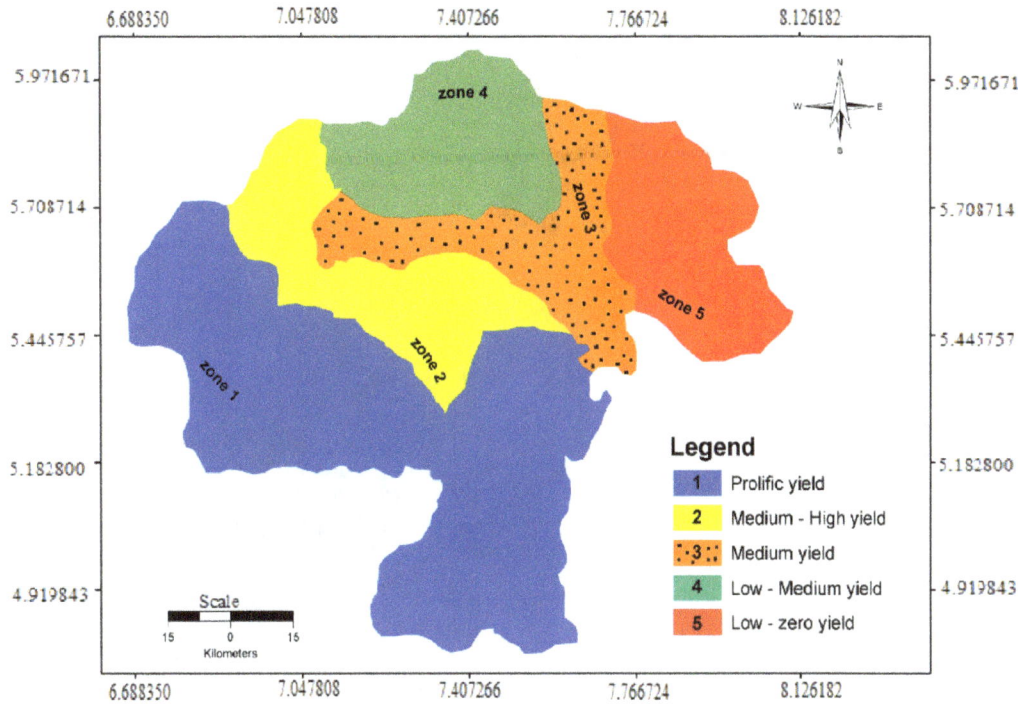

Figure 9. Groundwater prospect Model in the Imo River basin.

This output map is correlated with the groundwater data collected in the field, and represents the groundwater prospect map of the basin. This map integrates the analytical results of pumping test data and well status information in five groundwater areas. Area 1 is characterized by prolific yield; ≥ 25 m^3/h. Area 2 is medium to high yield (M-H), with medium yield being ≥ 11 m^3/h, and area 3 is the medium yield areas. Area 4 is low to medium (L-M), with low yield as ≤ 10 m3/h, whereas Area 5 is the area characterized by very poor to zero yield. This area accounts for a good number of dry or abortive wells. Ganapuram et al. (2008) used a similar classification to assess the groundwater potentials of the Musi basin, and qualitatively they classified the basin into very good, good, moderate, poor and nil based on hydro-geomorphologic conditions.

Area 2 (Medium-high yield) included parts of 6, 7, 8, 15, and 19, *Area 3* (Medium yield) include the northern parts of 3, 5, 6, 7, 8, 15, 19, 23 and 24, while *Area 4* (Low-medium yield) included the southern parts of 3, 5, 7, 8 and 15 county areas. This recommendation is a general guide, which does not eliminate site specific variations as may be applicable. Again, they do not in any way eliminate the mandatory site specific groundwater pre-drilling and down-hole geophysical studies for water well development in the basin.

Groundwater prospect increases from the north to the south of the basin following the characteristic decrease in shaliness from the north to the coastal plain sand of the

Benin Formation in the south. However, the proliferation of shallow private/commercial substandard water wells and the number of abortive, abandoned, and non-functional wells threatens sustainability of groundwater in the basin.

Information on standard water well development of all the 44 local governments shown in the GIS map contain attribute data and ID as contained in Table 1. Attribute data for each local government water well can be assessed by clicking on the particular local government area. A GIS map of this nature is very necessary. It will solve the problem of water well statistics, and remain an essential tool of government for decision making in water resource management. It will serve as a government working tool in the award of contracts for public water wells, and in post construction maintenance of the wells.

Justifying the GIS groundwater prospect model

The existing regional geology map of the basin (Figure 1c) is used to justify the GIS groundwater prospect model of this study. There is significant correlation between the regional geology and the groundwater prospect model as illustrated in Figure 10. Three groundwater prospect zones have been defined out of the five groundwater *Areas* classified earlier, based on the correlation. This correlation could be used to adjust or review boundary of geologic units in future studies. Overlapping boundaries

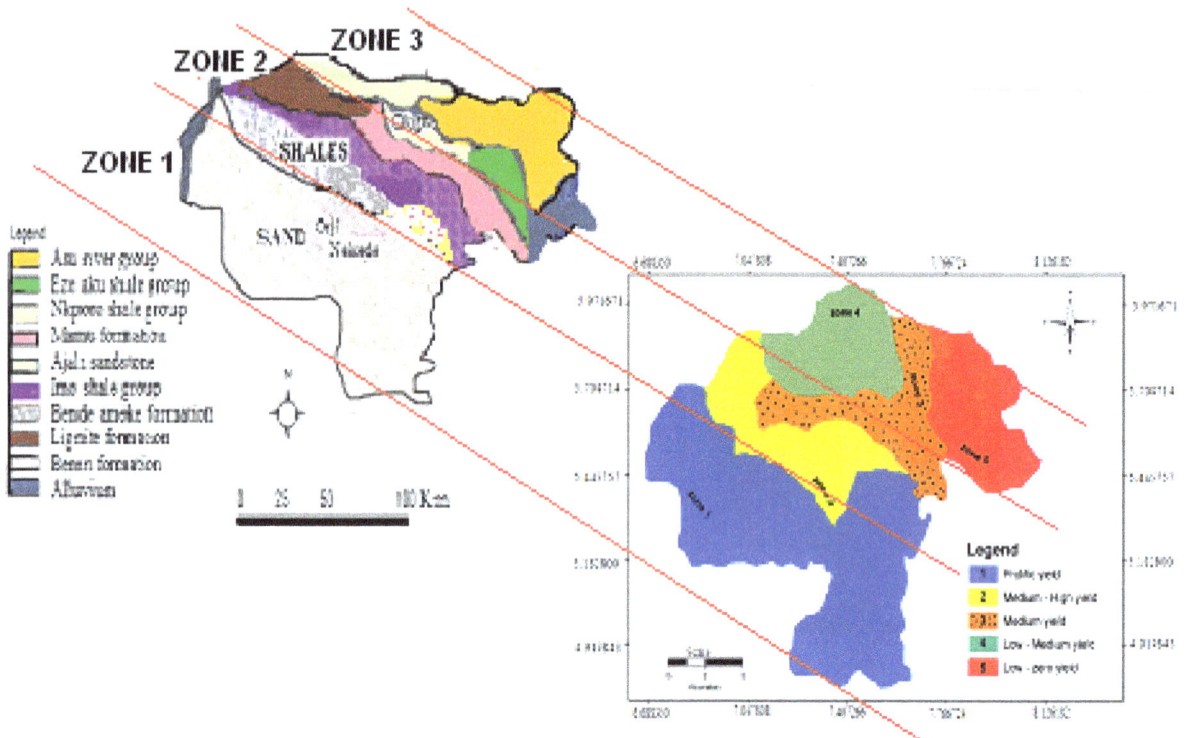

Figure 10. Correlation of Regional geology with the groundwater prospect model.

in the correlation can be described as transition areas. Zone 1, characterized by prolific yield consists of sand belonging to the Benin Formation. Zone two of medium to prolific yield is sandy with mixed shale, and the shaliness increasing to the north. Zone three consists predominantly of shale with minor sand units. This zone is characterized by poor to medium yield. This correlation has further confirmed that groundwater prospect is controlled by subsurface geology. It implies that spatial geophysics and geologic data integrated to GIS can provide comparable information about geology and hydrogeology of a region.

Other problems of groundwater development and supply in the basin

The lack of public water supply notwithstanding, private wells cannot replace public wells in communities. Less than three percent "privileged citizens" supposed to facilitate the maintenance of public water wells have their private shallow wells. They are therefore unconcerned with presence or functionality of public wells in their communities. As a result, most public wells are neglected; non functional, contributing to the inefficiency of public water supply, and low impact of government efforts towards public water supply in the area.

A summary of other major factors militating against

sustainable groundwater development in the basin is as follows:

1) Poor environmental quality causing groundwater pollution.

2) Improper or partial distribution of public wells due to lack of easily retrievable information on existing wells in the region, and favouritisms. Often a community may have two or more public wells whereas the next community does not have one.

3) Poor maintenance and often sabotage in the development and operational processes of public wells. Fund budgeted for water well development and maintenance may be diverted, resulting to the use of inferior materials and no proper supervision during the well construction.

4) Several non functional and abortive wells are not properly closed or sealed from interference; some become waste disposal pits. Such wells are transport pathways of contaminants to groundwater.

RECOMMENDATIONS

Water well partnership

Surface water development is thus recommended in the difficult, low to zero yield areas of zones 4 and 5, against the prevailing groundwater development. We are recommending two types of water well partnerships as innovative strategy to improve water supply, improve water quality, reduce incidents of well failure, and uphold a sustainable groundwater development system in the Imo River basin. First is the government-private partnership: In this case, government wells could be privatized even at immediately after completion to citizens under a mutual agreement. The partner citizens will be responsible to the day to day operation, distribution and management of the wells to the benefits of the host communities. Second is a private-private partnership: Here two or more individuals and organizations that have plans to develop their private wells could combine their resources. By so doing, rather than the proliferation of shallow substandard private/commercial wells, the partners now could develop standard water wells to serve their respective households and neighbourhood. Government may provide incentives to support water well partnerships.

Conclusion

Groundwater prospect GIS model of a river basin is a vital tool to sustainable development of groundwater resource. Comparing groundwater prospect GIS model with the basin's geology map provides better judgement on how to manage and conserve groundwater in a basin. Result of this study has confirmed that surface water development is most appropriate in the difficult, low to zero yield areas of zones 4 and 5, as, against groundwater development.

ACKNOWLEDGEMENTS

The authors are grateful to GEOPROBE Int'l Consultants Ltd, Imo State Water Development Agency (IWADA), Anambra-Imo River basin and Rural Development Authority, Abia State Water Corporation, and UNICEF Owerri for all the well data used. We thank Montclair State University New Jersey, U.S.A. for assistantship, and staff of GEOPROBE for field supports.

REFERENCES

Comely HH (1987). Cyanosis in infants caused by nitrates in well water. J. Am. Med. Assoc. 257:2788-2792.

Dobrin MB, Savit CH (1988). Introduction to geophysical prospecting, 4th edn. McGraw-Hills, New York

Ekwe AC, Onu NN, Onuoha KM (2006). Estimation of aquifer hydraulic characteristics from electrical sounding data: The case of middle Imo River basin aquifers, south-eastern Nigeria. J. Spatial Hydrol. 6:2.

Ekweozor CM, George IU (1990). First discovery of oil shale in the Benue Trough, Nigeria. Sci. Direct 69(4):502-508.

Ganapuram S, Vijaya Kumar GT, Murali Krishna IV, Ercan K, Cüneyd Demirel M (2008). Mapping of groundwater potential zones in the Musi basin using remote sensing data and GIS Advances in Engineering Software Elsevier- Science Direct. 40(7):506-518.

Gezahegn LB (2012). Delineation of groundwater potential zones of Upper Tumet catchment, Western Ethiopia using remote sensing and GIS. LAP Lambert Academic Publishing (2012-06-15), p. 92.

Ibe KM, Nwankwor GI, Onyekuru SO (2007). Assessment of Ground Water Vulnerability and its Application to the Development of Protection Strategy for the Water Supply Aquifer Owerri, South-eastern Nigeria. J. Environ. Monlt. Assess. 67:323-360.

Jasrotia AS, Singh R (2005). Groundwater pollution vulnerability using the drastic model in a gis environment, Evak-rui watersheds. India J. Environ. Hydrol. 13:11.

Khalek AA, Ahmed O (2008). Integration of Remote Sensing, Geophysics and Gis to Evaluate Groundwater Potentiality – A Case Study In Sohag Region, Egypt. The 3rd International Conference on Water Resources and Arid Environments, 1st Arab Water Forum.

Kumar PGN, Srinivas P, Jaya CK, Sujatha P (2010). Delineation of groundwater potential zones using Remote sensing and GIS techniques: A case study of Kurmapalli Vagu Basin in Andhra Pradesh. India Int. J. Water Res. Environ. Eng. 2(3):70-78. Academic Journals.

Kamaleshwar P, Ravindran K, Prabakaran B (2000). Groundwater Prospect Zoning Using Remote Sensing and Geographical Information System: A Case Study in Dala-Renukoot Area. J. Indian Soc. Rem. Sens. 28(4):249-263.

Kearey P, Brooks M, Hill I (2002). An Introduction to Geophysical Exploration. Minshull Geol. Mag. 140:366.

Kelly WE (1977). Geo-electric sounding for estimating aquifer hydraulic conductivity. Groundwater 15(6):420-425.

Mayilvagama MK, Mohana P, Naidu KB (2011). Delineating groundwater potential zones in Thurinjapuram watershed using geospatial techniques. Indian J. Sci. Technol. 4(11):1470-1476.

Nwachukwu MA, Feng H, Alinnor J (2010a). Assessment of heavy metal pollution in soil and their implications within and around mechanic villages. Int. J. Environ. Sci. Technol. 7(2):347-358.

Nwachukwu MA, Huan F, Maureen IA, Umunna FU (2010b). The Causes and the Control of Selective Pollution of Shallow Wells by Coliform Bacteria, Imo River Basin Nigeria. Water Qual. Expo. Health 2:75-84.

Nwachukwu MA, Huan F, Ophori D (2010c). Groundwater Flow Model and Particle Track Analysis for Selecting Water Quality Monitoring Well Sites, and Soil Sampling Profiles. J. Spatial Hydrol. 10:1.

Onyeagocha AC (1980). Petrography and depositional environment of the Benin Formation Nigeria. J. Min. Geol. 17:147-11.

Omosuyi GO (2010). Geoelectric assessment of groundwater prospect and vulnerability of overburden aquifers at Idanre, SW Nigeria. Ozean J. Appl. Sci. 3:1.

Reyment RA (1965). Aspects of the Geology of Nigeria - University of Ibadan Press.

Reynolds JM (1997). An Introduction to Applied and Environmental Geophysics. John Wiley & Sons Inc.

Singh AK, Ravi PS (2004). Integration of Thematic Maps Through GIS for Identification of Groundwater Potential zones. Proceeding of Map India Conference 2004© GISdevelopment.net.

Shahid S, Kumar NS (2008). GIS Integration of Remote Sensing and Electrical Sounding Data for Hydrogeological Exploration. J. Spatial Hydrol. 2:1.

Telford WM, Geldart LP, Sheriff RE (1990). Applied geophysics. Cambridge University Press.

U.S. Department of Transport (U.S. DOT) (2004). Federal highway administration: Application of geophysical methods to highway related problems. Pub. No FHWA-IF-04-021 (2004).

Ungemach P, Mostaghimi F, Duprat A (1969). Essais de determination du coefficient d-emmagasinement en nappe libre application a la nappe alluvial du Rhin. Bull. Int. Assoc. Sci. Hydrol. XIV(3):169-190.

6

Assessment of pollution levels of groundwater in parts of Imo River Basin, South Eastern Nigeria

Ijeh Boniface Ikechukwu[1] and Onu Nathaniel Nze[2]

[1]Department of Physics, College of Natural and Applied Sciences, Michael Okpara University of Agriculture, P. M. B. 7267, Umuahia, Abia State, Nigeria.
[2]Department of Geosciences, Federal University of Technology, Owerri, Imo State, Nigeria.

Twenty five samples of groundwater were obtained from various boreholes in the study area and subjected to physico-chemical analysis using standard laboratory techniques. The study was aimed at the assessment of the pollution status of the groundwater supply in parts of Imo River Basin and its environs using some pollution indications namely: electrical conductivity, phosphate (PO_4), total dissolved solids (TDS), nitrate (NO_3^-), sulphate and turbidity. The values of the physico-chemical parameter were correlated with the World Health Organisation (WHO) values. The result shows that the level of pollution is relatively high in Owerri. The highest concentration of electrical conductivity and nitrate were obtained at Owerri. The highest concentration of phosphate was observed at Mbaise area. The most saline part is the southern part proximate to the delta area. Moderate to higher values of TDS were observed at the densely populated areas of Owerri, Umuahia and Aba. Electrical conductivity in Owerri ranged from 101 to 181.6 NS/cm which exceed the WHO standard of 100 NS/cm. Phosphate levels exceeding the WHO standard of 5 mg/L were observed at the Mbaise axis. No borehole location exceeded the required maximum standard for TDS which show general stability.

Key words: Groundwater, physico-chemical analysis, pollution status, Imo River Basin, borehole.

INTRODUCTION

Water is essential for livelihood as well as socio-economic development of any community. Many communities in Nigeria, especially in the Imo River Basin area rely on surface and groundwater for both domestic and agricultural water supplies. It is estimated that approximately one third of the world's population use groundwater for drinking (Nickson et al., 2005). Groundwater pollution is a growing environmental problem, especially in developing countries. Many major cities and small towns in Nigeria depend on groundwater for water supplies, mainly because of its abundance, stable quality and also because it is inexpensive to exploit. However, the urbanization process threatens the groundwater quality because of the impact of domestic

and industrial waste disposal. This results in aquifer deterioration, since some of these waste products, including sewage and cesspool may be discharged directly into the aquifer system. Water soluble wastes and other materials that are dumped, spilled or stored on the surface of the land or in sewage disposal pits can be dissolved by precipitation, irrigation waters or liquid wastes and eventually seep through the soil in the unsaturated zone to pollute the groundwater. Once contaminated, it is difficult, if not impossible, for the water quality to be restored. Thus constant monitoring of groundwater quality is needed so as to record any alteration in the quality and outbreak of health disorders. Groundwater quality depends, to some extent, on its chemical composition (Wadie and Abduljalil, 2010) which may be modified by natural and anthropogenic sources. Rapid urbanization, especially in developing countries like Nigeria, has affected the availability and quality of groundwater due to waste disposal practice, especially in urban areas. Once groundwater is contaminated, its quality cannot be restored by stopping the pollutants from source (Ramakrishnaiah et al., 2009). As groundwater has a huge potential to ensure future demand for water, it is important that human activities on the surface do not negatively affect the precious resource (Sarukkalige, 2009). Poor environmental management creates havoc on the water supply, hygiene and exacerbating public health (Okoro et al., 2009). Tay and Kortatsi (2008) emphasize on the importance of groundwater globally as a source for human consumption and changes in quality with subsequent contamination can, undoubtedly, affect human health.

Geology and hydrogeology of the study area

The Imo River Basin lies between Latitudes 4° 38'N and 6° 01'N and between Longitudes 6° 53'E and 7° 32'E and covers an area of about 9100 km². The boundaries are defined by its surface drainage divides. There are two main sub- basins within the basin: The Oramirukwa—Otamiri sub- basin and the Aba River sub-basin. The estuary of the Imo River at the Atlantic Ocean forms the southern boundary. There are two prominent features at north-eastern and north-western boundaries; these are the Udi-Okigwe-Arochukwu and the Awka-Umuchu-Umuduru sedimentary cuestas, respectively (Uma, 1989).

The Imo River Basin is based on a bedrock of a sequence of sedimentary rocks of about 5480 m thick and with ages ranging from Upper Cretaceous to Recent (Uma, 1986). The deposition of these sedimentary rocks is related to the opening of the South Atlantic Ocean and the formation of the rift-like Benue Trough of Nigeria in the Mesozoic (225-65 M.Y.B.P.) (Schlumberger, 1985).

Generally, there are two different classes of formations underlying the Imo River Basin. About 80% of the basin consists in Coastal Plain Sand, which is composed of non-indurated sediments represented by the Benin and Ogwashi-Asaba Formations, and alluvial deposits at the estuary at the Southern end of the Imo River Basin. The remaining 20% is underlain by a series of sedimentary rock units that get younger southwestward, a direction that is parallel to the regional dip of the formations.

The Ajali Sandstone of Maastrichtian age is the oldest exposed formation in the basin, outcropping at its north-eastern fringe along a NW-SE band (2 to 4 km width). It consists of thick friable, loosely consolidated sandstones (Uma, 1989). Overlying the Ajali Sandstone conformably is the Nsukka Formation (Maastrichtian-Lower Paleocene), which extends to a relatively broader stretch of land than the former. It consists of alternating sequences of sandstones, shales and sandy shales. It

dips at about 6°, on the average, to the south-west. The Imo Shale of Paleocene-Lower Eocene age overlies the Nsukka Formation unconformably. It consists of a thick sequence of blue and dark grey shales with occasional bands of clay-ironstones and subordinate sandstones (Ekwe et al., 2006). Next in the depositional sequence is the Ameki Formation (Eocene), which consists of sand and sandstones. The lithologic units of the Ameki Formation fall into two general groups (Whiteman, 1982; Arua, 1986); an upper grey-green sandstones and sandy clay and a lower unit with fine to coarse sandstones, and intercalations of calcareous shales and thin shelly limestone. Next in the depositional sequence is the Ogwashi/Asaba Formation (Oligocene to Miocene), which is generally made up of clays, sands, grits and seams of lignite alternating with gritty clay. This formation is characterized by its up dip and down dip pinch outs within the Imo Basin. The Ogwashi/Asaba Formation is overlain by the Benin Formation (Miocene to Recent) which is the most extensive of all the formations, which covers more than half of the area of the basin. It consists of sands, sandstones, and gravels, with intercalations of clay and sandy clay. The sands are fine-medium-coarse grained and poorly sorted. The map of the study area is shown in Figure 1.

METHODOLOGY

Twenty-five groundwater samples were collected from wells located around the Imo River Basin of South Eastern Nigeria. The samples were stored in a sterilized 250 ml bottles and then taken to the laboratory for analysis. The electrical conductivity, total dissolved solids, Nitrate, sulphate, phosphate and salinity were determined using a HA-CH 44600-00 Condutivity/TDS meter at a temperature of 20°C. These samples were refrigerated and analyzed within 24 h. All plastics and glass wares utilized were pre-washed with detergent water solution, rinsed with tap water and soaked for 48 h in 50% HNO_3 then rinsed thoroughly with distilled- deionized water. They were then air-dried in a dust free environment. The pH was determined using a HACH sensor 3 pH meter. The turbidity was determined using a spectrophotometer. The result is presented in Table 1.

RESULTS AND DISCUSSION

Investigations of the pollution status of groundwater in the study area were conducted recently by collecting water samples from boreholes in different locations in the study area (Table 1). Water samples from 25 randomly selected boreholes in the study area were analyzed for chemical quality at the UNICEF Water Project, Owerri, and Imo State Environmental Protection Agency, respectively. The result was geo-processed to obtain groundwater quality maps showing the spatial variation of electrical conductivity, sulphate, phosphate, total dissolved solids (TDS), salinity, nitrate respectively. There specific parameter maps facilitate the rapid assessment of the extent of pollution of the various locations

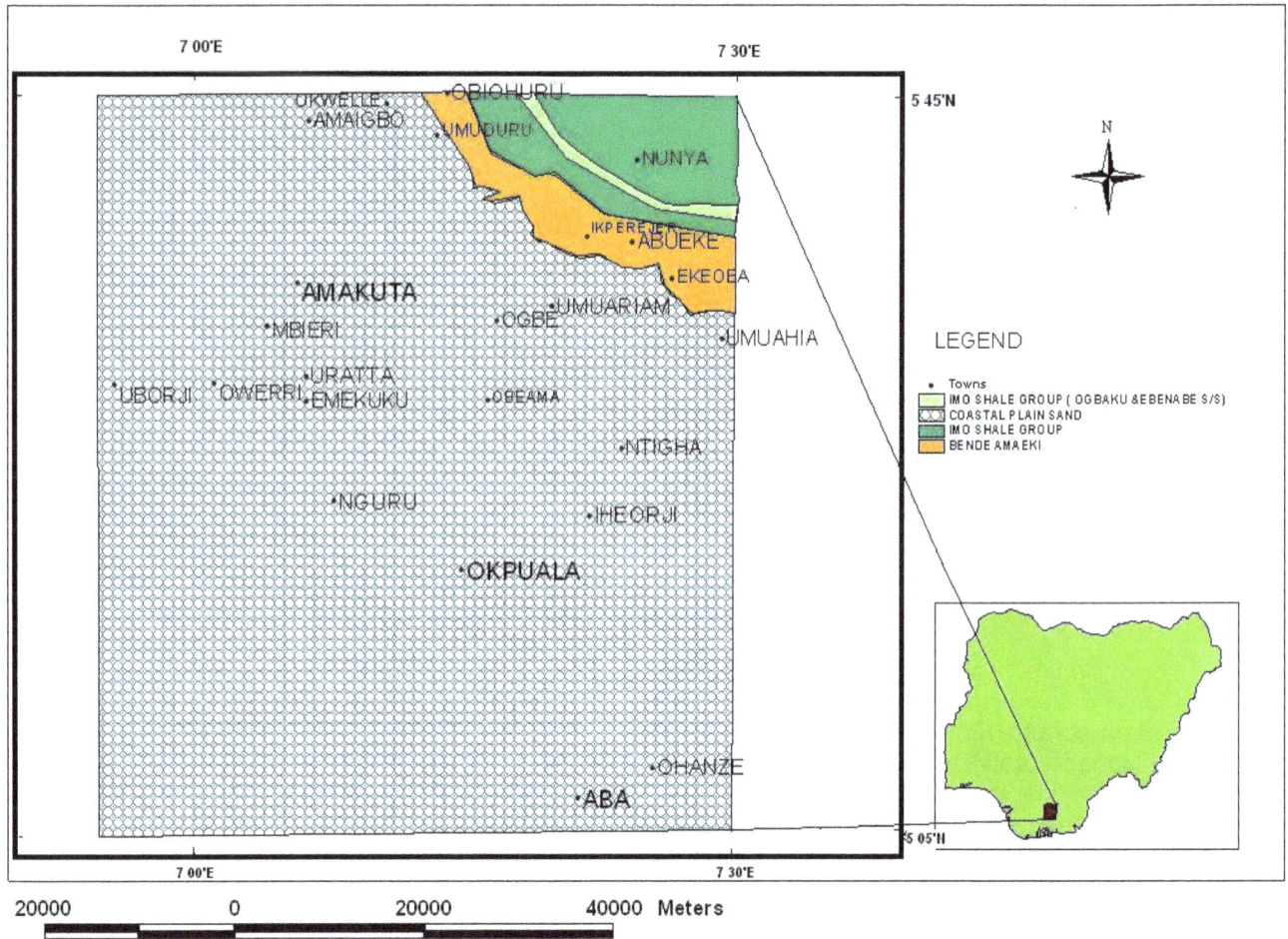

Figure 1. Location map of the study area.

within the study area in terms of their respective concentrations. Contour maps of the spatial variation of electrical conductivity, phosphate, sulphate, salinity, total dissolved solid, nitrate, and turbidity were also developed. The highest conductivity was obtained at BH35 in MCC Road, Owerri (181.6 µs/cm).

Electrical conductivity

The map of the spatial variation of electrical conductivity is shown in Figure 2. Electrical conductivity of water is used as an indicator of how salt- free, ion-free, or impurity-free the sample is; the purer the water the lower the conductivity (the higher the resitivity). The World Health Organization standard for acceptable electrical conductivity is 100 µs/cm. In addition to BH35m, other areas with electrical conductivity above the WHO standard are at BH39 (111.32 µs/cm), BH141 (148 µs/cm), BH42 (101.0) µs/cm), BH59 (153.0 µs/cm), BH65 (135.0 µ/cm) and BH66 (136.0 µs/cm). Except for Amauzi Obowo (BH59) and Avu Amaaku (BH66), all the other

locations with electrical conductivity above 100 µs/cm are located within Owerri Municipal. Pure water has an electrical conductivity of 5.5 µs/cm, a measure of the total dissolved solid (TDS), while rain water and ocean water have 5000 to 30000 µs/cm and 45,000 to 60,000 µs/cm respectively. Normal groundwater has a range of 100 to 2000 µs/cm (Offodile, 2002). It is interesting to note that in spite of the huge populat ion in Aba the electrical conductivity is not quite high. This is a possible indication that the hydro-geological factors that determine the rate of pollution are favorable in Aba, making it less vulnerable to pollution. The location on the map corresponding to Nkwo Obohia, which had the highest resistivity, shows lower electrical conductivity.

Phosphate

Figure 3 shows the map of phosphate concentration in the study area. Phosphorus is one of the key elements necessary for the growth of plants and animals. Phosphates are not toxic to people or animals unless

Table 1. Results of groundwater quality analysis of samples collected from selected boreholes in the study area.

S/N	CODE	LOCATION	pH	ELECTRICAL CONDUCTIVITY (µs/cm)	TDS (mg/l)	NITRATE (mg/l)	SULPHATE (mg/l)	PHOSPHATE (mg/l)	SALINITY	TURBIDITY (NTU)
		WHO STANDARD (WHO,2011)	6.5-8.5	100.0	250.0	50.000	250.00	5.00	50.0	50.0
1	BH27	UMUBIAM, ABOH MBAISE	6.80	98.2	49.0	0.002	0.02	16.05	0.5	0.1
2	BH28	UMUEZEALA-AMA	6.50	86.3	43.0	0.001	0.58	12.50	0.3	BDL
3	BH30	IFE, EZINIHITTE MBAISE	6.3	86.3	42.9	5.000	0.50	6.07	0.5	0.0
4	BH31	NKWOGWU, ABOH MBAISE	6.3	83.1	41.5	0.020	0.45	14.20	0.6	0.1
5	BH32	ASSUMPTA, CATH., OWERRI	6.8	44.8	22.6	1.200	0.80	2.40	BDL	BDL
6	BH33	OFOROLA, OWERRI WEST	6.5	94.1	47.0	0.100	BDL	BDL	16.5	BDL
7	BH34	URATTA, OWERRI NORTH	6.0	95.6	47.9	0.020	0.40	BDL	13.2	BDL
8	BH35	MCC RD. OWERRI NORTH	7.3	181.6	91.6	13.200	BDL	0.85	13.2	10.0
9	BH36	TIMBER SHADE OWERRI	6.5	21.1	36.5	13.500	2.00	BDL	3.3	BDL
10	BH37	UBOMIRI MBAITOLU	6.5	5.3	2.6	27.200	1.00	BDL	2.5	BDL
11	BH38	OKWU NGURU ABOH MBAISE	6.5	36.2	19.0	0.300	0.01	BDL	11.5	0.5
12	BH39	OWERRI GIRLS' SEC, OWERRI	7.0	111.3	58.7	1.300	0.80	BDL	7.42	BDL
13	BH40	MBIERI, MBAITOLU	6.9	81.0	44.2	0.150	3.00	BDL	1.65	BDL
14	BH41	ALADINMA OWERRI	7.1	148.6	77.7	0.130	4.00	BDL	2.48	BDL
15	BH42	UGAKWOCHE OBUBE, OWERRI	6.0	101.0	51.0	11.000	12.50	BDL	1.68	BDL
16	BH43	UMUCHOKU OBUBE, OWERRI	6.5	53.0	26.5	15.000	0.30	BDL	0.75	BDL
17	BH44	UMUAKPAA OBUBE, OWERRI	6.3	82.0	41.0	0.120	2.50	BDL	1.16	BDL
18	BH45	OBOAME, ABOH MBAISE	6.0	36.3	19.0	7.000	0.20	BDL	11.55	BDL
19	BH46	VILLA MARIA OWERRI	6.9	49.0	25.0	0.000	BDL	BDL	4.95	BDL
20	BH48	UMUGUMA AMAAKU, OWERRI WEST	7.0	16.7	8.3	32.800	1.20	BDL	2.50	0.3
21	BH49	UGORJI OWERRI WEST	6.7	20.8	10.5	35.200	4.00	BDL	6.60	2.0
22	BH50	OBINZE BARACK OWERRI WEST	6.8	36.5	18.0	39.600	3.00	BDL	3.30	5.0
23	BH51	OBINZE MAMI MKT	7.5	5.3	2.6	37.200	1.00	BDL	2.50	10.0
24	BH52	UZII PRIMARY SCH. OWERRI	7.2	66.6	33.3	50.600	0.01	BDL	3.30	0.1
25	BH53	UDO EZINIHITTE MBAISE	6.9	90.6	49.3	0.120	BDL	BDL	3.30	5.0

*BDL : Below detectable limit.

they are present in very high levels. Digestive problems could occur from extremely high levels of phosphate. The WHO standard for phosphate in drinking water is 5 mg/L. This standard is exceeded in the following areas: BH27 (16.05 mg/L); BH28 (12.50 mg/L); BH30 (6.07 mg/L); BH31 (14.20 mg/L). BH27, BH28 and BH31 are in Mbaise area, which is reported to be the most densely populated rural area in Nigeria (Onwuegbuche, 1993). They also engage in a lot of farming requiring the use of fertilizer. Much of the areas around Aba, Umuahia and Owerri have generally low concentration of phosphate (≤ 2 mg/L), perhaps, due to less farming in those urban centers.

Nitrate

Figure 4 shows the contour map of spatial variation of

nitrate concentration in the study area. Nitrate is an essential ingredient of plant nutrition. It is, however regarded as an indicator of pollution in public water supply (Offodile, 2002). The WHO standard for nitrate in drinking water is 50 mg/L. This standard is exceeded in BH53 (50.6 mg/L).

Total dissolved solids

Figure 5 shows the map of the spatial variation of the total dissolved solids (TDS) in the study area. The total dissolved solids (TDS) provide a rough indication of the overall suitability of water for whatever purpose. The WHO standard for TDS in drinking water is 250 mg/L. No borehole location exceeded the required maximum standard. The map indicates higher values around the densely populated areas of Owerri, Umuahia and Aba.

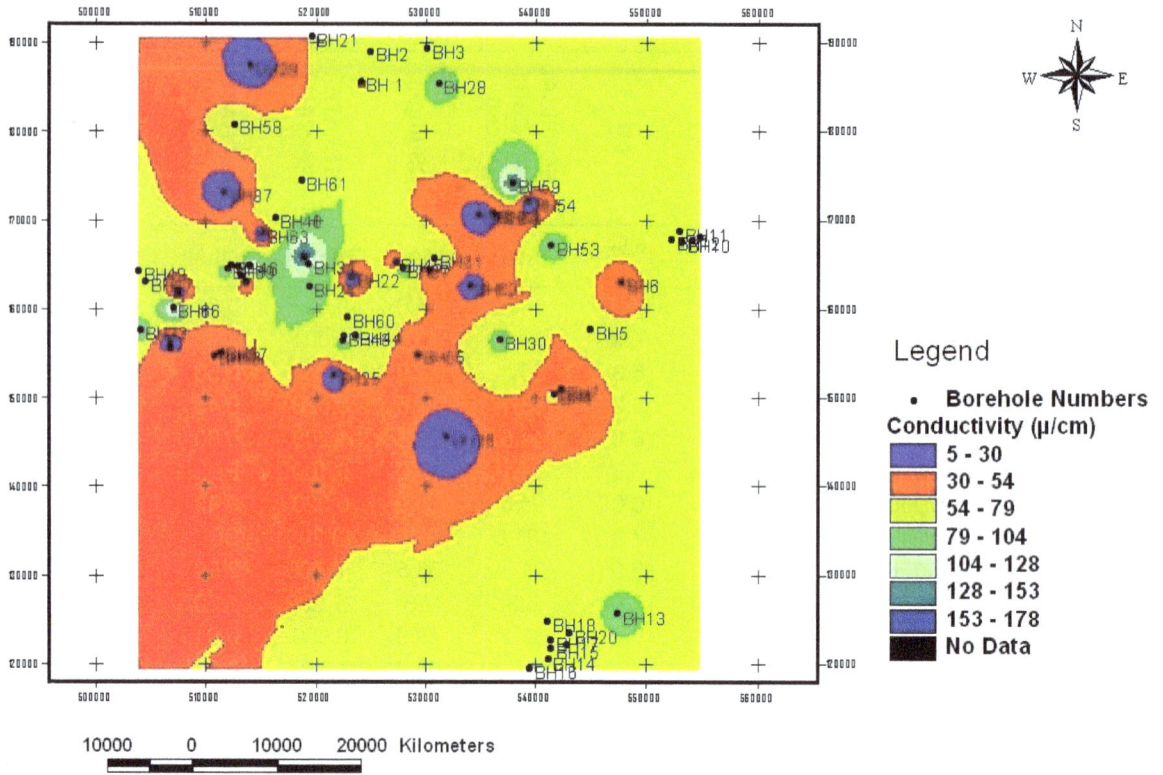

Figure 2. Spatial variation of electrical conductivity in the study area.

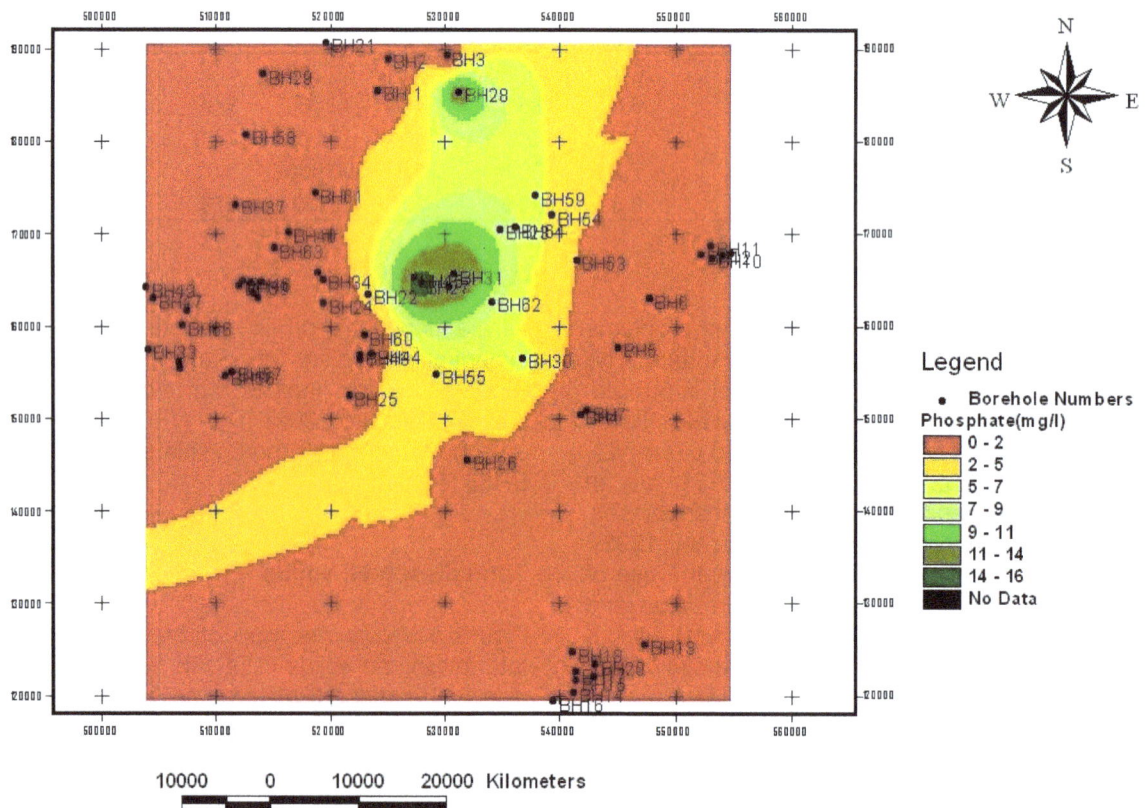

Figure 3. Spatial variation of phosphate concentration in the study area.

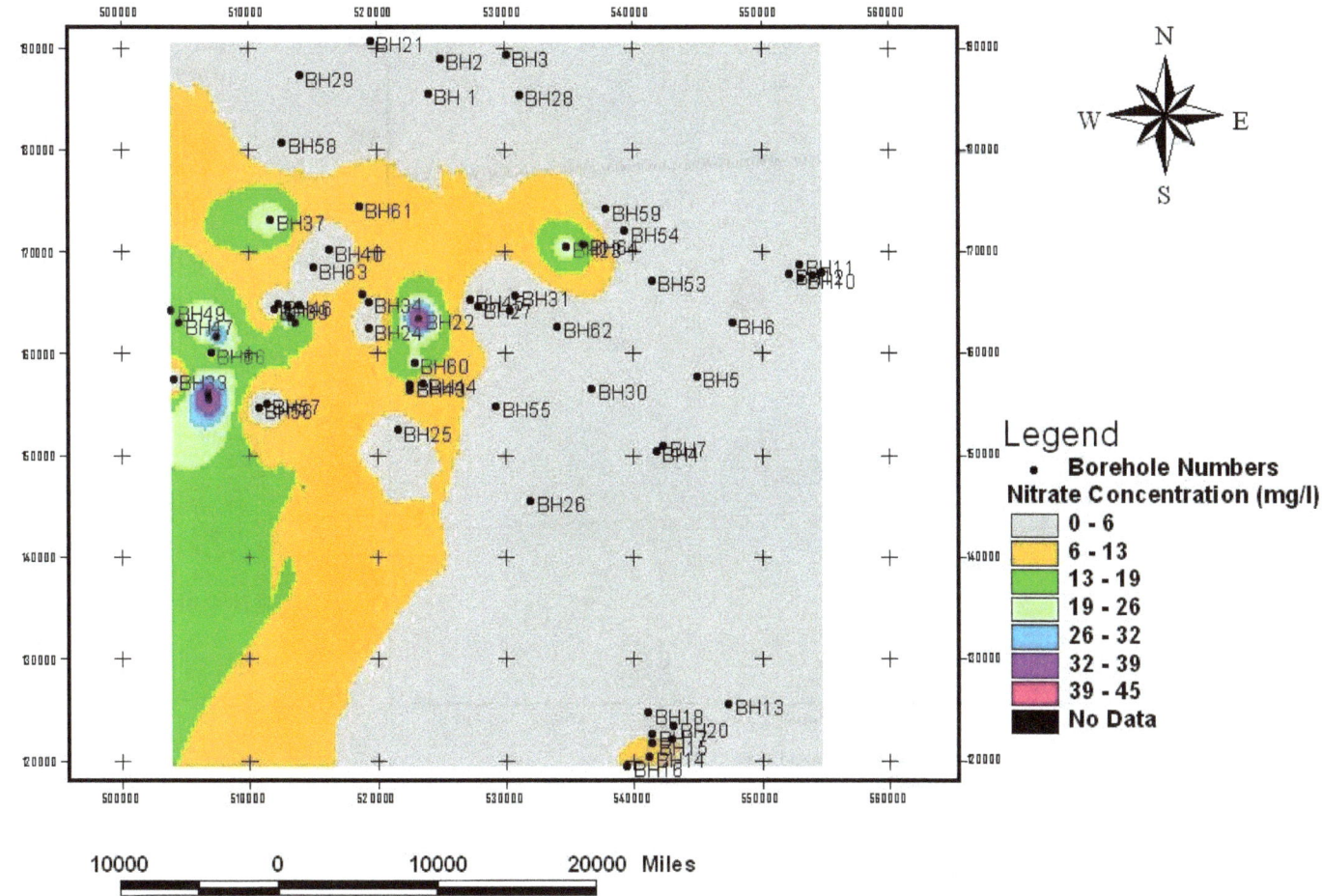

Figure 4. Spatial variation of nitrate concentration in the study area.

There are also higher values around Mbaise area (BH62). This may be due to increased pollution arising from increased domestic activity, fertilization and industrial waste.

Turbidity

Figure 6 shows the contour map of the turbidity in the study area. Turbidity is the amount of cloudiness in the water. This can vary from a river full of mud and silt where it would be impossible to see through the water (high turbidity), to a spring water which appears to be completely clear (low turbidity).Turbidity can be caused by silt, sand and mud, bacteria and other germs, and chemical precipitates. It is very important to measure the turbidity of domestic water supplies, as these supplies often undergo some type of water treatment which can be affected by turbidity. Turbidity was measured in nephelometric turbidity units (NTU), using a turbidity meter because of its accuracy. The map shows that most of the areas investigated are within acceptable WHO standard.

Sulphate

Figure 7 shows the map of the spatial variation of sulphate concentration in the study area. Sulphate occurs mostly as Calcium Sulphate (Gypsum). Sodium and Magnesium Sulphate are readily soluble in water while Calcium Sulphate is less so. Sulphur is useful to plants (Offodile, 2002). High levels of sulphate in drinking water can cause diarrhea (EPA/CDC, 1999). The WHO standard for Sulphate in drinking water is 250 mg/L. From the study no borehole was found to have excess sulphate. The map shows that the northeast quadrant of the study area and a bit of the southeast have generally less concentration of sulphate than the west, northwest and south of the study area.

Conclusion

From the groundwater quality analysis, the highest concentration of electrical conductivity and nitrate were obtained at Owerri. This may have resulted from the domestic and industrial activities typical of a hugely populated

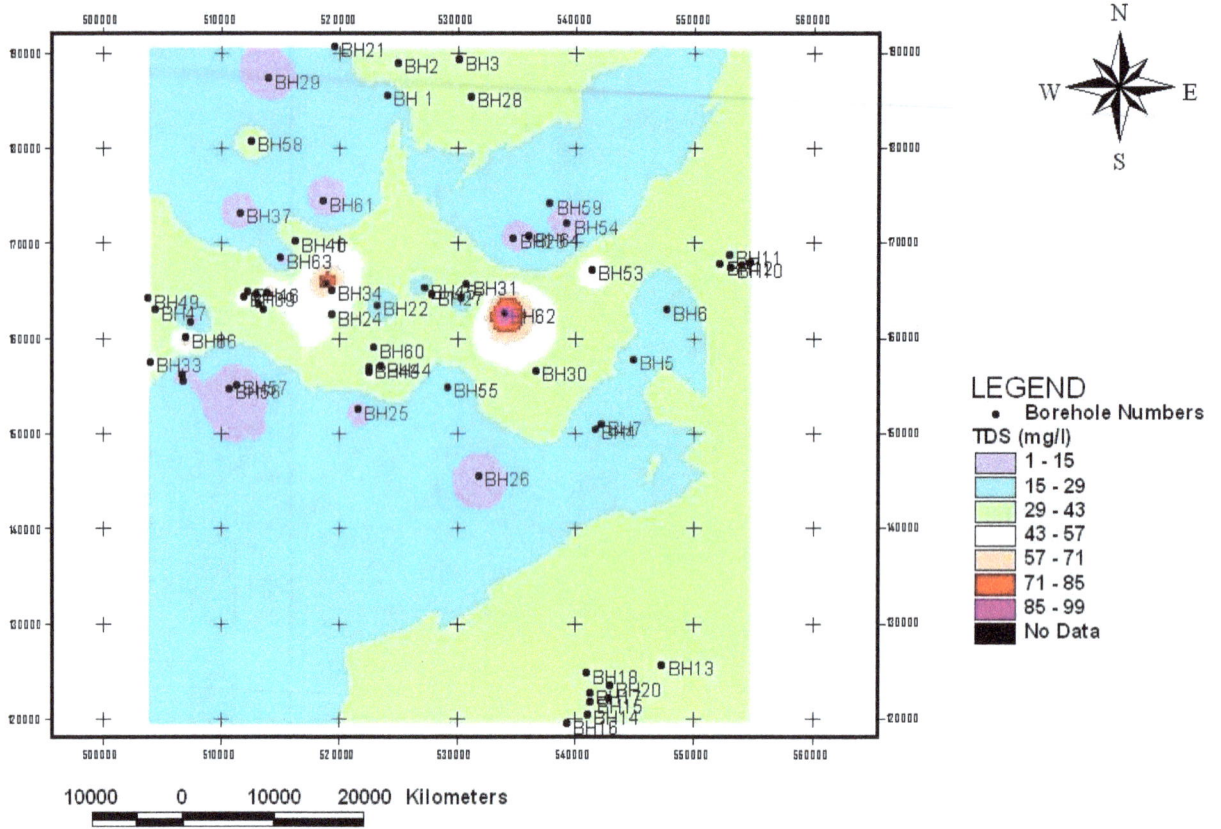

Figure 5. Spatial variation of totalled dissolved solid in the study area.

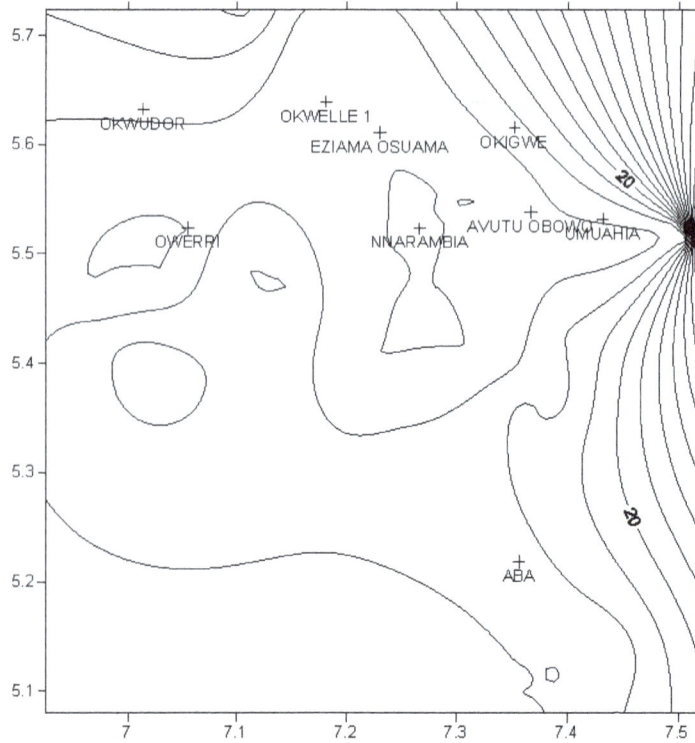

Figure 6. Contour map of turbidity in the the study area (C. I. = 5NTU).

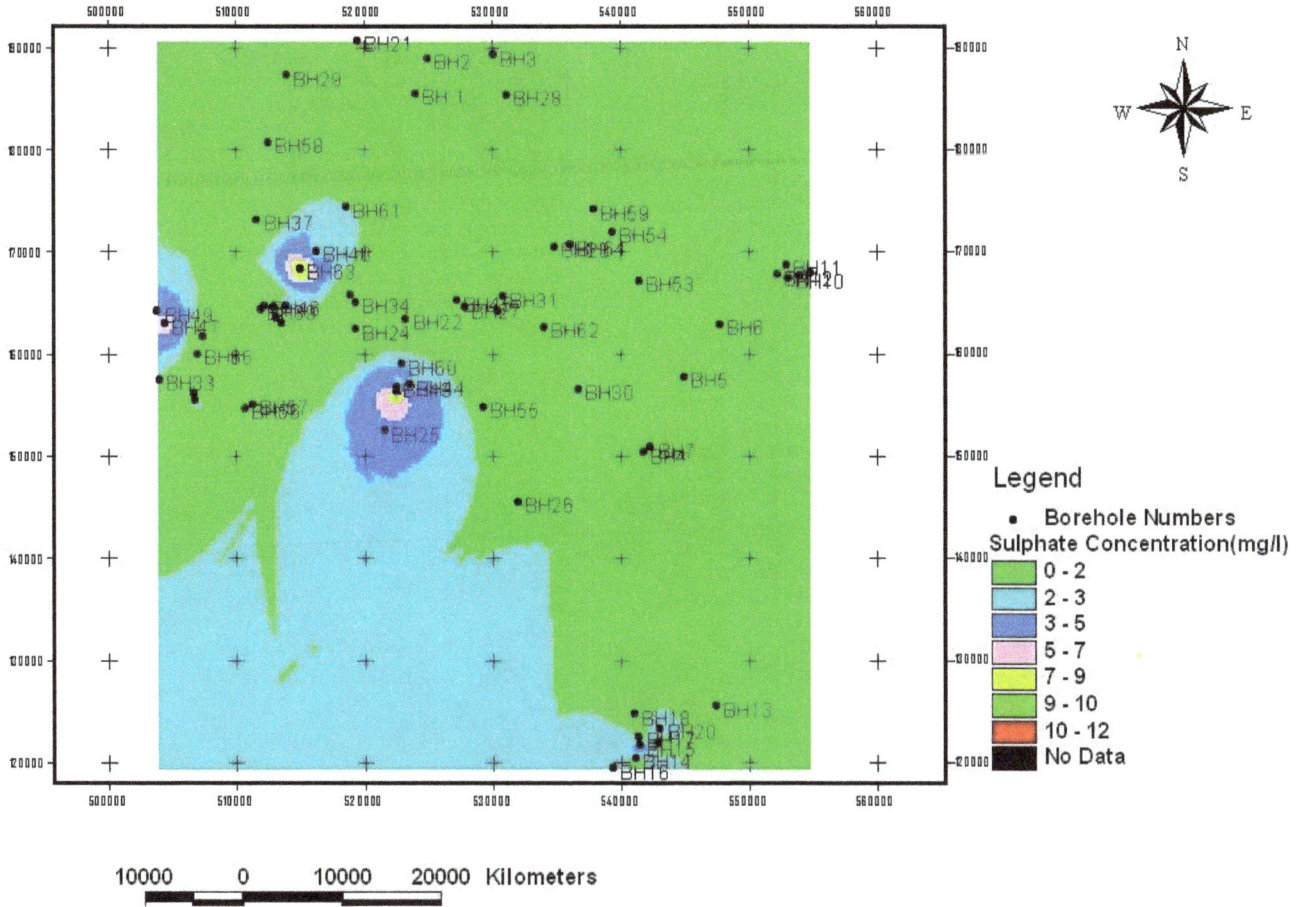

Figure 7. Spatial variation of sulphate concentration in the study area.

area as Owerri. The location on the map corresponding to Nkwo Obohia shows lower electrical conductivity. The highest concentration of phosphate was observed at Mbaise area, the most densely populated rural area in the study area (Onwuegbuche, 1993). Their extensive farming practice with the use of fertilizer can account for this unusual concentration of phosphate. The most saline part of the study area is the southwestern part, probably because of its proximity to the delta area. From the study no borehole was found to have excess sulphate or TDS, but higher values of TDS were observed at the densely populated areas of Owerri, Umuahia and Aba. There are also higher values around Mbaise area. This may be due to increased pollution arising from increased domestic activity, fertilization and industrial waste.

It is recommended that there should be environmental interventions through public health education by community based health workers, awareness and sensitization campaigns be carried out for improved household and community sanitation around Owerri and Nkwo Obohia. Wells located within 50 m from pollution source should be abandoned and future wells should be constructed beyond 250 m from pollution source.

Adequate solid disposal method should be adopted, phasing out open dumpsites to safeguard public health from water borne diseases.

ACKNOWLEDGEMENTS

The authors wish to acknowledge with gratitude the Anambra-Imo River Basin Authority, the Imo State Rural Water Supply Agency, and UNICEF Owerri for giving us access to their information resources. Special thanks to Engr. Emeka Udokporo of FLAB Engineering and Mr. A. O. Kanu of the Abia State Water Board, who made a lot of useful material available and helped in many other ways; also Mr. R. C. Oty of the Anambra- Imo River Basin Authority, and Mr. R. N. Ibe of the Ministry of Public Utilities, Owerri for their tremendous help in the work.

REFERENCES

Arua I (1986). Paleoenvironment of Eocene Deposits in the Afikpo Syncline. J. Afr. Earth Sci. 5(3):279-284.
Ekwe AC, Onu NN, Onuoha KM (2006). Estimation of aquifer hydraulic

characteristics from electric sounding data: the case of middle Imo River basin aquifers, south-eastern Nigeria. J. Spatial Hydrol. 6(2):121-132.

Environmental Protection Agency/Centre for Disease Control Study (1999). Federal Register Notice of Sulfate Health Effects from Exposure to High Levels of Sulfate in Drinking Water Study, p. 25.

Ijeh BI (2010). Assessment of Pollution Status and Vulnerability of Water Supply Aquifers in Parts of Imo River Basin, Southeastern Nigeria. Unpublished Ph.D thesis, Department of Geosciences, Federal University of Technology, Owerri, Nigeria. pp.122-142.

Nickson RT, McArthur JM, Shrestha B, Kyaw-Myint TO, Lowry D (2005). Arsenic and other dringking water quality issues, Muzaffargarh District, Pakistan. Appl. Geochem. 20(1):55-68.

Offodile ME (2002). Ground Water Study and Development in Nigeria, Mecon Geology and Eng. Services Ltd., 2nd Edition. pp. 303-332.

Onwuegbuche AA (1993). Geoelectrical investigations in the Imo River Basin Nigeria. Unpublished Ph.D thesis. Department of Physics, University of Calabar, pp. 2-62.

Schlumberger (1985). Well evaluation conference Nigeria: Schlumberger Technical Services Inc.

Okoro EI, Egboka BCE, Anike OL, Onwuemesi AG (2009). Integrated water resources management of the Idemili River and Odo River drainage basins, Nigeria. Improving Integrated Surface and Groundwater Resources Management in a Vulnerable and Changing World (Proceedings of JS3 at the Joint IAHS & IAH Convention, Hyderabad, India, September 2009). IAHS Pub. pp. 117-122.

Ramakrishnaiah CR, Sadashivaiah C, Ranganna G (2009). Assessment of Water Quality Index for the Groundwater in Tumkur Taluk, Karnataka State, India. E-J. Chem. 6(2):523-530.

Sarukkalige PR (2009). Impact of land use on groundwater quality in Western Australia. Improving Integrated Surface and Groundwater Resources Management in a Vulnerable and Changing World (Proceedings of JS3 at the Joint IAHS & IAH Convention, Hyderabad, India, September 2009). IAHS Pub. pp. 136-142.

Tay C, Kortatsi B (2008). Groundwater quality studies: A Case study of the Densu Basin, Ghana. W. Afr. J. Appl. Ecol. p. 12.

Uma KO (1986). Analysis of Transmissivity and Hydraulic Conductivity of Sandy Aquifers of the Imo River Basin. Unpl. Ph.D. Thesis University of Nigeria, Nsukka.

Uma KO (1989). An appraisal of the groundwater resources of the Imo River Basin, Nigeria. J. Min. Geol. 25(1&2):305-331.

Wadie AST, Abuljalil GAS (2010). Assessment of hydrochemical quality of groundwater under some urban areas with Sana'a Secretariat Ecletica quimica. www.SCIELO.BR/EQ. 35(1):77-84.

Whiteman A (1982). Nigeria: Its Petroleum Geology Resources and Potentials, Vol.2, Graham and Trotman Publ., London SWIVIDE.

World Health Organization (WHO) 2011. Guidelines for Drinking Water quality, Vol.1, 4th Ed; Recommendations, Geneva.

7

Hydrogeochemical Study of Surface Water Resources of Orlu, Southeastern Nigeria

S. I. Ibeneme[1] , L. N. Ukiwe[2], A. O. Selemo[1] C. N. Okereke[1], J. O. Nwagbara[1], Y. E. Obioha[1], A. G. Essien[1], B. O. Ubechu[1], E. S. Chinemelu[1], E. A. Ewelike[1] and R. N. Okechi[3]

[1]Department of Geosciences, Federal University of Technology, Owerri Imo State, Nigeria.
[2]Department of Chemistry, Federal University of Technology, Owerri Imo State, Nigeria.
[3]Department of Biotechnology, Federal University of Technology, Owerri Imo State, Nigeria.

Hydrogeochemical assessment of surface water Resources of Orlu and its environs was carried out in order to evaluate the chemical composition and quality profile with respect to the environmental setting of the study area. A total of six samples were collected, from different rivers and streams traversing different communities in the area, and analyzed for major cations and anions as well as trace elements like Fe, Mn, Zn, Pb and As. The major cation concentrations determined were in the order of $Ca^{2+}> Na^+> K^+> Mg^{2+}$ with mean values of 4.14, 3.78, 1.60 and 1.19 mg/L respectively and that of the major anion concentration of the surface water is of the order $HCO_3^{2-}> Cl^-> SO_4^{2-}> NO_3^-$ with mean values of 19.83, 11.67, 3.45 and 0.24 mg/L respectively. Two major water groups were identified based on characterization on the Piper Trilinear diagram. They include $Ca-(Mg)-Na-HCO_3$ and $Ca-Mg-(SO_4)-HCO_3$. This reflects diverse effects of bedrock lithologies, base exchange processes, precipitation and weathering. The pollution index (PI) value of 9.10 (which is greater than the critical value of unity) shows that the surface water bodies in the area are slightly polluted by the trace elements and as such not fit for domestic use as those parameters exceeded the maximum permissible level set by the Nigerian Industrial Standard (2007) and the World Health Organization (2006) Standard. Finally the electrical conductivity (EC) values of 13.0 to 97.0 µs/cm and total dissolved solids (TDS) values of 8.0 to 66.0 mg/L, alongside the estimated sodium absorption ratio (SAR) of 1.74 revealed a fresh water grade that is fairly suitable for agricultural purposes.

Key words: Hydrogeochemistry, pollution index, water facies, water quality.

INTRODUCTION

Water is an essential resource and forms the primary need of man in his environment. It is a vital component of life for both plants and animals. It is available in forms of rain and snow thereby making rivers, oceans, streams, lakes, springs etc (Ibeneme et al., 2013). Water, is most readily available to man in the flow of stream and river.

These flows show wide variation both in time and place. Throughout ages, mankind has been faced with the problem of providing suitable water of sufficient quality for his use. Water pollution is one of the most serious environmental problems facing the world today as most industrial wastes are channeled directly to water bodies thereby affecting the qualities of the surface water systems (Nganje et al., 2010). The availability of large volumes of water is not on its own enough guarantee that the water is potable. The presence of objectionable tastes,

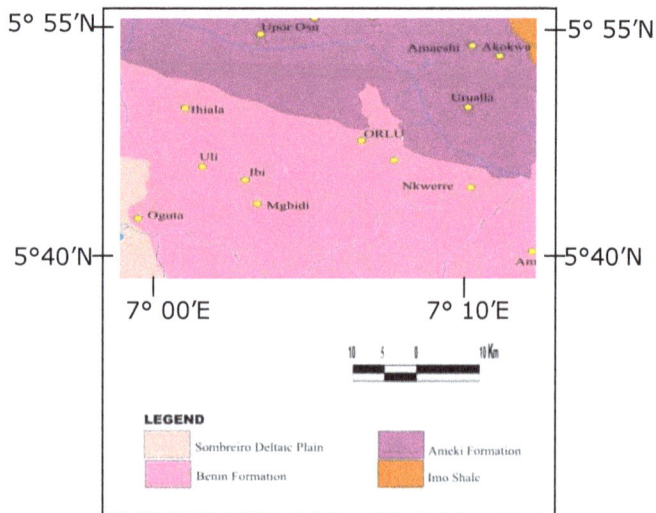

Figure 1. Geological map of the study area.

odour, colour as well as harmful substances in such water no matter how abundant it is, renders it unsuitable for domestic, industrial and agricultural uses (Okeke and Igboanua, 2003). Geochemical determination of surface water is one of the principal criteria for assessing the quality of water. Geology, particularly rock types, their weathered products and precipitation from rainfall contribute greatly to the chemistry of surface and ground water (Abimbola et al., 2002). The primary purpose of water analyses is therefore to determine the suitability of water for a proposed use; the main classes of use being domestic, agricultural and industrial purposes (Opara et al., 2005). This study is therefore carried out to ascertain the portability and suitability of the surface water resources of Orlu and environs for municipal use.

GEOLOGY AND PHYSIOGRAPHY OF THE STUDY AREA

The study area is located between latitudes 5°43'N to 5°51'N and longitudes 7°00'E to 7° 09'E (Figure 1). It is part of the Anambra sedimentary Basin, and is characterized by three major geologic formations with Bende-Ameki formation being the oldest. Bende-Ameki formation is overlain by Ogwashi-Asaba formation which in turn is overlain by the Benin Formation. The geology and hydrogeology of these units have been extensively studied by various authors (Reyment, 1965; Uma, 1989; Whiteman, 1982). The Benin formation comprises of a thick sequence of poorly consolidated to unconsolidated sandstones that are friable with sorting ranging from poorly to fairly sorted (Onyeagocha, 1980). Several grain sizes occur within the unit and the coarse and fine unit alternate along the vertical sequence. The thick sandy

units are frequently separated by thin and discontinuous clay streaks and lenses. The clay beds are thin (less than 1 m) and sometimes occur as lamination lining the bedding plane of the unconsolidated sandstone beds. The formation starts as a thin edge at its contact with the Ogwashi/Asaba formation in the north of the area and thickens southwards (Avobovbo, 1978). The Ogwashi–Asaba formation is identified within the Palaeocene Anambra Basin (Afikpo geosynyncline) (Oboh – Ikuenobe et al., 2005). The formation is characterized by alternation of clays, sands, grits and lignites (Bassey and Eminue, 2012). The formation occurs mainly in Benin, Asaba, Onitsha and Orlu areas, Reyment (1965) suggested Oligocene–Miocene age for this formation, but palynological results by the work of Cherie et al. (1978) assigned a Middle Eocene age to the basal part. The Ogwashi–Asaba formation is a surface lateral equivalent of the Agbada formation which occurs in the subsurface of the Niger Delta (Assez, 1989; Akpoborie et al., 2011). The Bende-Amaki formation of Eocene to Oligocene age consists of medium to coarse-grained white sandstone, which may contain pebbles, graygreen sandstone, bluish calcareous silt, with mottled clays and thin limestone. Considerable lateral variation in lithology has also been observed. The lower part of the formation consists of fine-coarse-grained lenses of sandstone with abundant calcarceous shales and thin shelly limestone. The Bende-Ameki formation overlies the impervious Imo shale group of Paleocene age, which is characterized by lateral and vertical variations in lithology (Akaninyene, 2012). The physiography is dominated by a segment of the NW-SE trending Awka-Orlu cuesta which rises to about 350 m above mean sea level and to about 200 m above the surrounding plains. These undulating ridges and plains are somewhat related to bedrock underlying the area. The drainage pattern in Orlu and environs is the dendritic type indicative of lithological, structural and topographic differences and also typical of alluvial rocks, which is typical of the geology of the area that consists mainly of sedimentary rocks (Figure 1).

MATERIALS AND METHODS

The study aims at determining the physico-chemical characteristics of the surface water resources in order to assess their quality and usability. Six water samples were collected from different rivers traversing different communities in the study area. *In-situ* measurements of temperature (°C), electrical conductivity (EC), total dissolved solids (TDS) were carried out in the field while samples were collected in clean new white plastic bottles (1.5 L) which were first rinsed with part of the water samples to avoid contamination. For unclear water sample, a filter membrane was used to filter off suspended particulate and then subjected to further detailed analyses. The TDS was measured using a specially calibrated TDS meter, with a small container where the water sample can be poured. The EC was measured using a WPACMD 400 m which can measure conductivity over the range of 10^{-2} to10^{-6} μ Siemens/cm^2 with an accuracy of 0.1 μ S/cm^2. The coordinates (longitude, latitude and elevation/altitude) of each sample location

Table 1. Results of geochemical analysis of water samples of Orlu and environs compared with NIS (2007) and WHO (2006) standard.

Parameter	Sw1	Sw2	Sw3	Sw4	Sw5	Sw6	WHO 2006	NIS 2007	Undesirable effects at higher levels
Temperature (°C)	28	29	21	27	21	20	25	Ambient	-
pH at 24°C	6.4	6.5	6.7	6.2	6.7	6.7	6.5-8.5	6.5-8.5	High pH-Corrosion; Low pH-Taste/Soapy feeling
Colour (apparent)	97.0	88.0	46.0	69.0	50.0	44.0	15	15	Appearance.
Turbidity (FTU)	12.0	25.0	38.0	16.0	6.0	23.0	5	5	Appearance.
Conductivity (μs/cm)	21.0	28.0	24.0	13.0	76.0	97.0	100	1000	-
TDS	11.0	15.6	16.2	8.0	38.0	66.0	500	500	Taste.
Hardness (mg/L)	8.9	8.0	15.0	19.0	14.0	16.0	-	-	High-Scale deposit and scum formation; Low-Possible corrosion.
Iron (mg/L)	2.03	1.67	0.14	0.07	0.52	2.63	0.3	0.3	Stain of Laundry and Sanitary wares.
Calcium (mg/L)	1.7	3.6	6.88	1.36	2.11	9.20	75	100	Scale Formation.
Magnesium (mg/L)	1.60	0.08	1.55	2.51	0.80	0.57	20	0.2	Hardness and Gastrointestinal Irritation.
Potassium (mg/L)	0.05	0.24	1.00	1.65	1.06	5.58	200	200	-
Sodium (mg/L)	2.5	6.8	2.30	1.70	3.30	6.10	200	200	Taste.
Sulphate (mg/L)	3.60	0.04	5.08	3.39	0.20	8.40	250	100	Taste and Corrosion.
Nitrate (mg/L)	0.15	0.13	0.15	0.18	0.40	0.40	10	50	Physiological problems.
Chloride (mg/L)	8.34	12.1	10.7	6.67	12.3	19.9	200	250	Taste and Corrosion.
Bicarbonate (mg/L)	12.8	17.8	21.9	13.4	16.5	36.6	500	500	-
Total hardness ($CaCO_3$) (mg/L)	4.4	4.9	24.0	3.4	5.44	23.0	100	150	High-Scale deposit and scum formation; Low-Possible corrosion.
Lead (mg/L)	0.42	0.22	0.05	0.32	0.06	0.08	0.01	0.01	Irritability, Acute Psychosis and reduced dermal sensitivity.
Arsenic (mg/L)	0.09	0.01	0.04	0.12	0.02	0.10	0.01	0.01	Toxic, Liver damage.
Zinc (mg/L)	2.10	2.02	1.18	2.12	1.38	1.35	0.01	0.01	Appearance and Taste.
Manganese (mg/L)	0.3	0.13	0.26	0.4	0.17	0.31	0.01	0.2	Stain of Laundry and Sanitary wares, Neurological disorder.

Sw1=Orashi River, Sw2=Njaba River, Sw3=Ezize River, Sw4=Ogidi River, Sw5=Ngakwu Stream, Sw6=Mgbede Stream.

in the study area was determined using the GARMIN GPSmap76CSx. These samples were labeled Sw1-Sw6 (Table 1). All the water samples were preserved in a refrigerator to exclude microbial activity and unwanted chemical reaction until analysis was completed within 2 days. The determinations of other physico-chemical properties of the water samples were performed within 2 days of sampling. HACH DR 2800 Spectrophotometer was used in the determination of different hydro geochemical properties such as Na, K, HCO_3, Cl, NO_3, SO_4, Fe, Mn, Zn, Pb and As. Analytical water test tablets prescribed for HACH DR Spectrophotometer 2800 using procedures outlined in the HACH DR 2800 Spectrophotometer manual were used for the examination of the water quality. Other analyses such as the determination of Mg and Ca concentrations were done by complexometric titration method.

RESULTS

The results of the analyses and the summary of the physico-chemical parameters in mg/L are presented (Table 1).

DISCUSSION

From the results, the total dissolved solids (TDS) for the surface water samples ranged from 8.0 to 66.0 mg/L with a mean value of 25.80 mg/L. The calcium content ranged from 1.36 to 9.2 mg/L with an average of 4.14 mg/L, while that of magnesium was from 0.08 to 2.51 mg/L. Magnesium ion (Mg^{2+}) concentration is generally low. The availability of magnesium ion in surface water systems of the area could be explained by occurrence of magnesium with calcium carbonate cement in detrital sedimentary formation. Similarly the sodium and potassium content on the other hand ranged from 1.70 to 6.10 mg/L and 0.05 to 5.58 mg/L respectively (Figure 2). Sodium must have entered the water system through natural system (that is, rainwater) (Egbunike, 2007). Other natural sources include weathering of feldspars (albite) and leaching of clay minerals (Spears and Reeves, 1975; Ogbukagu, 1986). For the anions, bicarbonate was the dominant anion in the surface water samples and ranged from 12.8 to 36.6 mg/L with a mean value of 19.83 mg/L. This was followed by chloride, with a range of 6.67 to 19.9 mg/L. The chloride concentrations are comparatively low because of the fact that chloride does not show any correlation with the components of pore water derived from mineral breakdown (Spears and Reeves, 1975). The concentration of rain water by evapotranspiration may be an important source of chloride in the area. Nitrate

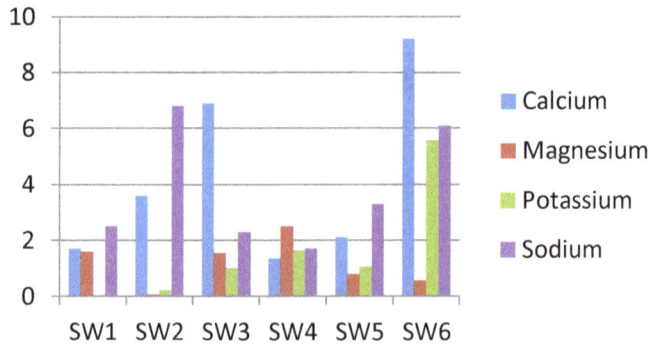

Figure 2. Major cations in the analyzed samples expressed as a bar chart.

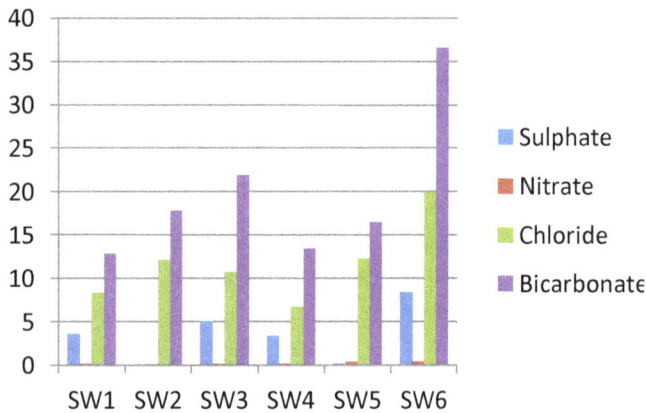

Figure 3. Major anions in the analyzed samples expressed as a bar chart.

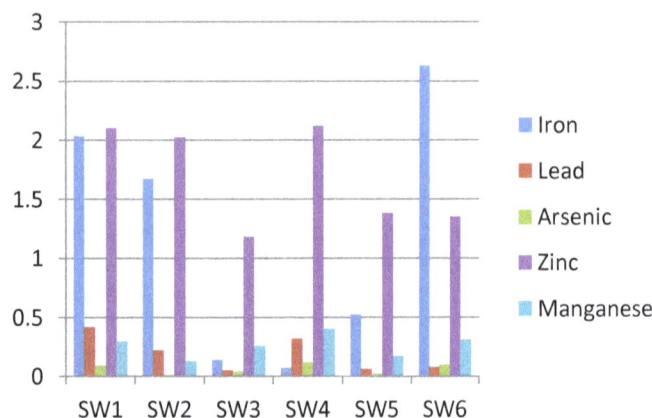

Figure 4. Trace elements concentration in the analyzed samples expressed as a bar chart.

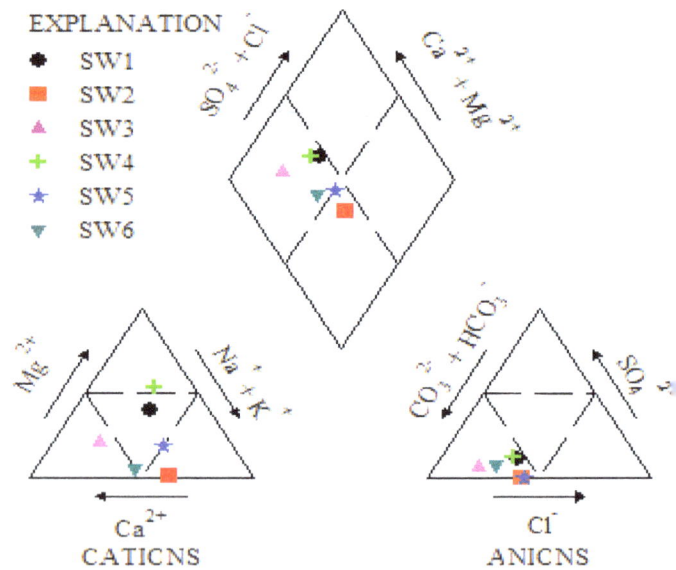

Figure 5. Piper trilinear diagram of the studied water samples.

probably the atmosphere. Sulphate concentration ranges from 0.04 to 8.4 mg/L with an average of 3.45 mg/L. The major source of sulphate in the area may result from gypsum and oxidation of sulphide ores. The pH ranged from 6.2 to 6.7 with a mean value of 6.53, which clearly showed the water to be slightly acidic.

The average concentrations of all the trace elements (Pb, Zn, Mn, Fe, As) (Figure 4) are higher than the NIS and WHO standard for drinking water (WHO, 2006; NIS, 2007) and this may be attributed to possible infiltration of petroleum products from the petrol stations, discharge from surface waste disposal sites, wastes from automobile workshops as well as the chemical composition of the bedding rock via which the water flows. The human health effects are linked to irritability, acute psychosis, reduced dermal sensitivity, highly toxic to infants and pregnant women (for Pb); liver and kidney damage (for As); neurological disorder (for Mn). Fe as well as Mn can cause stain of laundry and sanitary wares, while excess of Zn in water gives it unpleasant taste and appearance (NIS, 2007).

The water resources of the area were found to have two dominant hydrochemical facies according to the piper Trilinear diagram (Figure 5). These include Ca-(Mg)-Na-HCO_3 and Ca-Mg (SO_4)-HCO_3. According to Lohnert (1973) and Onyekuru et al. (2010), the former facies group has appreciable amount of $NaHCO_3$ which is an indication of cation exchange water. One of the characteristics of this water type is the higher carbonate hardness as compared to the total hardness. This in effect means that there is more HCO_3 than the available alkaline earth metal ions (Ca^{2+} and Mg^{2+}) in equivalent concentration (Lohnert, 1970). These excess bicarbonate ions then release the alkaline (notably Na^+) into the solution by exchange reaction with the cation exchangers

generally was low for all locations and had a range of 0.13 to 0.40 mg/L (Figure 3). Nitrate concentration falls within the accepted limit. The potential source of nitrate in the area may include legume, animal excrement, and

Table 2. Physico-chemical parameters of water resources of orlu for pollution index determination.

| Parameter | Surface water | | WHO 2006 | C_{ij} |
	Range	Mean (C_{ij})	(W_{ij})	W_{ij}
Temperature (°C)	21-29	24.33	25	0.9732
pH at 24°C	6.2-6.7	6.53	6.5	1.0046
Colour (apparent)	44.0-97.0	65.67	15	4.378
Turbidity (FTU)	6.0-38.0	20.00	5	4.000
Conductivity (µs/cm)	13.0-97.0	43.17	100	0.4317
Total dissolved solids	8.0-66.0	25.80	500	0.0516
Hardness (mg/L)	8.0-19.0	13.48	-	-
Iron (mg/L)	0.07-2.63	1.18	0.3	3.933
Calcium (mg/L)	1.36-9.2	4.14	75	0.0552
Magnesium (mg/L)	0.08-2.51	1.19	20	0.0595
Potassium (mg/L)	0.05-5.58	1.60	200	0.008
Sodium (mg/L)	1.70-6.10	3.78	200	0.0189
Sulphate (mg/L)	0.04-8.4	3.45	250	0.0138
Nitrate (mg/L)	0.13-0.40	0.24	10	0.024
Chloride (mg/L)	6.67-19.9	11.67	200	0.0584
Bicarbonate (mg/L)	12.8-36.6	19.83	500	0.0397
Total hardness ($CaCO_3$) (mg/L)	4.4-23.0	10.86	100	0.1086
Lead (mg/L)	0.05-0.42	0.19	0.01	19.00
Arsenic (mg/L)	0.01-0.12	0.05	0.01	5.00
Zinc (mg/L)	1.35-2.12	1.69	0.01	169.00
Manganese (mg/L)	0.13-0.40	0.26	0.01	26.00
Total				**234.16**
Mean				**11.71**

such as clay minerals. The Ca-Mg (SO_4)-HCO_3 type falls within normal alkaline water and is predominantly hydrogen carbonate sulphate (Piper, 1953). Assessing the water samples suitability for domestication pollution index computation using the method of Horton (1965) was utilized (Table 2). To determine the pollution index of the water samples, the formula below was used:

$$PI = \frac{\sqrt{[Max(Cij/Wij)]^2 + [Mean(Cij/Wij)]^2}}{2} \quad \text{.............. (1)}$$

The pollution index (Horton, 1965) obtained using the above relation is 9.10. This shows that the surface water resources of Orlu and its environs are slightly polluted by trace elements. The source of these pollutants could be from the open solid waste dumpsites disseminated in different localities (like Ugwu Mgbee and TESAC dumpsites from which leachate and other substances seep to join the Orashi River and Mgbede stream), effluent discharge from automobile workshops ("Mgbuka" in Amaifeke that empty their wastes directly into the Ogidi River) and the geology of the underlying bedrocks along the river channel (made up of predominantly sedimentary rocks). For agricultural purposes, sodium content of water was used for this purpose. The sodium absorption ratio

(SAR) estimated this. The two principal effects of sodium are reduction in soil permeability and hardening of the soil (Etu-Efeotor, 1981). For the study area, SAR lies between 0.79 to 3.55 with an average of 1.74, which revealed a class of water suitable for irrigation purposes (Wilcox, 1955).

Conclusion

All the physico-chemical parameters measured conform to the standard set by the regulatory bodies except pH and the trace element. The electrical conductivity (EC) values of 13.0 to 97.0 µs/cm, total dissolved solids (TDS) of 8.0 to 66.0 mg/L and sodium absorption ratio (SAR) of 1.74 revealed a class of water suitable for agricultural purposes. High concentrations of trace element as observed in the rivers sampled rendered the water unfit for domestic uses. Thus, the water resources of the area need to be treated for heavy metals prior to domestication.

REFERENCES

Abimbola AF, Odukoya AM, Olatunji AS (2002). Influence of bedrock on the Hydrogeochemical characteristics of groundwater in Northern part of Ibadan metropolis. W. Res. 13:1-6.

Akaninyene OA, Igboekwe MU (2012). Preliminary Lithologic Deductions for Michael Okpara University of Agriculture, Umudike, Abia State, Nigeria, using Vertical Electrical Sounding Method. Arch. Phy. Res. 3(4):292-302.

Akpoborie NA, Nfor AA, Etobro I, Odagwe S (2011). Aspects of the Geology and groundwater conditions of Asaba, Nigeria. Arch. Appl. Sci. Res. 3(2):537-550.

Assez LO (1989). In: Kogbe (ed). Geology of Nigeria. Rockview Publ., Jos, pp. 311-334.

Avobovbo AA (1978). Tertiary Lithostratigraphy of Niger Delta. Bull. Am. Assoc. Pet. Geol. 62:295-306.

Bassey C, Eminue O (2012). Petrographic and stratigraphic analyses of Palaeogene Ogwashi-Asaba formation, Anambra Basin, Nigeria NAFTA 63(7-8):247-254.

Cherie R, Onyike MS, Sowumi MA (1978). Some new Eocene pollen of the Ogwashi – Asaba formation, Southeastern Nigeria. Rev. Esp. micropateont. 10:285-322.

Egbunike ME (2007). Hydrogeochemical Analysis of Water Samples in Nando and Environs of the Anambra Basin of South Eastern Nigeria. Pac. J. Sci. Technol. 8(1): 32-35.

Etu-Efeotor JO (1981). Preliminary Hydrochemical Investigations of Subsurface Water in Parts of the Niger Delta. J. Min. Geol. 18(1):103-105.

Horton RK (1965). An Index Number System for Rating Water Quality. J. W. Poll. Cont. 37(3):12-16.

Ibeneme SI, Ukiwe LN, Essien AG, Nwagbara JO, Nweze CA, Chinemelu ES, Ivonye CA (2013). Assessment of the Chemical Characteristics of a Spring Water Source at Ife-Owutu, Ezinihite-Mbaise, Southeastern Nigeria. Am. J. Eng. Res. 2(10):282-290.

Lohnert EP (1973). Austauschwaesser In H Schneider (Hrsg) Die Grunwasserchliessung, Vulkan Verlag Essen 2nd Edition, pp. 138-144.

Lohnert EP (1970). Grundwasserchemismus und kationentaush im norddeutschen Flachland Z dt geol Ges, Sonderheft. Hydrgeol Hydrogeochem, Hannover, 139-159.

Nganje TN, Adamu CI, Ntekim EEU, Ugbaja AN, Neji P, Nfor EN (2010). Influence of mine drainage on water quality along River Nyaba in Enugu South-Eastern Nigeria. Afr. J. Environ. Sci. Technol. 4(3):132-144.

NIS (2007). Nigerian Indusrial Standard. 2, ICS 13.60.20, pp. 15-17.

Oboh-Ikuenobe FE, Obi CG, Jaramillo CA (2005). Lithofacies, Palynofacies, and Sequence Straigraphy of Paleogene strata in Southeastern Nigeria. J. Afr. Earth Sci. 41:79-101.

Ogbukagu IKN (1986). Water Supply of Njikoka and Awka Areas of the Anambra Basin, Nigeria. J. Afr. Earth Sci. 5(5):519-526.

Okeke OC, Igboanua AH (2003). Characteristics and quality assessment of surface water and groundwater resources of Awka Town, SE Nigeria. W. Res. J. 14:71-77.

Onyeagocha AC (1980). Petrography and Depositional Environments of the Benin Formation. J. Min. Geol. 17 (2):147-150.

Onyekuru SO, Nwankwor GI, Akaolisa CZ (2010). Chemical Characteristics of Groundwater Systems in the Southern Anambra Basin, Nigeria. J. Appl. Sci. Res. 6(12):2164-2172.

Opara AI, Onyekuru SO, Nwokei KI (2005). Hydrogeochemical study of surface and Groundwater Systems in Parts of Okigwe, Southeastern Nigeria. J. Sci. Eng. Technol. 12(2):6233-6242.

Piper AM (1953). A Graphic Procedure in the Geochemica Interpretation of water Analysis. Washington D.C, US. Geol. Surv. ISBN ASIN: B0007H2Z36.

Reyment RA (1965). Aspects of Geology of Nigeria University Press, Ibadan, Nigeria, pp. 24-29.

Spears DA, Reeves MJ (1975). The Influence of Superficial Deposit on Groundwater Quality in the Vale York. Q. J.Eng. Geol. 8:255-270.

Uma KO (1989). Appraisal of the groundwater resources of the Imo River Basin, Nigeria. J. Min. Geol. 25(1&2):305-315.

WHO (2006). Revision of the World Health Organization guide lines for drinking water quality: Report of the Final Task Group Meeting WHO, Geneva, p. 307.

Wilcox LV (1955). Classification and use of irrigation waters. U.S. Dept of Agric, Cire 967, Washington D.C.P. 19.

Whiteman A (1982). Nigeria, its petroleum Geology, Resources and Potentials. 2. Graham and Trotman Publ., London, pp. 234-241.

8

Transient numerical approach to estimate groundwater dewatering flow rates for a large construction site: a case study from the Middle East

GUHA Hillol, DAY Garrett and CLEMENTE José

Bechtel Power Corporation, Geotechnical and Hydraulic Engineering Services, U.S.A.

A three-dimensional, transient groundwater model was developed to determine the rate, volume, and number of pumping wells required to estimate the dewatering of three deep excavations at a large coastal construction site in the Middle East with a shallow groundwater table. There was limited site-specific hydrogeologic information for the site. A calibrated MODFLOW-based groundwater model was developed using the "the model-independent parameter estimation and uncertainty analysis" (PEST) software with pilot points and regularization mathematical techniques. Simulated heads were fitted against the monitoring well heads along extrapolated site groundwater head contours by estimating the hydraulic conductivity at each pilot point. Model-calibrated hydraulic conductivities obtained were within the range of medium to fine sand with silt values and matched closely with the subsurface material descriptions obtained through site geotechnical investigations. Dewatering of the three pits, each with approximate dimensions of 10 by 8 m and a depth of 20 m, was simulated through a series of sensitivity analyses to determine the number of wells, discharge rate, time duration to dewater the pits, and the volume of discharge water per pit to be diverted. Conclusions from the dewatering simulation estimations were as follows: (1) sensitivity analysis showed that the range of dewatering from each pit was dependent on the selected hydraulic conductivity and storage values, (2) storage was most sensitive to achieve the dewatered groundwater elevation depths, and (3) a one order-of-magnitude decrease of storage resulted in a shorter duration to dewater a pit. In summary, model simulations showed that site-specific pumping tests should be performed to optimize the design of a dewatering well system, specifically in low hydraulic conductivity soils where using large capacity wells is not feasible. The use of a numerical transient groundwater model is warranted for dewatering estimations as site-specific conditions are complex.

Key words: Middle East, construction dewatering, groundwater modeling, MODFLOW, PEST.

INTRODUCTION

Groundwater dewatering designs are often required prior to undertaking any subsurface geotechnical constructions (Ergun and Naicakan, 1993; Powers et al., 2007). The amount of dewatering is a function of the depth of the groundwater as it relates to the depth and size of the construction area (Powers et al., 2007; Preene, 2012). Understanding the geology, hydrogeology, and heterogeneity of the subsurface prior to undertaking dewatering is critical in successful geotechnical construction designs. Dewatering in a heterogeneous system requires development of a numerical groundwater model to better predict the number of well points or pumping wells required and also the time-variant nature of the dewatering process (Boak et al., 2007). We present a case study of a dewatering prediction rate estimate with the development of a numerical groundwater model of a large coastal construction site in the Middle East. The main objectives of this paper are to:

1. Develop a calibrated numerical groundwater model at and around three pits to include the site groundwater elevation shown in Figure 1 (determined from geotechnical boring logs and a limited number of groundwater observation wells) and
2. Estimate the rate, amount, and number of pumping wells needed to lower the groundwater elevation by approximately 20 m for each of the three pits with plan dimensions of approximately 10 by 8 m.

METHODOLOGY

Hydrogeological conditions

The study area is composed of Miocene and Pliocene sandy limestone, marl, gypsum, and beachrock formations. Evaporatic and low supratidal flats (Sabkha) are predominant and close to the study area. In the immediate study area geotechnical boring logs show interlayered medium to coarse sand with silt followed by thick fat clay layer at a depth of -21.5 m below ground surface. Groundwater elevation at the proposed dewatering area varied between 5 to 10 m below ground surface and is under unconfined condition, however, at greater depths groundwater is under confined conditions. Annual precipitation rates are very low along with high evaporation throughout the year result in a permanent water deficit. Although recharge is nearly zero due to high evaporation than precipitation; however, there are occasions when short heavy showers result in some recharge to the groundwater in similar hydrologic conditions (Memon et al., 1986; de Vries and Simmers, 2002; Kalbus et al., 2011).

Groundwater model development

A three-dimensional numerical groundwater model was developed using the pre-processor Groundwater Vistas, version 5 (ESI, 2007), utilizing the U.S. geological survey MODFLOW 2000 numerical model code for groundwater flow. The model included two layers, a horizontal grid dimension of 10 by 10 m aligned in East–West and North–South directions, resulting in a total of 328 and 200 columns. The model dimension encompassed an area greater than that of the pit so as to minimize any modeled boundary effects near the pit dewatering area. The length and width of the model domain was 1,990 by 3,267 m, respectively. The average model vertical depth was approximately 27.6 m (elevation depth of -21.5 m) at the three dewatering pits. The two layers in the model were simulated as an unconfined aquifer as there were no confining lithologic units separating the two layers. The four boundaries of the model were depicted using the general head boundary (GHB) conditions (Figure

2). The GHB conductance was determined by multiplying the hydraulic conductivity of the layer with the area of the finite difference grid dimension of each cell and dividing by the thickness of the layer at each of the cell nodes corresponding to the GHB. As there was no surface water features or water level monitoring well outside the model domain, the immediate groundwater head assumed outside the model domain was based on the surface elevation (assuming that groundwater was at the surface).

The model was divided into two layers to differentiate between the two distinct lithologies as determined from the geotechnical boring logs. Based on the boring logs, the top layer was denoted as medium dense to dense sand with silt, and the lower layer was denoted very dense sand with silt. The boring logs indicated that at a depth of approximately -21.5 m elevation, hard, fat clay exists at the pit area. As there were limited boring logs within the model domain, the existence and depth of the clay layer beyond the pit areas was not certain. For the purpose of model development, the clay layer was assumed to exist within the model domain. The clay layer was assumed as no-flow boundary within the model domain. The surface elevation of the model varied between 0.35 to 13 m, whereas, in the immediate project area, the range of surface elevation was 4.5 to 7 m. The thicknesses of model layer 1 varied between 0.5 to 9 m, and for layer 2, the thickness varied between 18 and 26 m. The model did not include any net recharge as it was assumed that groundwater infiltration was minimal and did not significantly impact the site.

Groundwater model calibration

The numerical groundwater model was calibrated under steady-state conditions, using the groundwater head contour distribution, as shown in Figure 1, as the target heads. The target heads were distributed along the contours in Figure 1. MODFLOW 2000 code was used to simulate the groundwater heads (Harbaugh et al., 2000). Prior to performing the calibration run, a total of 426 hydraulic conductivity pilot points were distributed within the model domain (Figure 2). These pilot points were distributed uniformly outside the immediate area of the project area but were increased in the area where the groundwater head contours are shown in Figure 1. The model-independent parameter estimation and uncertainty analysis (PEST) software with pilot point and regularizations (Doherty, 2010; Watermark Numerical Computing, 2010) was used to calibrate the model by fitting the simulated head against the monitoring well heads by estimating the hydraulic conductivity at each pilot point. Hydraulic conductivity was distributed among the pilot points using the Kriging interpolation method of geostatistics (Watermark Numerical Computing, 2010).

The regularization approach provides an advantage in estimating more parameters than there are observations to calibrate against. It is also a technique in minimizing the global objective function (ϕ_g). The global objective function can be defined as:

$$\phi_g = \phi_m + \mu\phi_r \quad (1)$$

where ϕ_m is the measurement objective function, ϕ_r is the regularized objective function, and μ is similar to a Lagrange multiplier, which in PEST is estimated through the Gauss-Marquadt-Levenberg optimization routine (Doherty, 2010).

The measurement objective function, ϕ_m, is defined as

$$\phi_m = (d - M(p))^t Q_1 (d - M(p)) \quad (2)$$

where d represents vector of field measurement, M is the modeled simulated values, Q_1 is the weight of the observation points, and

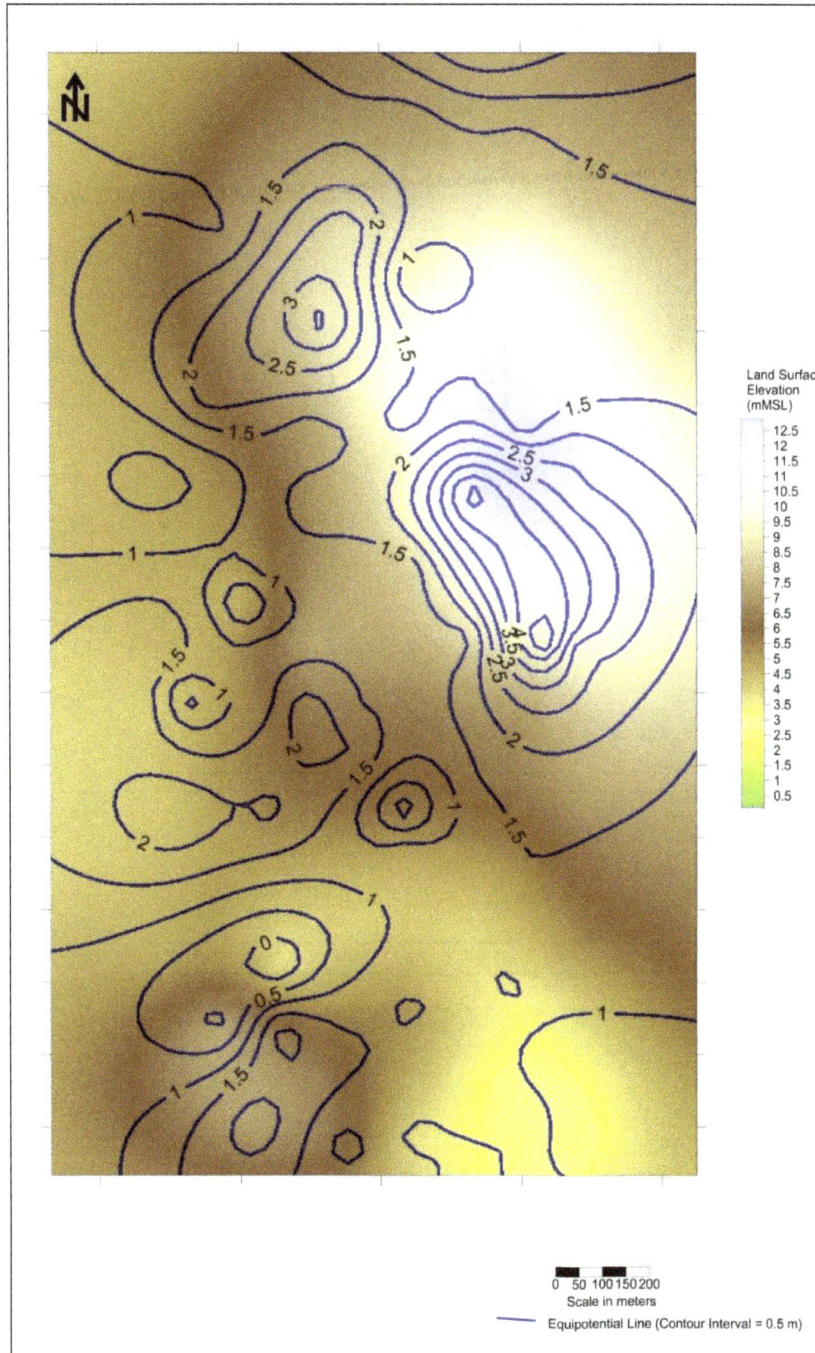

Figure 1. Site Observed Groundwater Levels

$d-M(p)$ is the difference between field and modeled data acting on parameter vector p.

The regularized objective function, ϕ_r, is defined as

$$\phi_r = (e - R(p))^t Q_2 (e - R(p)) \quad (3)$$

where e is the regularized observation values, R is a regularized operator acting on the parameter vector p, Q_2 is the weights assigned to the regularized observations, and $e-R(p)$ is the difference between regularization observation and regularized parameter vector values.

Figure 3 shows the simulated groundwater head distributions within the model domain under steady-state conditions. Figure 4 shows the simulated groundwater head distribution at and around the three-pit area. The distribution of the simulated groundwater heads was within 0.1 m of the measured heads (Figure 1) at the dewatering area and outside; that is, in areas to the northeast of the model domain, it varied between 0.5 and 1 m. The greater

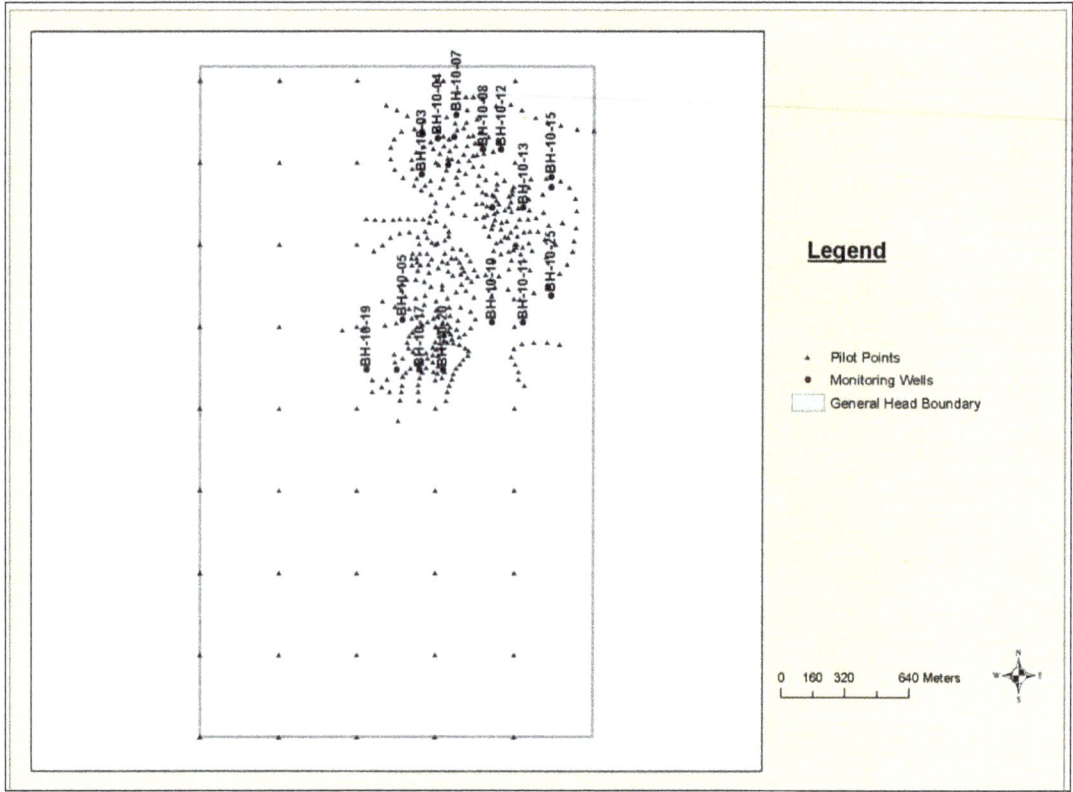

Figure 2. Model domain and pilot points distributions.

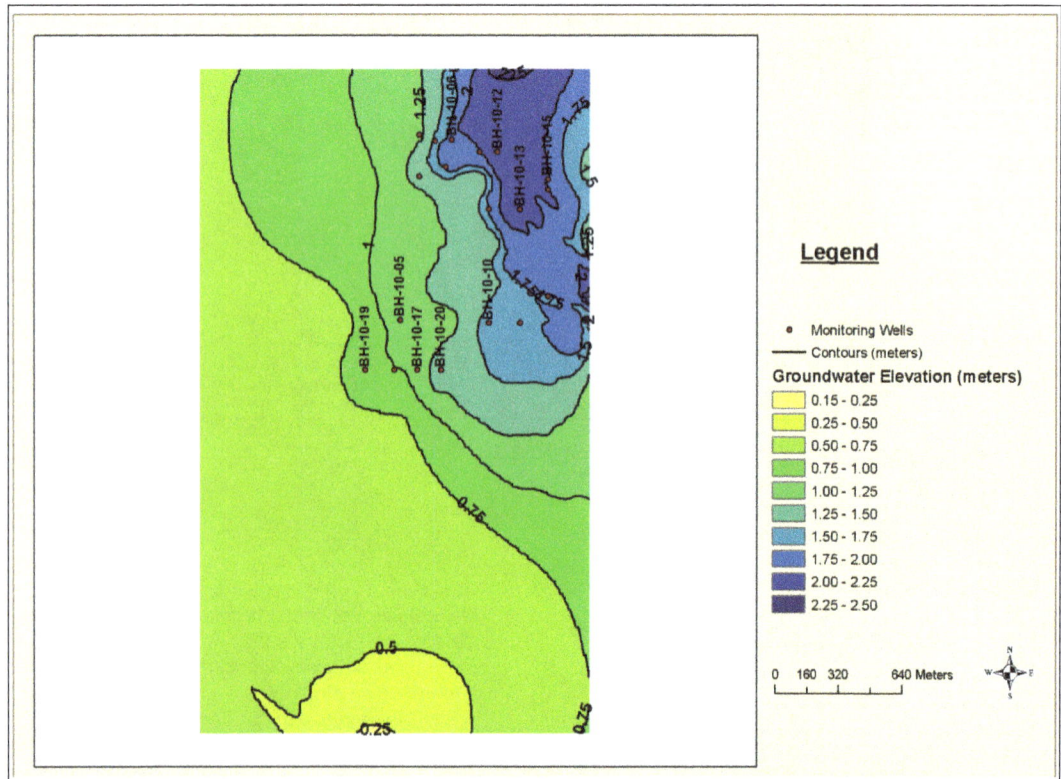

Figure 3. Groundwater head contours in the Calibrated Groundwater Model.

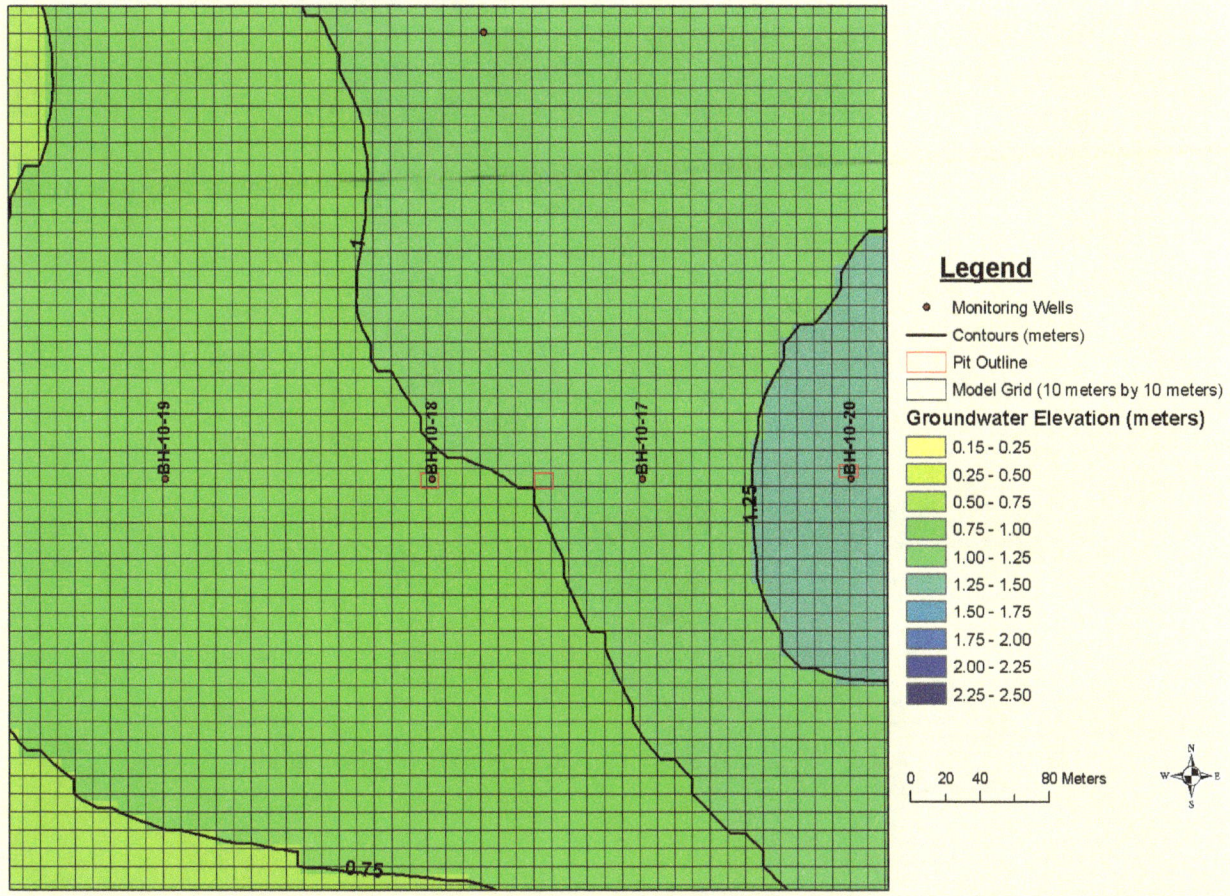

Figure 4. Groundwater head contours in the Calibrated Groundwater Model in the pit areas.

difference in heads between contoured heads (Figure 1) with that of the simulated head in the northeastern portion of the model domain was attributed to hydrogeologic conditions that may not be correctly conceptualized in the current groundwater model. However, the model-simulated heads did capture the mounding effects to the northeast. Some boundary effect is also responsible for the difference in the heads in the northeast portion of the model; however, the effects of the boundary did not have any impact at and around the three-pit area.

Figures 5 and 6 show the calibrated horizontal and vertical hydraulic conductivity distributions in layer 1 at the proposed dewatering zone. Similarly, Figure 7 depicts the calibrated horizontal hydraulic conductivity distributions in layer 2. The vertical hydraulic conductivity of layer 2 was 1.2×10^{-5} cm/s. The values of the horizontal hydraulic conductivity distributions were within the range of the medium to fine sand with silt values as provided in Table 4.6 of Fetter (1994). The estimated hydraulic conductivity values were determined inversely (no trial-and-error approach) by the model-independent parameter estimation and uncertainty analysis (PEST) suite of algorithms and the values are within the reported values of the lithologic units. The vertical hydraulic conductivity values were, on average, an order of magnitude smaller than the horizontal hydraulic conductivity values. Layer 1, which was described as medium dense to dense sand with silt, was characterized by horizontal hydraulic conductivities ranging from 1.1×10^{-4} cm/s to 1.5×10^{-3} cm/s at and around the three pits (Figure 5). The higher values of horizontal hydraulic conductivities than the vertical hydraulic conductivities is attributed to the

anisotropy of the soil types, where the dominant flow is along the horizontal direction which is typical in this kind of hydrogeological settings. In layer 2, the horizontal hydraulic conductivity values at and around the three pit areas were mostly within the range of 1.1×10^{-4} cm/s to 4.4×10^{-4} cm/s; in the immediate area of Pit 3, however, the value ranged from 1.3×10^{-3} cm/s to 4.6×10^{-3} cm/s (Figure 7).

RESULTS AND DISCUSSION

Pits dewatering estimate

Once the groundwater model was calibrated against the groundwater heads, the next step was to estimate dewatering rates for the three pits (Figure 4). Pits 1 and 2 were defined by a surface perimeter of approximately 10 by 8 m, and Pit 3 was approximately 10 by 7 m. Because the grid dimensions of the calibrated model were 10 by 10 m, finer grid discretization of 0.5 by 0.5 m was utilized at and around the three pits to numerically estimate the required dewatering at the pits. The grid was also expanded at selected areas beyond the vicinity of the pit. The maximum grid dimensions in the finer resolution model were 14 by 14 m. The total number of rows and

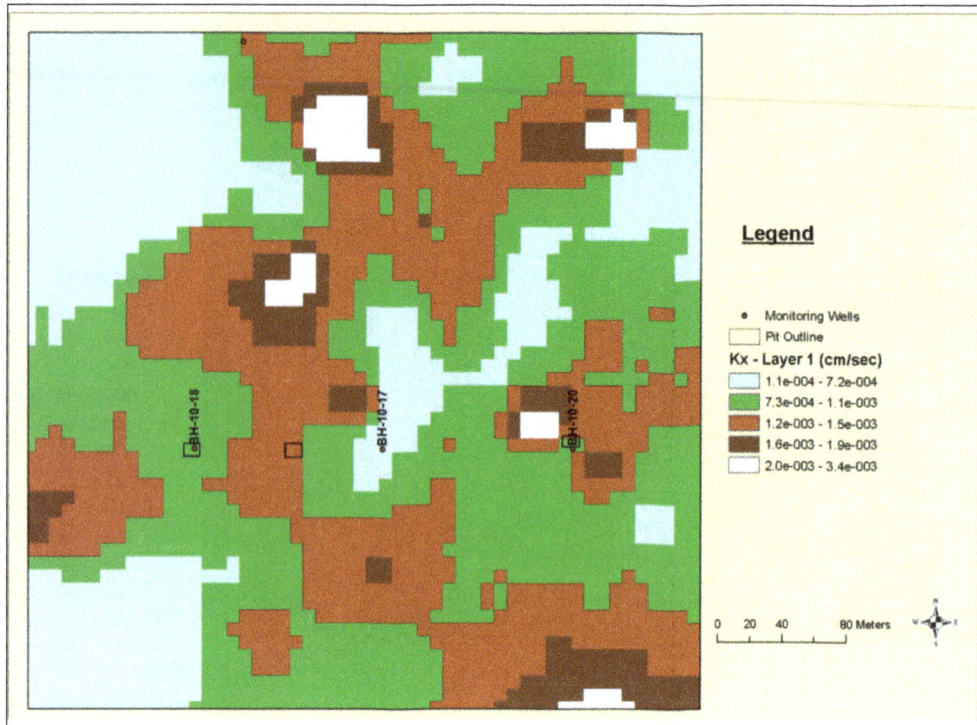

Figure 5. Distribution of Horizontal Hydraulic Conductivity Values in Layer 1 of the Groundwater Model at the pit areas.

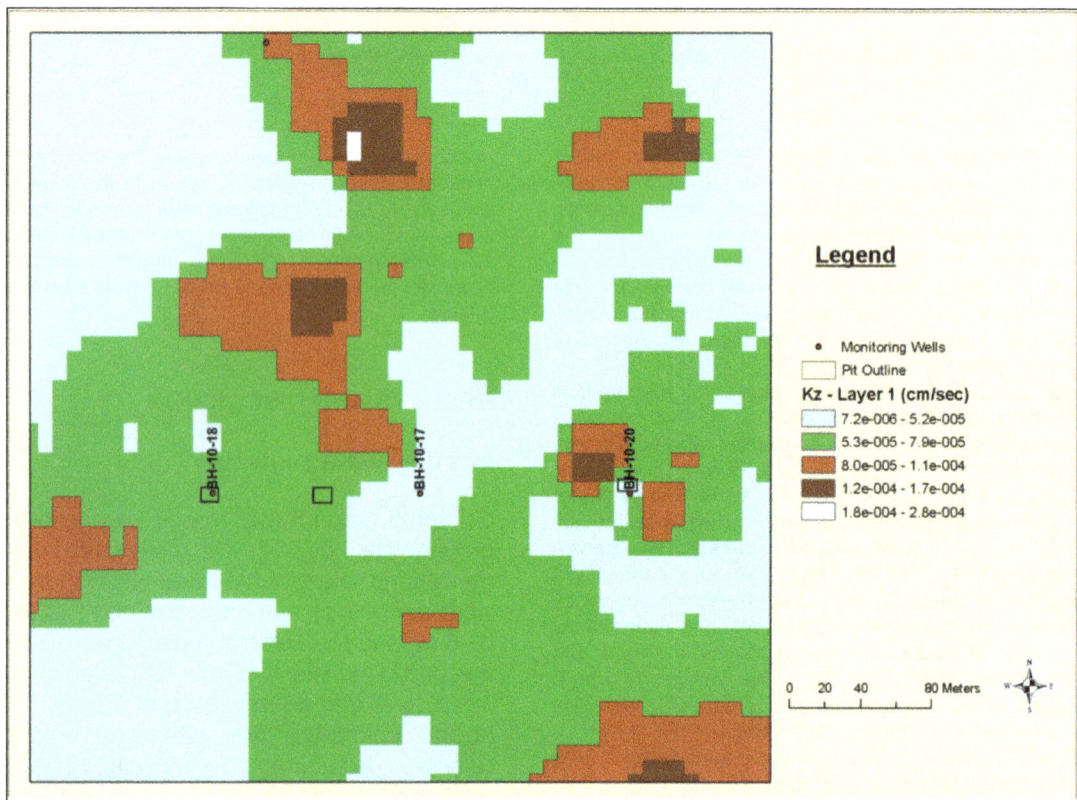

Figure 6. Distribution of Vertical Hydraulic Conductivity Values in Layer 1 of the Groundwater Model at the pit areas.

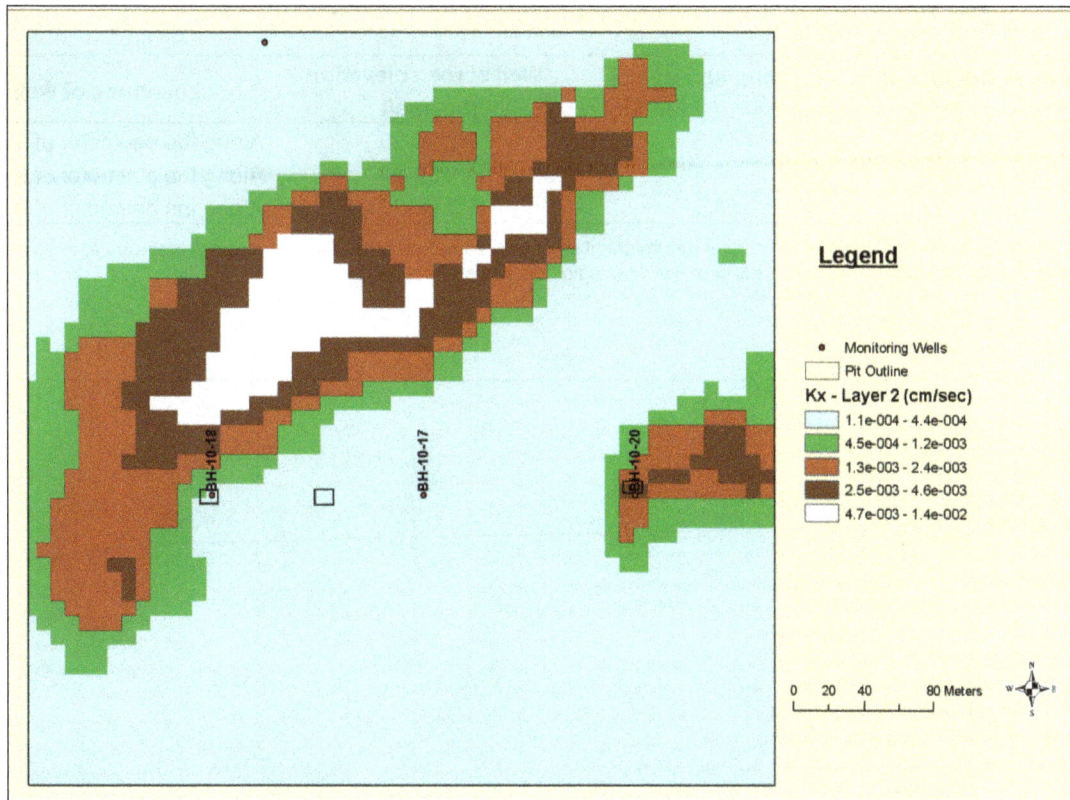

Figure 7. Distribution of Horizontal Hydraulic Conductivity Values in Layer 2 of the Groundwater Model at the pit areas.

columns in the model was 524 and 947, respectively. A steady-state simulation of the finer resolution model was run to compare the groundwater heads with those of the 10 by 10 m grid dimension model. The results of the heads compared satisfactorily within the model domain.

The three pits were dewatered in order, starting with pit 1, followed by pit 2 and then, ultimately, pit 3. It was assumed that once each pit was dewatered, concrete would be poured in each pit for construction (base and side walls). In the groundwater model, the pit area that had been dewatered was simulated as a no-flow zone for the next pit dewatering simulation. It was assumed that during the dewatering process, there was no rainfall at the site, and groundwater pumped from the pit was discharged offsite to avoid recharging the aquifer in the immediate area of the three pits.

Pit dewatering was simulated by pumping from wells along the perimeter of each pit. The wells were spaced at distances of approximately 1 m from each other. The number of wells, pumping rate, depth, and location of the wells are shown in Table 1. As dewatering is a transient process, aquifer storage terms were represented uniformly within each of the two layers. For layers 1 and 2, specific yield values of 0.21 and 0.18, respectively, were selected. These are average values for fine sand and silt as stated in Table 4.4 of Fetter (1994). The selection of the pumping rate of each well at 19.1 m^3/day

was arbitrary and was based on experience with dewatering in such geologic settings. However, the grain size analysis data show that the sediments at the site are tight, and thus, higher pumping rates may not be feasible. Prior to beginning any field dewatering activities, an aquifer test including step-test analysis for well pumping should be undertaken to estimate the most feasible rate of groundwater pumping.

Pit 1 dewatering rates

A total of 34 pumping wells, each pumping at 19.1 m^3/day, were distributed along the perimeter of the pit. Pumping proceeded for a total of 30 days before the groundwater head decreased to a level of approximately -18.5 m elevation. Figure 8 shows the time versus head and discharge in the location of the center of pit 1. The total water discharged at the end of the 30 days of pumping was approximately 8,500 m^3.

Pit 2 dewatering rates

Prior to simulating dewatering at pit 2, the area of pit 1 was simulated as a no-flow zone (to mimic the construction of the pit). The configuration and rate of the

Table 1. Pumping Well Information Used in the Groundwater Model.

Pit	Number of wells	Pumping rate of each well (gpm)	*Well screen elevation (m - msl)	Locations of wells
1	34	3.5	6.2 to -20	Along the perimeter of the pit
2	34	3.5	6.2 to -20	Along the perimeter of the pit
3	36	3.5	6.2 to -20	Along the perimeter of the pit

* The well screen elevation is based on aquifer depth identified in the groundwater model. For actual dewatering at the site, the well screen depth may vary depending on actual site hydrogeologic conditions encountered.

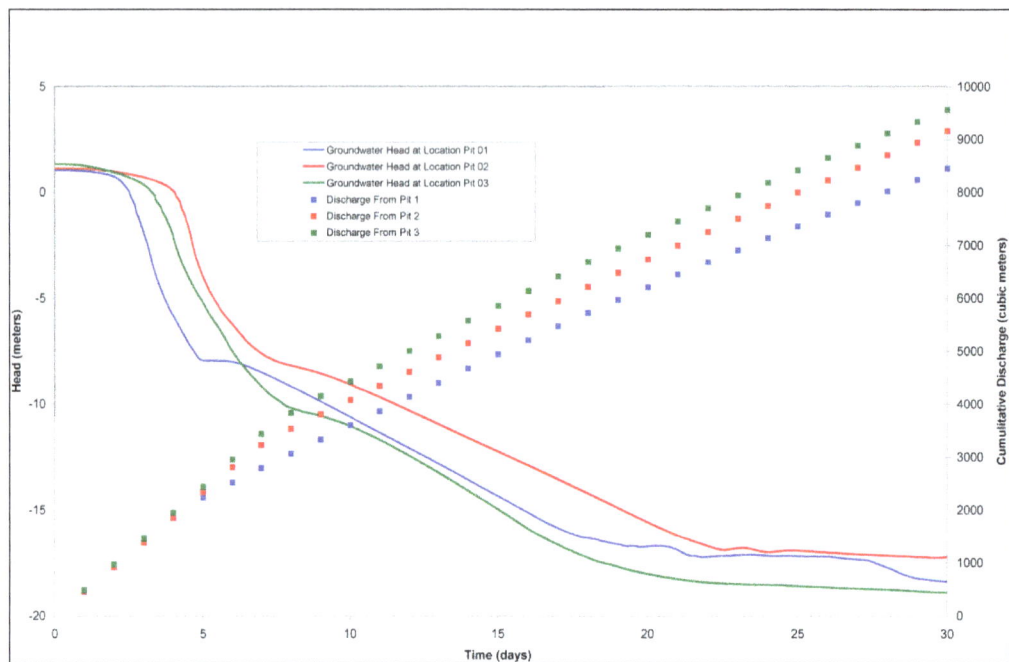

Figure 8. Groundwater Head and cumulative discharge with time for the three pits.

pumping wells were similar to those in the dewatering of Pit 1. Pit 2 was also dewatered for 30 days, with the groundwater head elevation decreasing to approximately -17.3 m at the target location in Pit 2 (Figure 8). The failure to reach the target elevation depth of -18.5 m after 30 days was attributed to a few dry wells and also to the lower hydraulic conductivities in the area of Pit 2. It is expected that sump pumps could be used in the bottom of the excavation to lower the groundwater head an additional meter to the target elevation. After 30 days of pumping, the total discharge from Pit 2 was approximately 9,200 m^3.

Pit 3 dewatering rates

Pit 3 was dewatered with a total of 36 pumping wells, each discharging at 19.1 m^3/day. The two additional wells in Pit 3 were the result of the slightly different pit outline

and the fact that the initial groundwater table was higher than the level of the other two pits. Figure 8 shows the relationship between time and groundwater heads and discharge at Pit 3. The total groundwater discharge from Pit 3 at the end of 30 days of pumping was approximately 9,600 m^3. Pits 1 and 2 were simulated as no-flow boundary conditions to represent post-construction conditions.

Sensitivity analysis

To determine the impact that the hydraulic conductivity and storage values have on dewatering estimates for the three pits, a series of sensitivity analyses were conducted. Pumping rates were similar to the rates used in the dewatering of Pits 1, 2, and 3 as already discussed under "Pit dewatering estimate" above. The sensitivity analyses included:

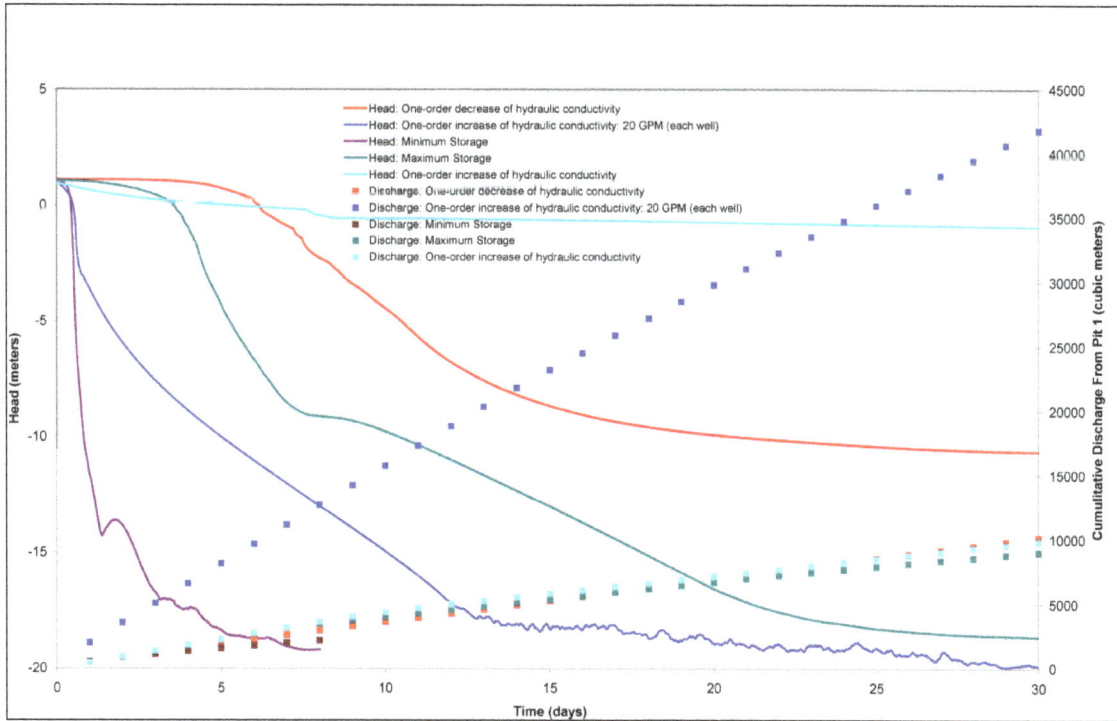

Figure 9. Sensitivity of Groundwater Head and cumulative discharge with time for Pit 1.

1. Increase of hydraulic conductivity (both horizontal and vertical) by one order of magnitude from the calibrated model values;

2. Decrease of hydraulic conductivity (both horizontal and vertical) by one order of magnitude from the calibrated model values;

3. Minimum storage (specific yield) values for fine sand (layer 1 in the model) and silt (layer 2 in the model) of 0.1 and 0.03, respectively; these values are shown in Table 4.4 of Fetter (1994) and

4. Maximum storage (specific yield) values for fine sand (layer 1 in the model) and silt (layer 2 in the model) of 0.28 and 0.19, respectively; these values are shown in Table 4.4 of Fetter (1994) Figures 9 to 11 show plots of time versus groundwater head and discharge for Pits 1, 2, and 3, respectively.

Pit 1 sensitivity analyses

Increasing the hydraulic conductivities by one order of magnitude resulted in very minimal groundwater dewatering (lowering of the groundwater elevation to -1.0 m) after 30 days of pumping. However, increasing the pumping rate of the individual wells from 19.1 to 109 m³/day lowered the groundwater head to an elevation of -18.5 m within 15 days of pumping; however, pit dewatering reached a depth elevation of -20 m until 30 days of pumping. Although the required simulated depth of dewatering was achieved within 15 days of the start of

pumping, it took another 15 days to achieve an extra 1.5 m (that is, to -20 m elevation) due to the groundwater fluctuations occurring very close to the fat clay below layer 2. The total groundwater discharge after 30 days with pumping rate of 109 m³/day from each well was approximately 42,000 m³ (Figure 9).

Decreasing the hydraulic conductivity by one order of magnitude resulted in a groundwater elevation at the pit after 30 days of pumping of about -10.5 m elevation. The total groundwater discharge after 30 days was about 10,000 m³. Due to lower hydraulic conductivities (one order-of-magnitude decrease from calibrated values), the model showed that it will take more time (that is, more than 30 days pumping at a rate of 19.1 m³/day) to lower groundwater heads at Pit 1 to a target elevation of -18.5 m.

Groundwater storage played a greater role in the estimate of groundwater heads with time during dewatering. Decreasing the groundwater storage (that is, specific yield) to 0.1 and 0.03 in layers 1 and 2, respectively, yielded faster pit dewatering. The model sensitivity run showed that Pit 1 achieved the target dewatering elevation of -18.5 m within 6 days of pumping. The total groundwater discharged after 6 days of dewatering was approximately 1,800 m³. When specific yield values were increased to 0.28 and 0.19 for layers 1 and 2, groundwater heads reached the dewatering target of -18.5 m elevation within 28 days. The total groundwater discharged after 28 days was approximately 8,600 m³.

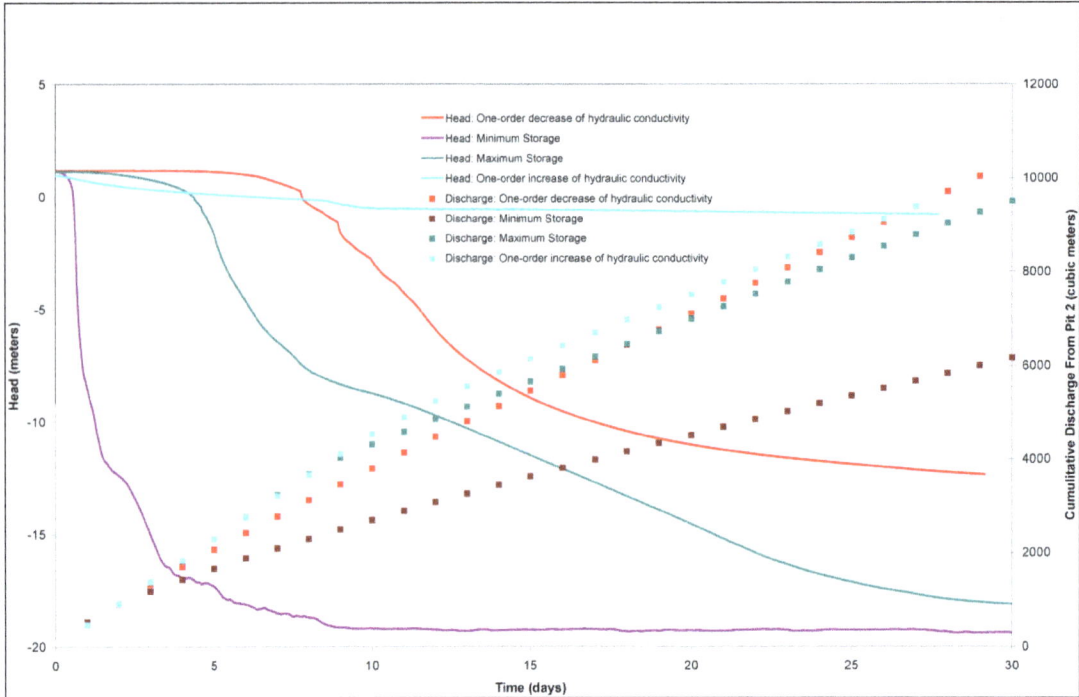

Figure 10. Sensitivity of Groundwater Head and Cumulative Discharge with Time for Pit 2.

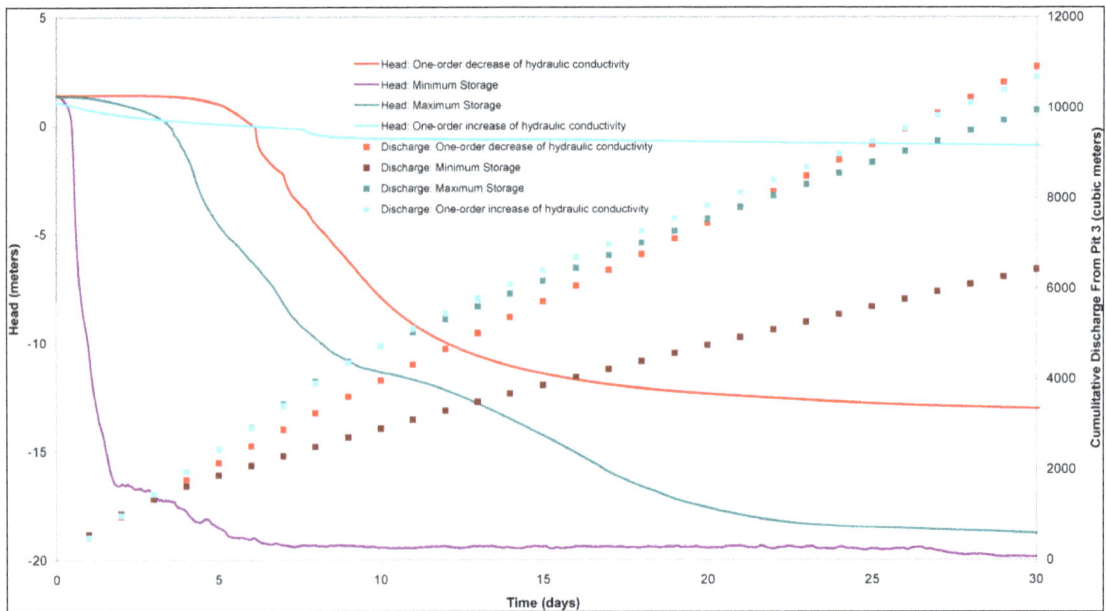

Figure 11. Sensitivity of Groundwater Head and Cumulative Discharge with Time for Pit 3.

Pit 2 sensitivity analyses

Pit 2 sensitivity analyses were conducted under conditions similar to the Pit 1 sensitivity analyses. The results of increasing and decreasing the hydraulic conductivity and storage values were similar to those in the Pit 1 sensitivity analyses (Figure 10).

Pit 3 sensitivity analyses

Pit 3 sensitivity analyses were conducted under conditions similar to the Pits 1 and 2 analyses. The results of increasing and decreasing the hydraulic conductivity and storage values were similar to the results from the Pits 1 and 2 sensitivity analyses (Figure 11). The sensitivity analyses show that a one order-of-magnitude increase of hydraulic conductivities above the calibrated hydraulic conductivity values will require a longer time to achieve the target dewatering elevation of -18.5 m, with approximately 34 to 36 wells pumping at a rate of 19.1 m^3/day. However, if the pumping rates for the individual wells are increased to 109 m^3/day, then it is possible to achieve the target dewatered elevation depth for the pits, as shown in the sensitivity analysis of pit 1. Storage plays a large role in dewatering estimation; a slight increase or decrease of storage impacts the time it takes to reach the groundwater level dewatering target. The lower the storage values for the sediments, the faster the groundwater dewatering target elevation is attained.

Assumptions

The following assumptions were used to evaluate the groundwater dewatering heads in the model:

1. The groundwater at the site represents uniform density;
2. There is no tidal influx affecting the dewatering pits;
3. The hydraulic conductivity and storage used in the model represent the values at the site;
4. There is no recharge during the dewatering processes;
5. Lithologic and groundwater information in the model at distances away from the pit areas is not ascertained from the available geotechnical field investigations and
6. Pumping rates used in the calculation during dewatering are constant. Initial water levels prior to dewatering of the pits are similar to those shown in Figure 1.

Conclusions

A groundwater model was developed for the purpose of estimating both the dewatering rates and the volume of groundwater that required removal from the three pits to achieve a construction dewatering groundwater elevation of -18.5 m. The model was calibrated against estimated maximum observed groundwater heads at the area of the pits. The estimated range of hydraulic conductivities was based on the lithologies from the geotechnical borehole logs. Dewatering of the three pits was evaluated by performing model simulations, starting with Pit 1, then Pit 2, and lastly, Pit 3. For Pit 1 dewatering, a total of 34 pumping wells, each pumping at 19.1 m^3/day for 30 days, were simulated until the target elevation was reached. The pumping wells were distributed along the periphery of the pit. The same well configuration was used to simulate dewatering at Pit 2. For Pit 3, 36 pumping wells were used because the pit outline (perimeter) is slightly different than those of Pits 1 and 2. In addition, the preconstruction dewatering groundwater heads at Pit 3

are slightly higher than the heads at Pits 1 and 2. The approximate total amount of water diverted to lower the dewatered groundwater elevation for each pit to -18.5 m was about 9,500 m^3. This estimation was based on model simulation values.

A series of sensitivity analyses were conducted to determine the effects of increasing and decreasing aquifer hydraulic conductivities and storage values on the dewatering estimates. The sensitivity analyses show that a one order-of-magnitude increase of hydraulic conductivity above the calibrated values would require more than 30 days of pumping to achieve the target dewatering depth. However, if the pumping rate was increased from 19.1 to 109 m^3/day for each of the 34 wells, the target depth of -18.5 m would be achieved within 15 days. Decreasing the hydraulic conductivity by one order of magnitude shows that the groundwater target depth is not achieved within the 30 days. Complete dewatering will require more time if the pumping rate from each well remains at 19.1 m^3/day.

Storage (or specific yield) values have a significant impact on the dewatered groundwater elevation depths. Decreasing the specific yield to 0.1 and 0.03 for layers 1 and 2 caused the groundwater elevation at the Pits to reach the target dewatered elevation within 6 days. Increasing the specific yield to 0.28 and 0.19 for layers 1 and 2 showed that the time to achieve the target elevation was approximately 28 days. The sensitivity analyses show that the range of total water dewatered from each pit depends on the selected hydraulic conductivity and storage values and varies from approximately 9,000 to 27,000 m^3 (for a one order-of-magnitude increase of hydraulic conductivity with a 109 m^3/day pumpage from each well) to achieve the target dewatered elevation of -18.5 m.

The model results show that it is possible to dewater the three pits using groundwater wells; however, specific information of dewatering pumping rates needs to be deciphered from site-specific pumping tests (step and constant rate discharge tests) to determine the hydrogeologic properties. The actual amount of water discharged will also depend on groundwater level conditions at the site. The current groundwater model is a planning-level tool to estimate dewatering potential at the three pits; however, for project design, site-specific information such as pumping tests should be performed to better estimate the site's dewatering capability. The number of wells used for dewatering simulations is not expected to be the exact number that is required for the actual dewatering process. For a required inflow into an excavation, there is more than one potential solution; a greater number of smaller wells, for example, achieve the same results as a smaller number of larger, higher-volume wells.

It is therefore recommended that a pumping test (step and constant rate discharge tests) be performed at any one of the pit areas. The tests would further determine how much water can be diverted from each well and would fine-tune the well pumping rate. The rate of flow into a pumped well or well point depends on the area and permeability of the ground immediately outside the well and on the hydraulic gradient causing the flow. Evaluation of the pumping test would provide a refined estimate of average hydraulic conductivity and storage of the pumping domain. The groundwater model would then be refined to better estimate the dewatering rates.

If the site step tests and pumping tests show that each of the individual wells can be pumped at rates greater than 19.1 m^3/day, then the groundwater model would be run with the new pumping rate. The model would help determine the required number of wells needed and the placement of the wells before groundwater dewatering is initiated. Additional dewatering methodologies should also be considered, including a series of well points, sumps, and/or cutoff walls and depends on the hydraulic conductivity and storage values determined from site-specific pumping tests.

REFERENCES

Boak R, Bellis L, Low R, Mitchell R, Hayes P, McKelvey P, Neale S (2007). Hydrogeological impact appraisal for dewatering abstractions. Science Report – SC040020/SR1. U.K. Environmental Agency. ISBN: 978-1-84432-673-0.

Doherty J (2010). Addendum to the PEST Manual. Watermark Numerical Computing, Revision: September 2010.

de Vries JJ, Simmers I (2002). Groundwater recharge: an overview of processes and challenges. Hydrogeol. J. 10:5-17.

Environmental Simulations, Inc. (2007). Guide to Using Groundwater Vistas Version 5, Reinholds, PA.

Ergun MU, Naicakan MS (1993). Dewatering of a Large Excavation Pit by Wellpoints, Third International Conference on Case Histories in Geotechnical Engineering, St Louis, Missouri, 1993, Paper No. 5.12.

Fetter CW (1994). Applied Hydrogeology, Third Edition, 1-691.

Harbaugh AW, Banta ER, Hill MC, McDonald MG (2000). MODFLOW-2000, The U.S. Geological Survey Modular Ground-water Model-User Guide to Modularization Concepts and the Ground-Water Flow Process: U.S. Geological Survey Open-File Report 00-92, 1-121.

Kalbus E, Oswald S, Wang W, Kolditz O, Engelhardt I, Al-Saud MI, Rausch R (2011). Large-scale Modeling of the Groundwater Resources on the Arabian Platform. Int. J. Water Res. Arid Environ 1(1):38-47.

Memon BA, Kazi A, Powell WJ, Bazuhair AS (1986). Estimation of Groundwater Recharge in Wadi Al-Yammaniyah, Saudi Arabia. Environ. Geol. Water Sci. 8(3):153-160.

Powers PJ, Corwin AB, Schmall PC, Kaeck WE (2007). Construction Dewatering and Groundwater Control: New Methods and Applications, Third Edition. John Wiley & Sons, Inc. ISBN: 9780471479437.

Preene M (2012). Groundwater Lowering in Construction: A Practical Guide to Dewatering, Second Edition. CRC Press, 673 pp.

Watermark Numerical Computing (2010). PEST: Model-Independent Parameter Estimation, User Manual: 5th Edition.

9

Groundwater resources of the Niger Delta: Quality implications and management considerations

H.O. Nwankwoala[1] and S.A. Ngah[2]

[1]Department of Geology, College of Natural and Applied Sciences, University of Port Harcourt, Nigeria.
[2]Institute of Geosciences and Space Technology, Rivers State University of Science and Technology, P. M. B. 5080, Nkpolu-Oroworukwo, Port Harcourt, Nigeria.

The Niger Delta region of Nigeria is of great socio-economic importance because of its huge and abundant reserves of crude oil and natural gas. The phenomenal increase in both industrial and social activities within the area in recent times has led to increase in groundwater abstraction rates. Groundwater development in the Niger Delta region has been carried out with little hydrogeological considerations. This is probably, due to strong political and other extraneous pressures. Consequently, borehole drilling has been indiscriminately and randomly sited and located, resulting to a myriad of problems. As a result, boreholes have failed where there are abundant groundwater storage, and many a time, have been drilled in, obviously, most unpromising locations. Salinity, bacteriological contamination, presence of undesirable iron and manganese as well as the high acidity of the groundwater and consequent corrosiveness pose serious distribution problems in the area. Efforts to solve the problems have been unsuccessful and supply of potable water remains grossly inadequate. Recent development plans proposed for the Niger Delta would call for high water demand. There is therefore an urgent need for a meaningful approach to the study of groundwater resources of the region. It is therefore hoped that many of the observations made in this paper will not only form a guideline for meaningful collection of data for quantitative analysis in future, but will also help in the understanding of the nature of groundwater resources of the region.

Key words: Groundwater resources, quality, aquifers, groundwater management, Niger Delta.

INTRODUCTION

The Niger Delta has an aerial extent of 75,000 km^2 and is located between latitude 4°30' and 5° 20' N and longitude 3° and 9°E. It is the second largest delta in the world with a coastline spanning about 450 km terminating at the Imo River entrance (Awosika, 1995). The region spans over 20,000 km^2 and it has been described as the largest wetland in Africa and consists mainly of freshwater swamps, mangrove swamps, beaches, bars and estuaries. This difficult terrain made it a region mostly forgotten by the rest of Nigeria, until the advent of petroleum in the area in the late fifties.

The pressure on water supplies and precious ecosystems in the coastal areas of the Niger Delta is very high and can increase in the future if urgent management measures are not put in place (Nwankwoala, 2011). Groundwater is the only source of water for water supply for both domestic and industrial uses in the Niger Delta and its demand will increase astronomically within the foreseeable future with increase in population, improved standard of living and more expansion and growth of the oil and gas industry (Oteri, 1983; Akpokodje, 2005; Nwankwoala and Udom, 2011).

Though the Niger Delta produces over 80% of Nigeria's petroleum, it is still very much a neglected part of the country. Beside efforts made by oil prospecting companies in the process of oil exploration and production, the region has not been studied in many areas and respects. Groundwater resources development of the Niger Delta is one such area where no serious effort has been made to investigate its nature, distribution and occurrence. Geologic considerations are rarely or inadequately incorporated into the design of boreholes (Amajor, 1989). Generally, groundwater has not attained a high level of development in the Niger Delta partly as a result of difficult environmental condition, low level of general underdevelopment of the region, inadequate finance; and partly perhaps as a result of deliberate neglect of the area by successive governments.

The Niger Delta is a large and ecologically sensitive region in which various water species (including surface and groundwater, saline and freshwaters) are in dynamic equilibrium (Abam, 1999). Because of the very nature of the region, groundwater constitutes the predominant, if not the only source of water supply in the area and unless a determined effort is made to understand the nature of the groundwater in the region, serious problems would be encountered in the area of water needs of the region, in future. Although, detailed stratigraphic analysis of the various geologic/geomorphic and aquifer systems (Etu-Efeotor and Akpokodje, 1990) have yielded a better understanding of the groundwater potentials of the Niger Delta region, they have not addressed the important and crucial question of over-abstraction and its associated consequences, particularly the possibility of large scale saltwater intrusion and general issue of sustainable supply of potable water in the region. This paper therefore appraises groundwater resources of the Niger Delta region, the quality impairments and management considerations.

GEOMORPHOLOGIC/GEOLOGIC SETTING

The geomorphology of the Niger Delta has been described by many researchers (NEDECO, 1954, 1959, 1961; Allen, 1965; Weber, 1971). The topography of the area is essentially flat, sloping very gently seawards. The area is low lying (usually does not exceed 20 m above sea-level) and is drained and criss-crossed by network of distributaries. The Niger Delta constitutes an extensive plain exposed to periodical inundation by flooding when the rivers and creeks overflow their banks. A prominent feature of the rivers and creeks is the occurrence of natural levees on both banks, behind which occur vast areas of backswamps and lagoons/lakes where surface flow is negligible.

Although various types of morphological units and depositiona environments have been recognized in the area (coastal flats ancient/modern sea, river and lagoonal beaches, sand bars, flooc plains, seasonally flooded depressions, swamps, ancient creeks and river channels), the area can be sub-divided into five majo geomorphological units (Figure 1) namely:

i) Active/abandoned coastal beaches
ii) Saltwater, mangrove swamps
iii) Freshwater swamps, back-swamps, deltaic plain alluvium anc meander belt
iv) Dry deltaic plain with abundant freshwater swamps (Sombreiro Warri deltaic plain) and
v) Dry flat land and plain.

Along the coastline lies a long coastal saline belt of active anc abandoned beaches built by ocean currents and tides. This area is comparatively higher than the adjacent areas and its width varies from 1 to 10 km. Parallel to, and north of the coastal saline belt o the beaches, is a stretch of mangrove swamp with an approximate width of 10 to 25 km. North of the mangrove swamp is the freshwater swamp which is in turn succeeded inland by dry areas that are not prone to periodical flood inundation.

Consequently, the present knowledge of the geology of the Nige Delta was derived from the works of the following researchers (Reyment, 1965; Short and Stauble, 1967; Murat, 1970; Merki 1970) as well as the exploration activities of the oil and gas companies in Nigeria. The formation of the so called proto-Nige Delta occurred during the second depositional cycle (Campaniar Maastrichtian) of the southern Nigerian basin. However, the moderr Niger Delta was formed during the third and last depositional cycle of the southern Nigerian basin which started in the Paleocene.

The geologic sequence of the Niger Delta consists of three mair tertiary subsurface lithostratigraphic units (Short and Stauble, 1967 which are overlain by various types of Quaternary deposits. Fron bottom to top, the tertiary units are the Akata, the Agbada and the Benin Formations (Table 1).

Hydrologic factors in the Niger Delta

The hydrologic conditions in a region are important ir understanding the groundwater situation, in that they determine the availability of water input into the basin for groundwater storage, the rate of groundwater recharge, and the movement of the water in the groundwater system, for extraction purposes. The factors o precipitation, runoff, evapo-transpiration and infiltration play major role. The process of water movement in the ground is more relatec to the factor of geology.

The amount of rainfall that takes place in a basin woulc determine the water input into the basin (Gobo, 1988). The annua rainfall in the Niger Delta is high and varies from 500 mm pei annum at the coasts, to about 300 mm at the northern part of the delta (Etu-Efeotor and Odigi, 1983). Evapo-transpiration is 100C mm, leaving an effective rainfall of 2000 mm. Of this effective rainfall, 37% or 750 mm is known to recharge the subsurface aquifers while the remaining 1250 mm flows directly into the streams (Akpokodje et al., 1996). This recharge which is 75% of the total precipitation is on the high side of the range commonly reported for unconsolidated sediments (Vecchioli and Miller, 1973 Legeette and Graham, 1994). This therefore ensures that the region is adequately supplied with water. Besides, rain may fall at any time of the year, even during the peak of the dry season further ensuring an all year round water input into the region. Basically, the Niger Delta water resources are drawn from the Eastern littoral hydrological and the Niger South hydrological zones. Infiltration and percolation processes from these broad recharge networks flow southwards into the underlying aquifers of the Benin

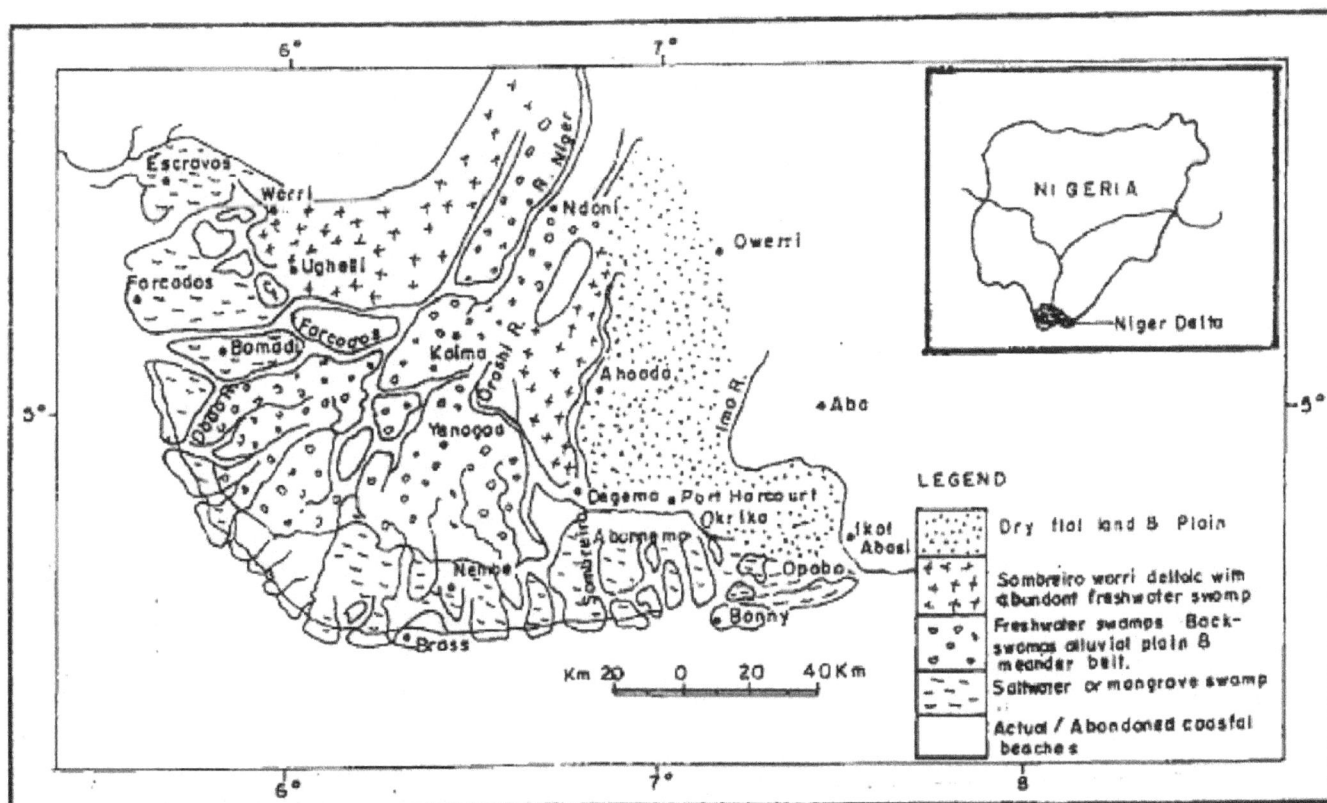

Figure 1. Major morphological units of the Niger Delta (modified after Akpokodje, 1987).

Table 1. Geological units of the Niger Delta (after Short and Stauble, 1967).

Age	Geological Unit	Lithology
Quaternary	- Alluvium (general) fresh water back swamp meander belt. - Mangrove and salt water. - Back Swamps - Active/Abandoned Beach ridges - Sombreiro Warri Delataic plain.	- Gravel, sand, clay, silt, sand, clay, some silt, gravel. - Medium-fine sands, - Clay and some silt.
Miocene	Benin Formation (coastal plain sand)	Coarse to medium grain sand with subordinate silt and clay lenses; Fluviatile marine
Eocene	Agbada Formation	Mixture of sand, clay and silt, fluviatile marine.

Formation beneath the Continental Shelf (Ngerebara and Nwankwoala, 2008). Both the structural and stratigraphic setting of the Niger Delta favour hydraulic gradient flow towards the coast, and hence into the Continental Shelf. This forms the basis of most freshwater aquifers located within the Continental Shelf. The general rainfall pattern in the Niger Delta ensures a permanent supply of water to the region.

The extent to which this rainfall reaches the ground to supplement the underground storage, greatly depend on the rate of infiltration, runoff pattern and the rate of evapo-transpiration. None of these factors had been studied quantitatively, but the following observations have been made (Ngerebara and Nwankwoala, 2008) in a more tentative way:

1) The rate of infiltration has generally been favoured by the prevailing flat nature of the Niger Delta which reduces runoff, by increasing retention time of rainwater on the land surface. Besides, the soil in most cases, except where surface clay and swamp prevail, is unconsolidated, porous and permeable, thus permitting quick infiltration of water underground.

2) The region is adequately drained by several streams and rivers. But more important is that these surface water bodies, flow through raised channels due to heavy load of sediments which settle within the channels. Consequently, the streams and rivers overlie most of the important aquifers in the area and feed them throughout the whole year. Most of the streams and rivers are therefore influent. The aquifers do not have to wait until flooding periods to be fed by

surface streams and rivers. The above prevailing conditions means that there is no period of the year when groundwater storage would have to be diminished by base flow but rather is ensured of a continuous discharge wherever the boundary with adjacent streams and rivers are continuous.

3) Evapo-transpiration measurements made on local scales by agricultural establishments indicate that water losses especially in areas of unconfined aquifers with water table near ground surface, may be high, due to high evapotranspiration during the dry season. This is in respect of the dense vegetation of the region and the dry condition that prevails during the dry season. But because of the huge reservoir prevailing under the Niger Delta, the more or less constant water supply in the region and the occurrence of clay lenses which often cut off aquifers from one another, the effect of evapotranspiration even where water is nearer the surface does not demand serious consideration in the Niger Delta.

Geological factors in the Niger Delta

The geology of the Niger Delta is well known and has been discussed by several authors (Reyment, 1965; Short and Stauble, 1967; Merki 1972; Weber and Daukoru, 1975; Avbovbo, 1978; Agagu, 1979; Whiteman, 1982; Doust and Omatsola, 1990; Owolabi et al., 1990; Koledoye et al., 2000, 2003). The influence of geology on the groundwater resources of the Niger Delta constitutes the most important factor besides that of climate in the region. Geology has been observed to be responsible for the complex groundwater distribution, extractability and quality in the Niger Delta. Unfortunately, the present knowledge of the true geological condition prevailing within the groundwater domain of the Niger Delta is limited.

Three major formations comprise the modern Niger Delta overlain by various types of Quaternary deposits (Table 1, Figure 1). These are the Akata Formation, which is predominantly shale and clay; the Agbada Formation which is generally fluviatile and fluviomarine, and the Benin Formation, constituting a continental deposit of sand and gravel (Murat, 1971). The depositional pattern which accompanied the accumulation of sediments during the formation of the delta, gave rise to structural traps (growth faults and roll-over anticlines) in the Agbada Formation. This constitutes the petroleum containing reservoirs in the Niger Delta. The Agbada Formation while suitable for petroleum accummulation, is too deep to be relevant to groundwater storage. There arises therefore, the major difference between the region where the petroleum geologist is prospecting for oil, that is, the Agbada Formation, and that, where the hydrogeologist is searching for water – the Benin Formation, in the Niger Delta.

Huge financial investments by oil companies have revealed the geology of the Agbada Formation in detail. Understandably, investigating the Agbada Formation, petroleum geologists had deliberately ignored the upper lying Benin Formation. Hence, the present knowledge of the Benin Formation is limited, compared with that of the Agbada Formation. An evaluation of the hydrogeology of the Niger Delta for petroleum exploration may appear to be ill-conceived. However, there is strong evidence that meteoric water from gravity-induced flow has recharged deep enough into the subsurface to possibly play a role in the distribution of hydrocarbons. According to Dickey et al. (1987) and Amajor and Gbadebo (1992), the extremely sandy nature of the upper Benin Formation and the abundant growth faults in the underlying Akata Formation have permitted meteoric water to penetrate very deep into the subsurface.

The controlling effect of geology on groundwater occurrence in the Niger Delta is no longer in doubt. The sedimentation pattern as well as stratification determines both the quality and quantity of water in the region. Its investigation is the first step towards a meaningful groundwater study of the region. The Benin Formation

therefore needs detailed investigation.

Groundwater occurrence in the Niger Delta

The main body of groundwater in the Niger Delta is contained in mainly very thick and extensive sand and gravel aquifers. Three main zones have been differentiated. These are: a northern bordering zone consisting of shallow aquifers of predominantly continental deposit, a transition zone of intermixing marine and continental materials and a coastal zone of predominantly marine deposits (Etu-Efeotor and Odigi, 1983; Amajor, 1989; Etu-Efeotor and Akpokodje, 1990). A distinct trend in aquifer properties have been observed following this division. Akpokodje et al. (1996) have summarized the hydrostratigraphic units of the Benin Formation as four well defined aquifers in the upper 305 m that vary in thickness to over 120 m. The aquifers vary from unconfined conditions at the surface through semi-confined to confined conditions at depth. The aquifers are separated by highly discontinuous layers of shales, giving a picture of an interval that consists of a complex, non-uniform, discontinuous and heterogeneous aquifer system. Although, majority of groundwater supply wells abstract water from these aquifers, there is evidence that industrial and municipal groundwater supply wells produce water from deeper aquifers in the Benin Formation.

Aquifers at the northern border of the Niger Delta are more continental in character, being composed of river loads coming from the hinter land. They are also encountered at shallower depths, so that in most cases, an average depth of 60 m had been all that was required to be drilled, to obtain very pure freshwater and in huge quantity. Clay materials, except a few metres found within the top soil, do not occur at depth. The sand is coarse to very coarse generally, and gravel layers are commonly encountered. The borehole performance in this section has generally been so good and the water quality so excellent, that sinking of wells at the northern borders of the Niger Delta has always been taken for granted. Very good examples of such regions are Port Harcourt, Ogoni, and Elele areas of Rivers State, eastern Niger Delta.

Moving coastwards from the northern borders of the Niger Delta, one comes across a transitionary zone of swamp lands. Two types of swamp lands are observed – the mangrove swamp lands and the freshwater swamp lands. The mangrove swamp lands are associated with tidal inlets and they are therefore more prominent in those areas where estuaries penetrated farther inland, such as the western and eastern zones flanking the pro delta. On the other hand, fresh swamp land persists more within the front of the delta where the dense network of streams and rivers combine to empty into the sea.

A common feature of the transition zone is the presence of clay embodiments within the aquifers. These clay lenses are erratically distributed laterally and vertically within the region. In several cases, strata logs of wells drilled less than 200 m apart, have been known to vary, and under such prevailing circumstance, prediction of aquifer performance in the region is difficult. However, the freshwater swamp lands which constitute the front of the delta, continue to indicate many features of continental environments, until very close to the coast. The aquifers are still shallow, consisting of predominantly sand and gravel materials, but clay intercalations become more prominent, than within the northern zones. Lignitic materials are also present in the aquifer and the presence of vegetative matter strongly point to sedimentation under shallow water condition.

Within the mangrove swamp lands, very strong evidences of marine conditions are indicated. Thicker lenses of marine clay are encountered and saline conditions are still well noticed. There is no doubt that these areas are protected from the dynamic zones of the deltaic front. This makes it possible for marine conditions to penetrate further inland, creating a more complex transition zone.

This is the case in the freshwater swamp lands. There is an intermixing of continental and marine sediments resulting in a very complex aquifer system. Generally, it has been necessary to drill beyond the 200 m depth before a good water yielding aquifer could be obtained and saline water intrusion problem plague the region. It is here, however, that artisan conditions, due to the interbedment of sand aquifers within clay aquicludes occur. But such confined aquifers are generally too deep seated to result in flowing wells (Ngah, 2009; Nwankwoala, 2011).

Within the sand bars and beaches of the coastal lands, boreholes still need to go deeper to reach quality good water aquifers. Beneath the coastal sands that form the surface deposits of this last zone, marine conditions predominate, until at depth where deep seated aquifers empty into the sea. Aquifers within the Niger Delta generally produce and perform better during the rainy season. They dwindle in yield during the dry season. At the coastal lands, rains feed and maintain phreatic aquifers during the rainy season. But with the incoming of the dry season, such aquifers dry up, and wells sunk into them commonly go without water at that season.

RESULTS AND DISCUSSION

Groundwater quality status

Table 2 shows the groundwater quality in the Niger Delta. The quality of groundwater in the Niger Delta closely follows the sedimentation pattern. As a result, three distinct zones are recognized. The continental deposits of the Northern border produce the best quality water in the region – fresh, pure and commonly uncontaminated groundwater. Within the transition zones, however, the complex sedimentary environment greatly influences the water quality. Most remarkable are the freshwater swamp lands where quality degradation associated with the breakdown of organic matter derived from vegetation buried in the sediments, are encountered. These generally take the form of high carbonate acidity and introduced hydrogen sulphide, commonly identified by the bad smells of some water samples from the area.

Iron contamination is also another feature of groundwater quality within the transition zones. The intensity of its occurrence has been observed to be higher within the freshwater swamp zones (Etu-Efeotor, 1981; Etu-Efeotor and Odigi, 1983; Odigi, 1989). But in the mangrove swamp land areas, notably Buguma, Bonny and Abonnema in Rivers State, cases of iron contamination have been encountered. They are commonly in the form of ferrous iron which generally remains in solution when water samples are freshly collected. On standing the samples, the ferrous iron comes in contact with oxygen of the air and is oxidized into its ferric equivalent which is generally brownish in colour.

The source of the iron contamination is not quite known but it is suggested to have been emplaced by iron fixing bacteria associated with sedimentary environments of decaying vegetative matter. According to Allen (1965) and Oomkens (1974), the Quaternary glaciation was accompanied by eustatic lowering of the sea level such that the paleo- strandline was at the present edge of the continental shelf. This geologic event would have exposed the sediments and created paleo-soils rich in iron oxides. The subsequent rise in sea level would have incorporated the paleo-soils into the geologic record.

Salinity trends in the Niger Delta

Salinity problems are encountered in the Niger Delta. But this is a case prevalent within the mangrove swamp lands and the coastal aquifers. Because of the importance of saline problem in the groundwater development of the Niger Delta, it is necessary to know the main types of saline pollution and where they would be expected. Generally, two types of saline gradient are noticed. There is a vertical gradient which changes with respect to distance from the sea. Vertical salinity gradient develops within the estuarine areas. They arise from the penetration of saline water inland through creeks and estuaries. They are therefore saline conditions that originate from the infiltration of salt water from creeks into underlying sediments. Such saline pollution however, applies only where the aquifer is uninterrupted in depth. A thick or extensive aquiclude can and does exclude further penetration of salt water under such circumstances (Nwankwoala, 2011; Ngerebara and Nwankwoala, 2008).

The WHO (2008) appears to be silent on specifications of limit for this parameter. Salinity tends to increase in a southerly direction. Least values of salinity are more common in the hinterland. The salinity values of groundwater in the study area appear to be generally tolerable. The salinity depths in deep boreholes in Bonny Island suggest saltwater intrusion into submarine freshwater aquifers (Ngah and Nwankwoala, 2013b). Naturally, the coastal aquifers drain into the ocean and are in contact with the ocean at the coastline where under natural conditions fresh water is discharged into the ocean. Excessive abstraction of groundwater appears to have resulted in a decreased seaward flow of fresh groundwater causing saline water to enter and penetrate inland through submarine outcrops. This phenomenon will progressively displace the freshwater thereby increasing the salinity depth (Ngah, 2009). Figure 2 is the map of the Niger Delta showing salinity limits/vegetation of the coastal zones of the Niger Delta.

Human activities at times enhance such saline intrusion in the area. This was the case at Isaka near Port Harcourt where the dredging of the Port Harcourt harbor admitted saline water into the aquifer in the area. Also, influence of land reclamation in Borokiri area of Port Harcourt induced saline water (Nwankwoala and Udom, 2008). It requires further deep drilling beyond a thick lens of clay before an uncontaminated aquifer will be encountered. Within sands on islands and beach ridges at the coast, a unique condition of salinity prevails in which a cone of freshwater overlies a saline layer. Saline

Table 2. Groundwater quality data in parts of the Eastern Niger Delta (Ngah and Nwankwoala, 2013a).

S/No.	Borehole Location	Geomorphic zone	Depth (m)	Iron (mg/l)	Chloride (mg/l)	Salinity (mg/l)	Conductivity (us/cm)	pH	Alkalinity (mg/l)	Total hardness (mg/l)
1.	Kanana	SWS	186	0.0	34	500	125	6.7	0.12	38
2.	Kala Degema	"	40.96	0.0	18	150.0	-	7.3	0.38	6
3.	Krakrama	"	75	2.5	48	700	100	6.8	-	16
4.	Abalama	"	60.96	2.0	36	10.74	35	6.7	30	11
5.	Buguma	"	60.96	0.0	10.64	17.55	-	6.2	6.5	17.0
6.	Okrika Mainland	"	320	0.0	35.5	58.5	19	6.2	10.40	20
7.	Ibuluya-Dikibo	"	180	0.0	32.0	18.00	20.00	4.5	10	21
8.	Bolo I	"	91.44	0.0	99	14.00	16.00	7.0	0.12	31
9.	Bolo II	"	91.44	0.0	97	16.00	40	6.8	0.1	35
10.	Kalio-Ama	"	82.88	0.5	10	331	58.0	6.4	0.1	61
11.	Abam-Ama	"	128.02	0.0	6.38	100	23.0	5.9	21	Artesian
12.	Okujagu	"	30.0	0.8	62	100	-	4.5	2	30
13.	George-Ama	"	109.73	1.8	31.5	60.25	29	5.1	-	22
14.	Isiokpo	CPS	70.1	2.0	38.94	64.4	20	5.6	15	10
15.	Aluu	"	60.96	0.2	10	50	23.00	6.1	2	6
16.	Umuoji	"	81	10.0	32	80	-	8.2	-	68
17.	Ogbakiri	"	78.03	0.01	24.2	51.9	4.5	6.2	0.5	60
18.	Ndele	"	72.5	0	10.6	100	13.5	7.5	19.0	9
19.	Omerelu	"	70.1	0.4	20	20	20	6.5	0.6	8
20.	Ubima	"	70.1	0.0	18	50	18	5.2	2	4
21.	Elele	CPS	60.96	0.0	29.8	49	15	6.0	3.0	5.0
22.	Ibaa	"	60.96	0.0	40	68.25	12	6.2	-	2.0
23.	Obelle	"	81.0	0.0	48	63	6.7	6.1	-	0
24.	Rumuewho	"	54.86	0.3	10	50	-	6.2	-	38
25.	Egwi	"	61.28	0.15	3.4	65	-	6.5	-	28
26.	Rumuoyo	"	57.3	0.30	24	73	-	6.4	-	8
27.	Ulakwo	"	67.06	0.0	13	60.2	100	7.4	-	34
28.	Opiro	"	138	0.0	16	100	12.5	6.2	6.0	10
29.	Rumuokochi	"	91	10	12	60	100	6.2	-	20
30.	Umuechem	"	132	0.30	11	100	36	6.7	3.0	6.0
31.	Kalibiama	CBR	281	0.0	10	202.0	440.0	8.4	-	488
32.	Bonny	"	304	0.8	5	186	140	5.9	-	78
33.	Oloma I	"	91.46	0.8	810	210.0	600	6.5	90	72
34.	Oloma II	"	82.88	0.0	250	220.0	380.0	6.2	20	454
35.	Illoma Opobo	"	19.8	0.6	330	60	280	7.2	-	84
36.	Gbokokiri	"	176.8	0.40	300	192.0	-	6.6	-	18
37.	Ikuru	"	190	0.38	351	242.0	-	6.7	-	70
38.	G.R.A. P.H.	"	170	0.00	26	50	7	5.3	2	6
39.	Creek Road	"	170.0	0.02	390.5	643.5	150.0	5.8	35	135
40.	Potts Johnson	"	180.0	0.02	401	661	160.0	5.8	18	135

CPS = coastal plain sands, FWS = freshwater swamp, SWS = saltwater swamp, CBR = coastal beaches and ridges, SWP = sombreirowarri deltaic plain.

water pollution becomes imminent, if such pool of freshwater is not intelligently extracted.

Aquifer vulnerability to contamination

Groundwater contamination may be defined as the induced degradation of natural water quality by the introduction of inorganic and/or organic compounds. It is usually more serious than surface water contamination because it is more difficult to detect in a timely manner, moves more slowly and requires special expertise to predict the path and rate of contaminant movement. Aquifers can be polluted by a combination of the following factors: agricultural activities, petroleum leakage and

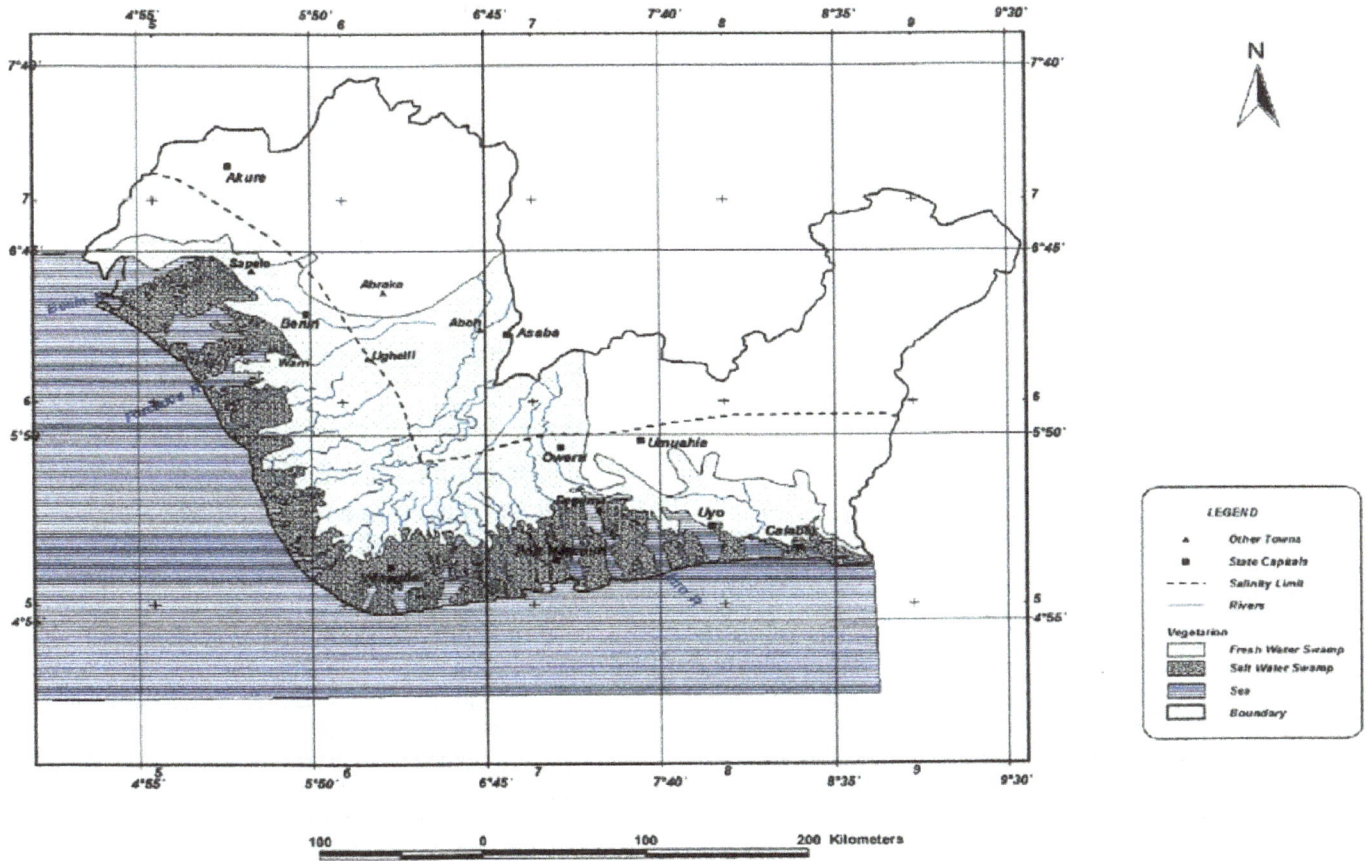

Figure 2. Map of the Niger Delta showing salinity limits/vegetation of the coastal zone (after Ogba and Utang, 2010).

spills, land disposal of solid wastes, sewage disposal on land, saltwater encroachment, deep well disposal of liquid wastes, mining activities, spread of urbanization to recharge areas, and seepage from industrial waste. Aquifers are vulnerable to pollution from these sources to the extent they are in proximity with such activities.

The first aquifer system is extremely vulnerable to pollution from surface sources. Consisting essentially of loose to poorly consolidated sandy materials, this aquifer is capped by laterized soil especially in the northern part of the study area. The laterite thins out in a southern direction and has become non-existent from the fresh water swamp to the coastline (Ngah, 2009). Where the laterite is thin or nonexistent, poor land use practices including use of fertilizers in farming and land disposal of solid wastes including refuse dump can result in the contamination of groundwater sources. For instance, the decomposition of refuse produces leachate, a highly polluting substance which can seriously degrade groundwater quality if allowed access to it. Moreover, the network of pipelines that convey petroleum products in the Niger Delta are buried in the first aquifer. Undetected leakage of hydrocarbon from the pipelines over a long period of time can pollute the aquifer.

Aquifer systems in the Niger Delta are separated by fairly thick clay/shale layer which by their nature, are impermeable, serving as a barrier to vertical (down-ward) percolation of contaminants from the first aquifer. The attenuation properties of clay also lead to an overall reduction of potency of contaminants and possible elimination through cation exchange. For this reason, it can be assumed that the aquifer systems below the first are shielded from pollution from surface sources.

Pollution of second aquifer system by surface sources can only be possible through vertical leakage if the overlying layer is an aquitard. Poor borehole completion practices may also lead to contamination of second aquifer system by the first if the well design does not exclude the first aquifer from contributing to the yield of the second.

In the coast, the shallow aquifer systems are unprotected from the dynamic coastal activities. Saltwater is likely to have mixed with shallow fresh water rendering it unfit for human use. Moreover, exploitation of the shallow freshwater aquifers will reduce the thickness and effective head of the freshwater wedge leading to saltwater intrusion into fresh water aquifers. The deeper aquifer systems are fed by recharge from distant (up-country) outcrops of the aquifers and through vertical leakage from the overlying aquitards. The deeper

Table 3. Activities that could contaminate groundwater aquifer and the likely aquifer to be affected in the Niger Delta (Ngah and Nwankwoala, 2013b).

S/No	Activity	Aquifer system				
		1	2	3	4	>4
1	Agricultural activities	x				
2	Petroleum leakage and spills	x				
3	Sewage disposal on land	x				
4	Saltwater encroachment	x	x	x	x	x
5	Deep well disposal of liquid wastes	x	x	x	x	
6	Mining activities	x				
7	Spread of urbanization to recharge areas	x				
8	Seepage from industrial waste lagoons	x				

aquifers are considered safe and free from pollution with zero vulnerability everywhere except in the coastal areas. Here, unplanned exploitation and/or over-pumping of groundwater could increase the already deep salinity depth thereby reducing the effective column of fresh water. Boreholes will therefore need to go even deeper to exploit freshwater. Table 3 outlines the activities and practices that could lead to the degradation of groundwater quality in the Niger Delta.

From the Table 3, it is evident that the first aquifer system is highly vulnerable to contamination and many activities of man impact directly on this aquifer. Where the lateritized soil cover or clayey overburden is thin or completely absent and the water table is shallow, the risk of contamination is much higher. Awareness of this situation amongst the stakeholders needs to be emphasized.

Groundwater management options in the Niger Delta

Groundwater deserves serious attention in the Niger Delta. This is because there are no alternative sources of water supply in many parts of the Niger Delta. This is particularly the case in those regions where estuaries penetrate very far inland, leaving many areas surrounded by saline creeks and inlets. Moreso, the network of deltaic tributaries create problem of isolation in the region. Such a condition has not favoured the growth of towns and cities. Little riverine communities scatter all over the delta and there is no likelihood that this situation would change in the very near future. The need therefore for the establishment of huge water development schemes would not arise for a very long time in the area. What would be needed would be water development projects tailored to expansion through increase in boreholes.

Furthermore, with particular reference to irrigation, waterlog problem prevalent in the Niger Delta, could be further aggravated through pumpage from the surface streams. Irrigation schemes, if planned in the area, would

therefore be better served by boreholes through which groundwater levels are lowered, rather than raised in order to improve the waterlog condition in the area.

Whether for domestic, industrial or irrigational purposes, it appears evident that groundwater constitutes the most economical, practical and sensible source of water supply in the Niger Delta. It would have to be harnessed in order to meet the water needs of the several development projects now planned for the region. Important steps need to be taken in order to utilize the groundwater resources of the region more efficiently and usefully in future. Among the steps are:

(1) A deliberate effort to promote more understanding in the profession of hydrogeology. Several practicing Engineers today in Nigeria's water industry dismiss hydrogeology as irrelevant in the process of supplying the nation with more water. Consequently, wells are drilled haphazardly and are pumped without regard to the characteristics of the producing aquifer. While such a state of affair may constitute no problem under our present level of development, there is no doubt however that things might not continue in a similar way for long.

(2) The need for a Water Resources research body, specifically created to coordinate studies on both ground and surface water resources of the region. It is only such a body that can absorb the demand in money, material and men that such a complex region as the Niger Delta demands.

(3) There should be laws governing the abstraction of groundwater in the region. Today, no laws prohibiting individuals from drilling water wells anyhow, anywhere and at anytime exists. Over-pumpage in coastal aquifers can be dangerous where no regulation exists. This should be addressed with all seriousness.

(4) There is a need for accurate data collection, monitoring and on longer time scale to be able to detect and document effects of water degradation and conversely show the effects of remedial activities, despite the superimposition on natural climatic variability. Documentation, storage and dissemination of knowledge

are important. Through the development of awareness, knowledge and capacity at the national and local level, it is envisioned that the overall knowledge gap will diminish- a step towards sustainable development and management of water resources.

THE WAY FORWARD

Quite unfortunately, in spite of the fundamental role groundwater plays in human well being, as well as that of many ecosystems, groundwater basins are difficult to govern and manage, partly because of poor information, and also because of poor visibility of the resource, the need for reliable data and information in support of water resource planning is central to any strategy. Technology, knowledge transfer and sound research cooperation should receive sufficient attention at the regional scale for any meaningful solution of regional groundwater problems.

More importantly, any sustainable suggestion towards the improvement of water resources management in the Niger Delta must include but not limited to the following:

i) Re-introduction of long-term hydrological observations and investigation of new data collection on water use, irrigation and agriculture lands, water sediment deposits, industrial demands, urban development, recharge, hydraulic properties as well as groundwater/surface water interaction;

ii) Protection of groundwater resources to safeguard long-term use and balance the demands of economic development with ecosystem conservation;

iii) Regional studies of hydrogeology, hydraulic properties, regional flow system and water quality that cross state boundaries;

iv) Greater integration of the relevant information systems, e.g hydrology, hydrogeology, water quality, land use, sediment transport etc;

v) Since the responsibility of protecting our groundwater resources is a collective responsibility with everybody as a major stake holder, an aggressive public awareness programme comparable to that for HIV/AIDS is recommended for water users, planners and policy decision makers at all levels.

REFERENCES

Abam TKS (1999). Dynamics and quality of water resources in the Niger Delta.Proc IUGG 99 Symposium HSS Birmingham) IAHS Publ. No. 259:429-437.

Agagu OK (1979). "Potential geo-pressured geothermal reservoirs in the Niger Delta subsurface." Nig. J. Sci. 13:201-215.

Akpokodje EG (2005). Challenges of sustainable groundwater resources management in the Niger Delta. Paper presented at the National Stakeholders Workshop on Joint Management of Shared West African Coastal Aquifer Resources, Organized by the Federal Ministry of Water Resources in collaboration with UNESCO/IHP-IAH, held at the Hotel Presidential, Port Harcourt Nov. 21-22.

Akpokodje EG (1989) Preliminary Studies on the Geotechnical Characteristics of the Niger Delta Subsoil. Engr. Geol. 26:247-257

Allen JRL (1965). A review of the origin and characteristics of Recent Alluvial sediments of the Niger Delta. Sed. 5:89-191.

Amadi UMP, Amadi PA (1990). Saltwater migration in the coastal aquifers of Southern Nigeria. J. Min. Geol. 26(1):35-44.

Amajor LC (1989). Geological appraisal of groundwater exploitation in the Eastern Niger Delta; (ed. C.O Ofoegbu) Groundwater and Mineral Research of Nigeria: Braunschweig/Weisbaden, FriedrVieweg and Sohn, pp. 85-100.

Avbovbo AA (1978). Geologic Notes: Tertiary lithostratigraphy of the Niger Delta. AAPG Bull. 62(2):295-306.

Awosika LF (1995). Impacts of global climate change and sea level rise on coastal resources and energy development in Nigeria. (ed. J.C Umolu) Global Climate Change: Impact on Energy Development DAM TECH Nigeria Limited, Nigeria

Dickey P, George GO, Barker C (1987). Relationships among oils and water compositions in the Niger Delta. Am. Assoc. Pet. Geol. Bull. 71(10):1319-1328.

Doust H, Omatsola E (1990). "Niger Delta" (ed. J. D.Edwards and P.A Santogrossi), Divergent/Passive Margin Basins. AAPG Memoir 48:201-238.

Etu-Efeotor JO (1981). Preliminary hydrogeochemical investigations of subsurface waters in parts of the Niger Delta. J. Min. Geol. 18(1):103-107.

Etu-Efeotor JO, Akpokodje EG (1990). Aquifer systems of the Niger Delta. J. Min. Geol. 26(2):279-285

Etu-Efeotor JO, Odigi MI (1983). Water supply problems in the Eastern Niger Delta. J. Min. Geol. 20(1&2):183-193.

Gobo AE (1988). Relationship between rainfall trends and flooding in the Niger-Benue River Basins. J. Met. 13:132-139.

Murat RC (1970). Stratigraphy and paleogeography of the Cretaceous and Lower Tertiary in Southern Nigeria. In: Dessauvagie TTJ, Whiteman AJ (Eds.) AfrGeol, University of Ibadan Press, Ibadan, Nigeria. pp. 251-266.

Ngah SA (2009). Deep aquifer systems of eastern Niger Delta: Their hydrogeological properties, groundwater chemistry and vulnerability to degradation. Unpublished PhD Thesis, Rivers State University of Science and Technology, Port Harcourt, Nigeria. 247 pp.

Ngah SA, Nwankwoala HO (2013b). Salinity Dynamics: Trends and vulnerability of aquifers to contamination in the Niger Delta. Comp. J. Environ. Earth Sci. 2(2):18-25.

Ngerebara OD, Nwankwoala HO (2008). Groundwater potentials in the offshore Niger Delta environment, Nigeria. Electr. J. Environ. Hydrol. 16:28. http//www.hydroweb.com

Nwankwoala HO (2011). Perspectives on fresh and saline groundwater interactions in the coastal aquifer systems of the Niger Delta. Int'l J. Sustain. Dev. 4(1):81-91.

Nwankwoala HO, Udom GJ (2008). Influence of land reclamation on the status of groundwater in Borokiri Area of Port Harcourt, Niger Delta, Nigeria. Int. J. Nat. Appl. Sci. 4(4):431-434.

Nwankwoala HO, Udom GJ (2011). A preliminary review of potential groundwater resources of the Niger Delta. Int. J. Appl. Environ. Sci. 6(1):57-70.

Odigi MI (1989). Evaluating groundwater supply in Eastern Niger Delta, Nigeria. J. Min. Geol. 25:159-164

Oteri AU (1983). Delineation of saline intrusion in the Dungeness shingle aquifer using surface geophysics. Quart. J. Eng. Geol. 16:43-51.

Reyment RA (1965)." Aspects of the geology of Nigeria." Ibadan Univ. Press, Nigeria.

Short KC, Stauble AJ (1967). Outline geology of the Niger Delta. Bull. Am. Assoc. Pet. Geol. 54:761-779.

Vecchioli J, Miller EG (1973). "Water resources of the New Jersey part of the Ramapo River Basin" U.S. Geol. Surv. Water Supply Paper 1973.

World Health Organization, WHO (2008). Guidelines for Drinking Water Quality, Recommendations, Geneva 2:67.

10

Neotectonic belts, remote sensing and groundwater potentials in the Eastern Cape Province, South Africa

K. Madi[1] and B. Zhao[2]

[1]Geology Department, University of Fort Hare, Private Bag, X1314, alice, 5700, Eastern Cape, South Africa.
[2]TWP Projects (PTY) LTD, P.O Box 61232, Marshalltown 2107, Johannesburg, South Africa.

The Eastern Cape Province is surrounded by three major neotectonic belts (south, east and north). Each one of these neotectonic belts has its own particularities. This study aims at characterizing and targeting potentially high yield aquifers in the neotectonic zones in the Eastern Cape Province. The methods used in this study include a comprehensive literature review, examination of digital elevation models and extraction of lineaments through remote sensing. The results indicate the following: 1) the eastern neotectonic belt has a higher density of lineaments oriented NW-SE, which makes it the second important neotectonic belt; these lineaments correlate with the normalized difference vegetation index indicative of a good circulation of groundwater; 2) the surface topography is not uniform, high elevations in the east are related to the uplift that took place in the Quaternary, most vector gradients are oriented E-W; 3) in the south the Eastern Cape great lineament oriented E-W is now considered a neotectonic domain because of many seismic epicenters that occur therein, its geomorphologic shape in a graben type form is a favourable structure for groundwater catchment. This project has given fundemental highlights to characterize the major and potential good yield aquifers in the Eastern Cape Province in south Africa.

Key words: Neotectonics, groundwater, lineament, remote sensing, hot spring, uplift.

INTRODUCTION

Three neotectonic belts almost surround the Eastern Cape Province in South Africa (south, east and north near the country of Lesotho (Figure 1). If the eastern neotectonic belt was qualified as the high level neotectonic domain through remote sensing in terms of abundance of lineaments, a careful examination on the term "neotectonics" would undoubtedly indicate that the true neotectonics occurs in the northern part of the Eastern Cape province near the country of Lesotho. First, this region is located in the Kokstad-Koffiefontein seismic belt that runs in an east-west direction; second, seven hot springs (Figure 2) are present in this remarkable seismic belt of South Africa, near the border with the country of Lesotho.

Another important thing is that in the Kokstad-Koffiefontein seismic belt (Figure 3), in the town of Smithfield near Aliwal North, the local power station might have been built on an active fault, and the Orange River tunnel intersected with much pain a huge aquifer that flooded the place.

In the Eastern Cape northern neotectonic belt, Olivier (1975) found that in the Orange-Fish tunnel flooding occurred after an abnormally well developed fissure-zone was intersected approximately 550 m south of the shaft 2. The inrush of water was associated with the collapse of the tunnel roof, as part of the roof major fissure was exposed by the fateful blast. He noted however that influence of earthquakes on the pattern of semi-diurnal tidal

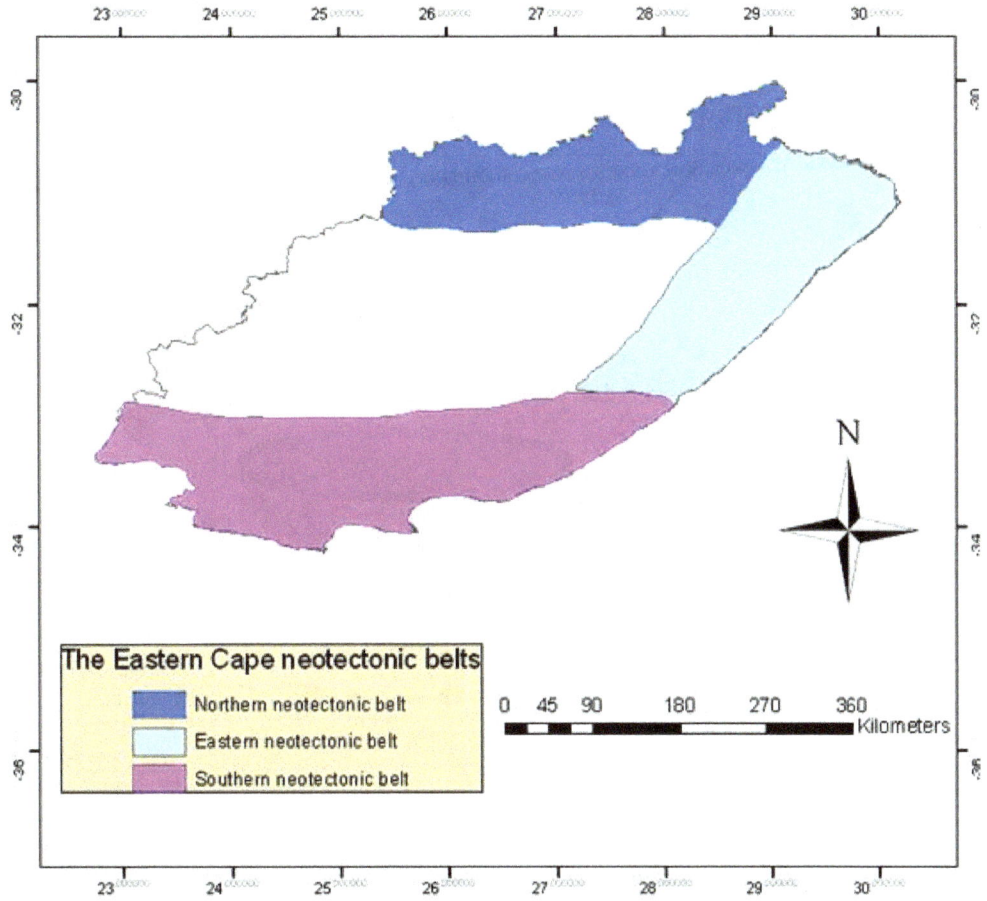

Figure 1. The Eastern Cape neotectonic belts.

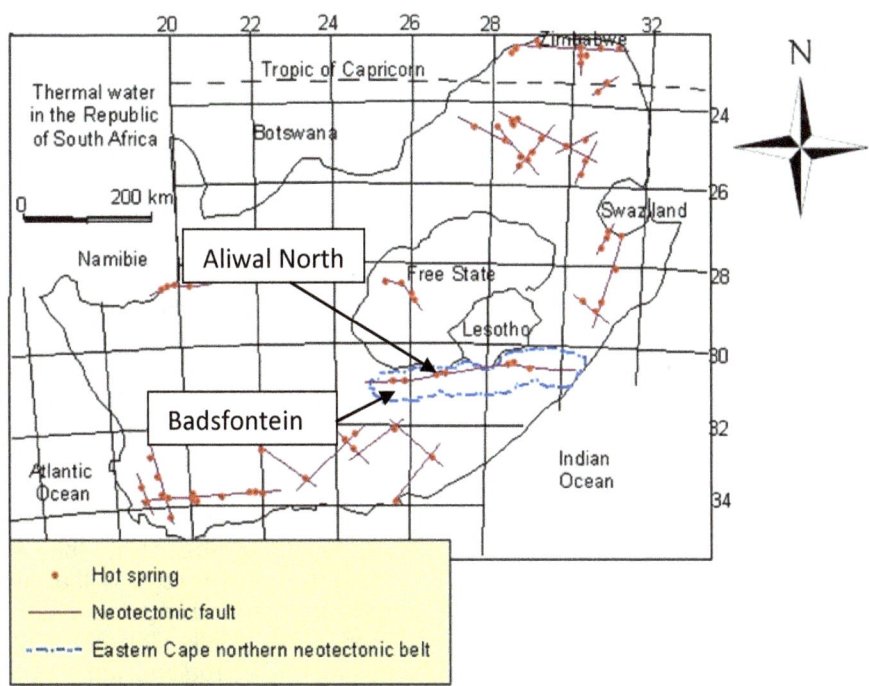

Figure 2. Hot springs in the Republic of South Africa.

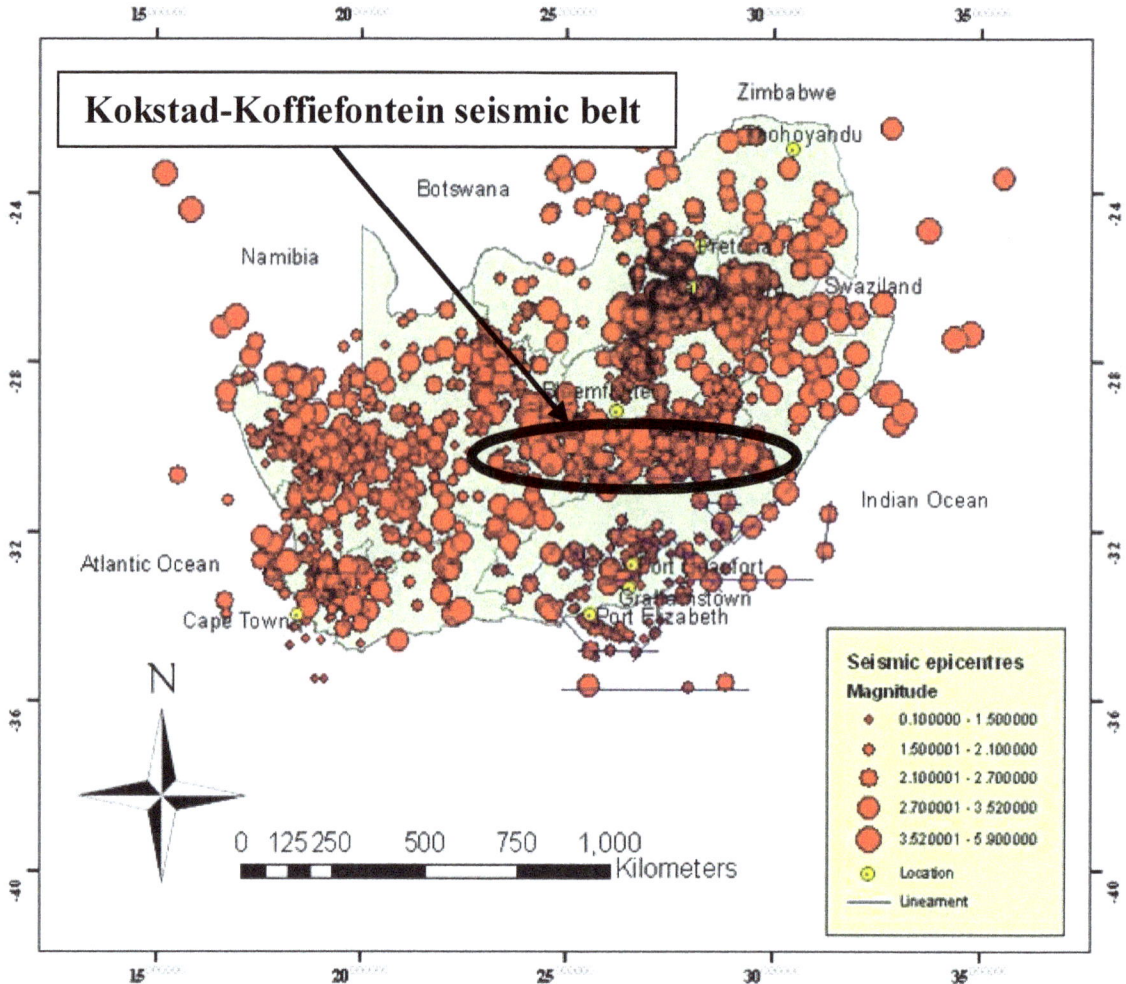

Figure 3. Seismic epicenters of new events in South Africa. Source of data: IRIS Earthquake Browser.

fluctuations is of special interest as regards the flooding problem. The flow of the central thermal spring, on the Badsfontein increased markedly for a period of at least 3 months after a local earthquake was felt during August 1956 (Whittingham of Geological Survey, personal communication, 1970).

During the Ceres-Tulbagh earthquake disaster of 1969 in the Western Province, tremors affected the tunnel area; displacements of 3, 4.6 and 7.6 cm on the graphs in some boreholes after some earthquakes of magnitude 6.5 and 6.2 in 1969 and 1970 were noted. This neotectonic belt is for the above mentioned facts, one of the zones of potentially high yield wells in the Karoo aquifers in the Eastern Cape Province.

In the southern neotectonic belt, the Coega Bavianskloof Fault (CBF) in the Cape Province has reactivated fault scarps that are, in some places, between 2 to 4 m high. The Worcester fault lies south of the CBF, with similar strike and orientation as the CBF, and extends toward the Ceres cluster (Bejaichund et al.,

2009); a striking neotectonic activity is the one that occurs in the eastern neotectonic belt, which is mainly characterized by the spectacular uplift from Swaziland to Amatole (Ciskei) in a NNE-SSW trend parallel to the coast line. Artyushkov and Hofmann (1986) mentioned that intensive crustal uplift (Figure 4) began in South Africa in the Oligocene period affecting most of the continental areas after a long period of relative stability. They also indicated that this neotectonic uplift formed most of the positive topographic features on the continent. In southern Africa, the uplift took place in the early Miocene (up to 300 m) and in the late Pliocene and Pleistocene (up to 900 m) with no stretching or shortening of the crust; this is indicative of a plume material in some regions that ascended from below and rapidly spread along the base of the lithosphere and eroded the mantle lithosphere in vast areas beneath the continents. In regions with hot asthenosphere, a strong weakening of the mantle lithosphere which allows its erosion can be associated with the high temperature of

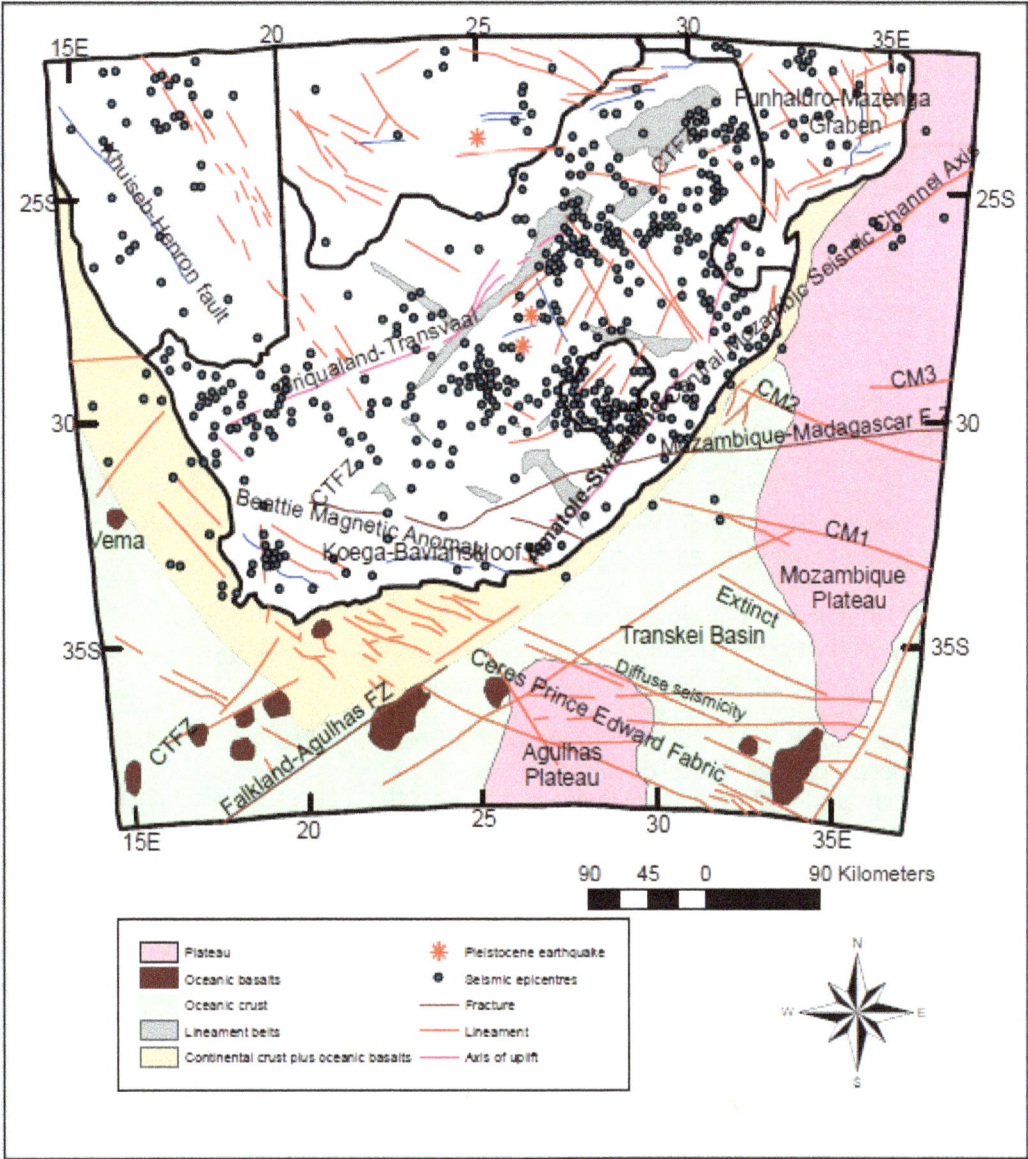

Figure 4. Seismic epicentres in relation to the post-Karoo tectonic framework of South Africa and nearby regions (Andreoli et al., 1996).

the plume material. In regions where the asthenosphere is at moderate temperature, weakening of the mantle lithosphere can result from infiltration of volatiles from the plume material.

Dobson et al. (2010) proposed three competing evolutionary models for the uplift: (1) the major phase of uplift occurred in the late Cretaceous, (2) the major phase of uplift occurred at ~30 Ma, and (3) ~ 900 m of the modern topography being generated rapidly 100 m/Ma in the Plio-Pleistocene (c. 3 Ma). This uplift is controlled by the NNE-SSW trending Miocene-Pliocene Griqualand-Transvaal related to the subsidence of the Kalahari basin (T. Partridge, Personal Communication, 2007) and Amathole-

Swaziland (formerly Ciskei-Swaziland) uplift (Figure 4) axes that stretch across almost the whole South Africa – Swaziland region (Partridge and Maude, 2000), and the uplift may be related to horizontal compressive stress. This uplift would have caused most of the rivers to flow toward the Atlantic Ocean, and for Eastern Cape towards the Indian Ocean (Esterhuizen, 2008); it is associated with numerous thermal springs and spas. Maouche et al. (2013) for instance reported that Quaternary and Pliocene travertines, deposited from hot springs, can reveal much about neotectonic and hydrothermal activity from their studies in the Guelma Basin in Algeria.

Remote sensing has been used in many disciplines as

a cost effective tool through study of satellite images, especially in groundwater, gas and oil exploration. Geological interpretation derived from remote sensing has been extensively used for the purpose of identification of lineaments and fractured zones along which flow of groundwater may take place.

Mohamed (2010) indicated that the surface lineaments are in parallelism with the subsurface basement fault. Ölgen (2004) stated that Earh scientists have been interested in linear features on the earth's crust since the early period of each observation. Elmahdy and Mohamed (2012) highlighted that lineaments are features that represent pass ways for groundwater accumulation, groundwater discharge and seawater intrusion into coastal and inland aquifers. Burnett (2011) pointed out that lineaments and surface dips are useful for locating trapped groundwater. Contes and Carla (2011) used remote sensing in Puerto Rico, and found that geomorphic data agrees with lineaments as faulting and fracturing in addition to linear bedding control features. Moreover, Ali et al. (2012) mentioned that remote sensing has been used in geology for lithological discrimination of different rock types and delineation of geological and structural features.

Satellite images provide quick and useful baseline information on the parameters controlling the occurrence and movement of groundwater like geology, lithology/ structural, geomorphology, soils, land use, lineaments (for example, Mogaji et al., 2011). Raghou highlighted that the use of satellite remote sensing for groundwater exploration is an established procedures which has a long pedigree, and is a powerful means to target potential groundwater resources. This study was conducted in the Eastern Cape neotectonic belts using satellite imagery, and was aimed at extracting lineaments in order to find zones of high density lineaments that can be used for groundwater exploration. High density lineament zones can also be used while considering environmental issues; nuclear wastes dump sites can be placed in zones not affected by high seismicity and high density lineaments in order to avoid groundwater contamination.

METHODOLOGY

Satellite images (Landsat 4-5 TM) were chosen because they offer good scenes that can be exploited from remote sensing in order to extract linear features. They were downloaded for free from Glovis (Global Visualization View), USGS website by a selection of rows and paths. The scenes were already processed only with systematic correction (Level 1G) due to processing constraints. The systematic correction (Level 1G) provides radiometric and geometric accuracy. Different scenes covering the three neotectonic belts (south, east and north) plus the central inactive zone were used for lineament extraction in ENVI 4.8. Enhancement of images was done in different stages as follows:

1. Linear interactive stretching
2. Smoothing using Median filter 3x3
3. Edge detection using Sobel filter

Images were processed using the Sobel operator with non editable kernels. Georeferenced processed images were then exported to Arc Map for lineament editing. After editing the lineaments, rose diagrams were produced using the Rose Plot 4.0, some scenes were then chosen due to their lineament density for Normalized Difference Vegetation Index (NDVI) in regions where major focus can give directions for groundwater exploration. In order to have a clue on the water flow direction, samples of Digital Elevation Models from each neotectonic belt were selected with associated grids, the southern neotectonic belt with grid 3327, the eastern neotectonic belt with grid 3228, and the northern neotectonic belt with grid 3028. Digital Elevation Models acquired from the National Geo-Spatial Information in the Western Cape (www.cdsm.gov.za) were in excel format (X, Y and Z) with coordinates in WGS84 datum Hartebeestoek 94 reference system. Samples for each one of these grids were plotted using the Surfer 10 software in order to depict the possible predominant water flow direction and to see the elevation in meters; first the surface topography was plotted followed by the contour lines and finally the vector gradients. They were then overlayed in order to produce a graphic representation in 3D.

RESULTS

Lineament extraction procedure adopted for this study

Smoothing filter

Smoothing filters are also called low-pass filters because they let low frequency components pass and reduce the high frequency component. The impulse response of a normal low-pass filter implies that all the coefficients of the mask should be positive. One has to bear in mind that low-pass filtering blurs the image and removes speckles of high frequency noise; on the other hand, larger masks will definitely result in noise blurring effect. The parametric low-pass filter is given by a 3x3 kernel where the coefficients are determined by a factor b; when b is equal to 1 the parametric low filter is equal to a mean filter. In Figure 5 smoothing was applied on the image taken from the northern neotectonic belt near the town of Matatiele.

$$c(b) = \left(\frac{1}{a+b}\right)^2 \begin{bmatrix} 1 & b & 1 \\ b & b^2 & b \\ 1 & b & 1 \end{bmatrix}$$

Stretching

The methods of Gonzalez and Woods (1992) that were adopted for this study indicate that an image pixel distribution can be monitored, a high contrast image contains a wide distribution of pixel count covering the entire amplitude range, whilst image that have a low contrast has pixel amplitudes confining in a relatively narrow range. An example of a stretching that was applied on an image can be seen in Figure 6. The distribution of pixels on the entire amplitude signifies the

Figure 5. (Smoothed image) Map showing the neotectonic zone around Matatiele, the strike-slip fault in blue and the hot spring.

image was processed with success.

Convolution morphology (Sobel operator)

Before deciding on which filter can be used for automatic extraction of lineaments, satellite images underwent different filters for testing, such as Laplacian, Robert and Sobel. It was found that the Sobel operator was convenient; on the processed image ridges appear white, while valleys appear darkers.

Lineament densities characterization for some scenes

The lineament extraction was done in two stages, after smoothing the image was exported to ArcMap for editing the lineaments that could be easily depicted, then images was convoluted in ENVI 4.8 with the Sobel Operator using a 30% darkening with non editable kernel, and reexported to ArcMap to complete the editing of lineaments. This method was adopted for all the scenes. Scenes were selected according to their path and rows in the neotectonic belts.

The 170/081 scene is characterised by scattered lineaments, though located in the northern neotectonic zone with the presence of the Aliwal North hot spring, no neotectonic fault could be perceived. It takes deep geophysical investigation to highlight the presence of deep structures below.

The scene 169/081, which is also located in the Kokstad-Koffiefontein seismic belt is also characterized by high altitudes that can reach 1600 km at some points. On the satellite image it clearly appears that structures which are seemingly new rivers are deflected by the long fault below; the short fault has beheaded the new rivers; the hot spring might be aligned in a fault system parallel to the strike-slip fault, or it is possibly related to an E-W neotectonic fault that extends to Aliwal North. Many lineaments are found in the western part of the scene, correlating very well with the east part of the previous scene (170/081). These faults are oriented SW-NE and are good indication for groundwater exploration. The satellite image showing the strike-slip faults can be seen in the Figure 5.

The scene Path 168, Row 082 around the town of Port Saint Johns near the East coast is characterized by the presence of three major features: lineaments, faults and dolerite dykes. This area is located near the zone where the Amatole-Swaziland uplift axis took place in the Quaternary. Most lineaments and the few remarkable dolerites are quasi oriented NW-SE, except for the faults that trend in an almost NE-SW at the extreme coast (Figure 7), which is a sample of how lineaments are extracted from all the images.

The scene (Path 169, Row 082) is characterized by increase in lineaments along the east coast, this is almost similar to the area of Port Saint Johns, since this area is also located in the vicinity of the Quaternary Amatole-Swaziland axis of axis; dolerites and lineaments in this neotectonic zone indicate potential high yield wells, the increase in lineament intensity along the coast (Figure 8) might be the consequence of the reactivation along the Agulhas Falkland Fracture Zone in the Indian Ocean combined with the uplift inland. Unlike the previous scenes, the area covered in the scene Path 169, Row 083 this area around the city of East London is characterised by the presence of faults, lineaments, dolerites, which show a general trend in a NE-SW direction,

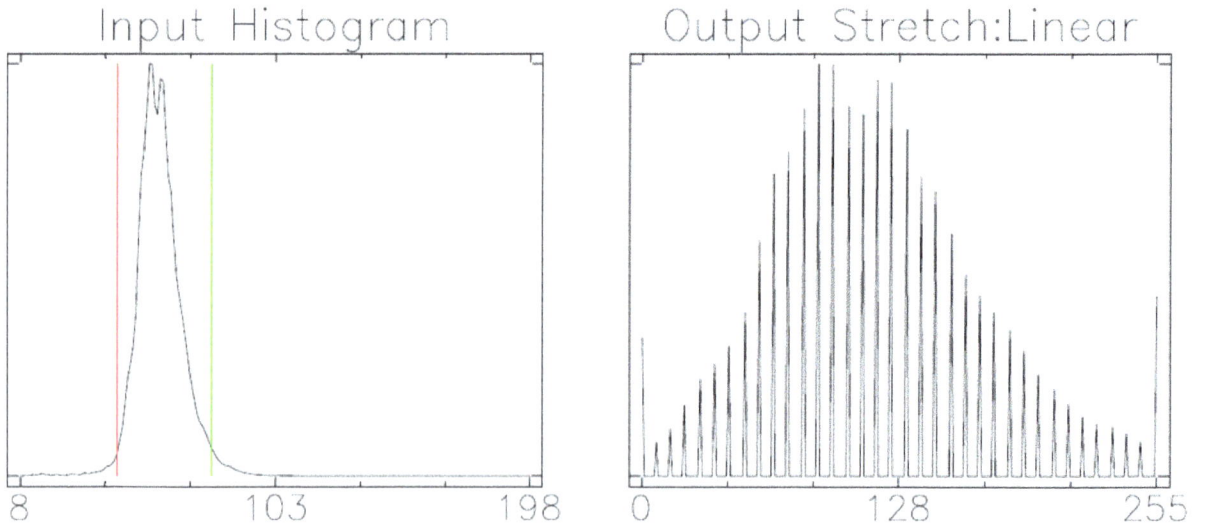

Figure 6. On the left hand side, image before stretching; on the right hand side, an output histogram showing amplitude distribution over the entire range.

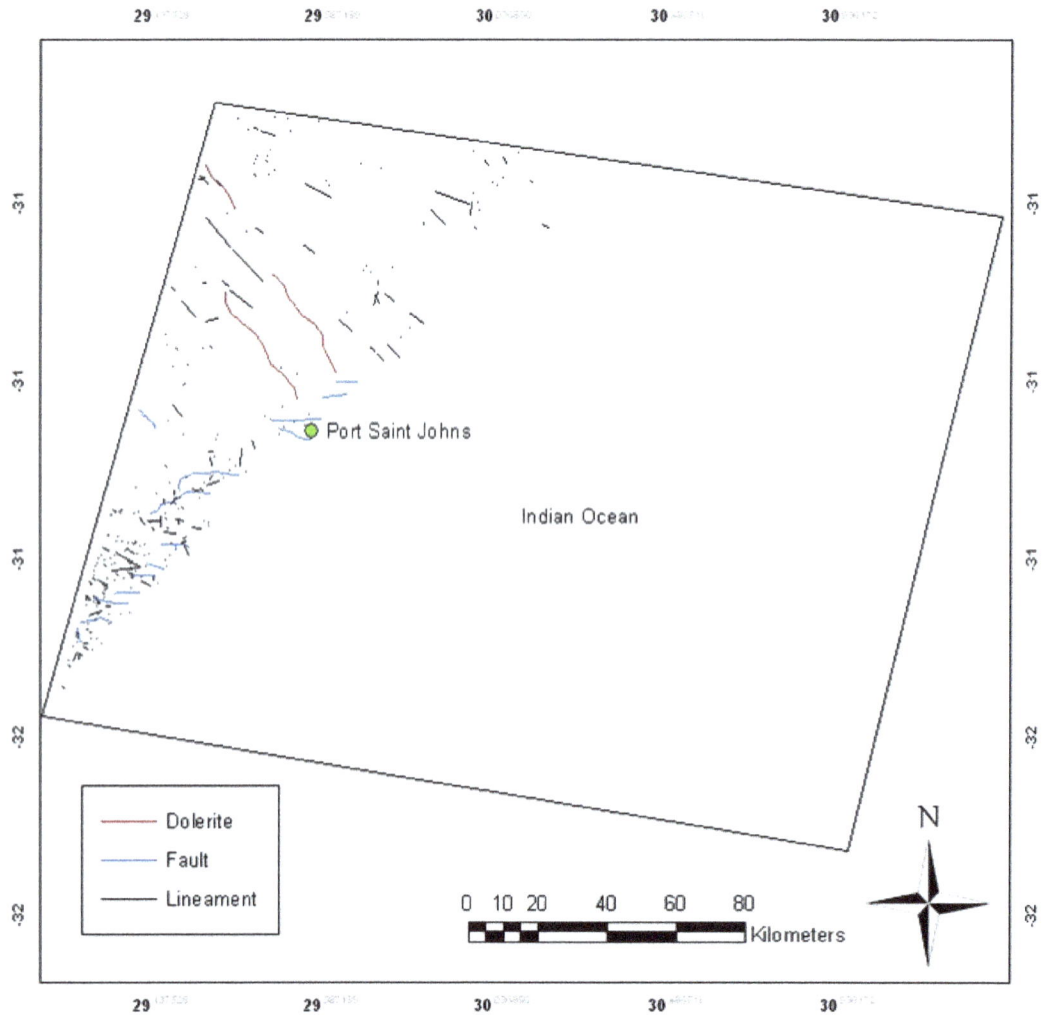

Figure 7. Faults, lineaments and dolerites from scene 168/082.

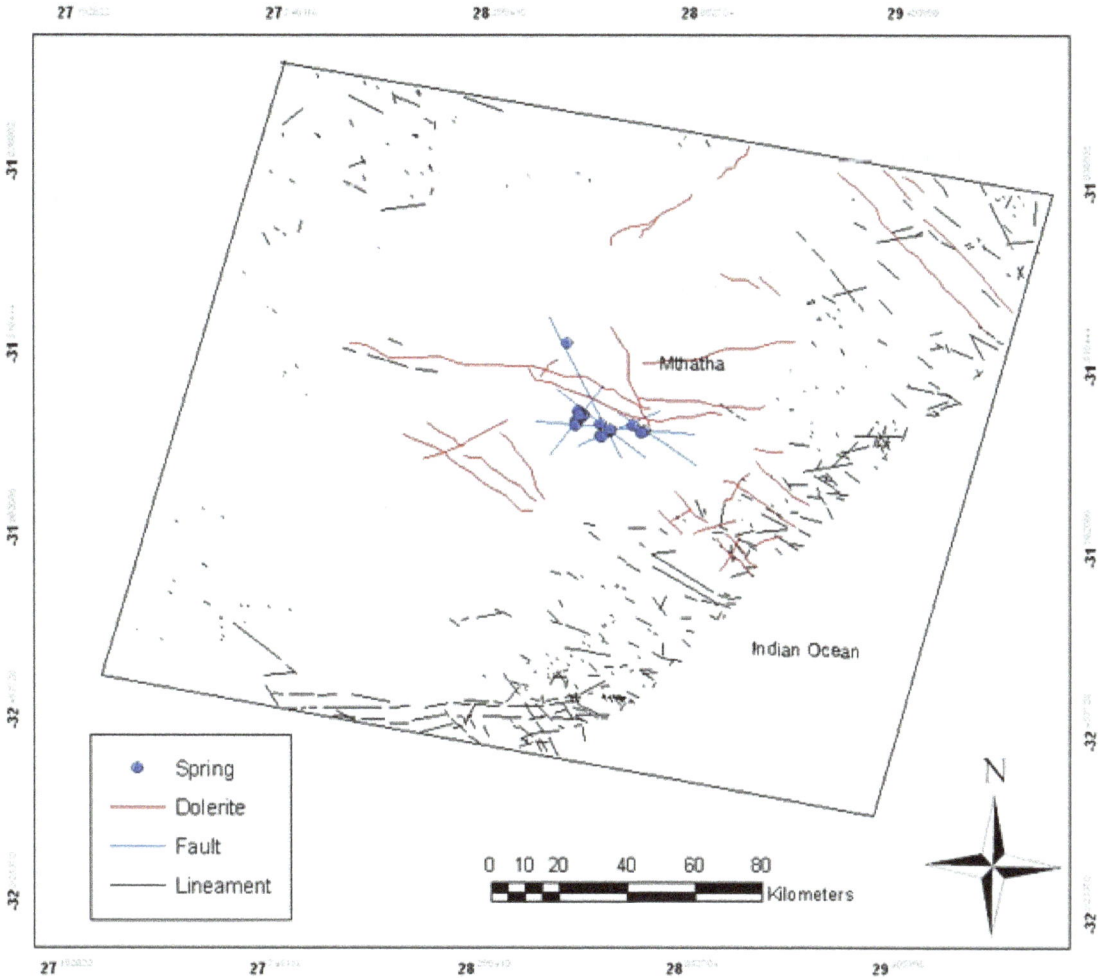

Figure 8. Lineaments extracted from satellite image 169/082.

though few lineaments are oriented SE-NW.What seems to appear to be lineaments as proper surface discontinuity at the geological term sensu stricto, are only negative weathering of dolerites, this can be seen on the satellite image (Figure 9). If there is a negative weathering of dolerite illustrated by a very long linear structure, this negative feature can only be initiated if there is a zone of weakness below the earth surface; dolerite dykes emplacement in the Karoo Basin are unequivocally consequence of the opening of the Gondwana, during this Gondwana opening, extensional fractures were generated and the doleritic magma emplaced itself following these openings to form what are known today as the Karoo dolerite dykes. This is to say that if there is negative weathering of dolerites, this negative weathering is a subsequent event, a negative weathering can only occur where there is already preexisting zone of weakness (fault or fracture below the Earth surface), where water can percolate and weather little by little the dolerite on the surface.

This scene (Path 170, Row 083) located almost in the southern neotectonic belt, is characterised by a big lineament extending from east to west. If more lineaments were observed in the two previous scenes, it clearly appears that the more the examined area is far from the coast, the less the intensity of lineaments (Figure 10). The big lineament intersects both the Fort Beaufort and the Grahamstown fractures, these two intersections are good potentials for groundwater exploration. This big lineament, which can be called the Eastern Cape Province Great Lineament (ECGL) can also be seen on a DEM that was used as an ancillary data, this great lineament has seismic epicenters (Figure 11), and is considered to be a current neotectonic domain.

This great lineament appears as conspicuous linear feature north of the town of Grahamstown; this linear feature crosses the province and stretches from east to west. Field observations indicate that this linear feature is similar to a graben, and is approximately located at the contact Cape-Karoo Supergroups. It was observed NW of the town of Grahamstown at latitude 33.17215 and longitude

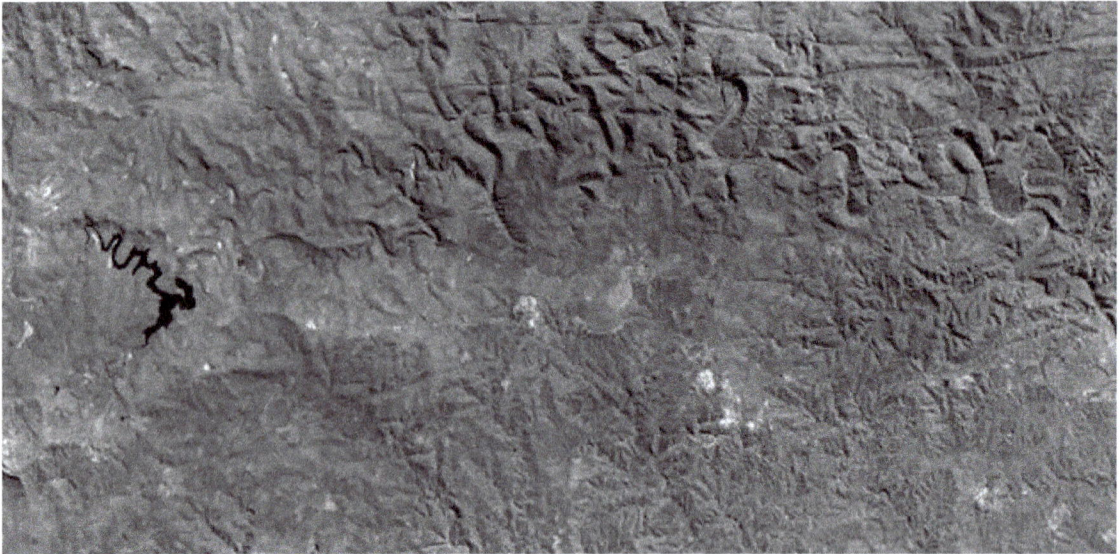

Figure 9. Negative weathering of dolerites depicted as lineaments in a E-W direction.

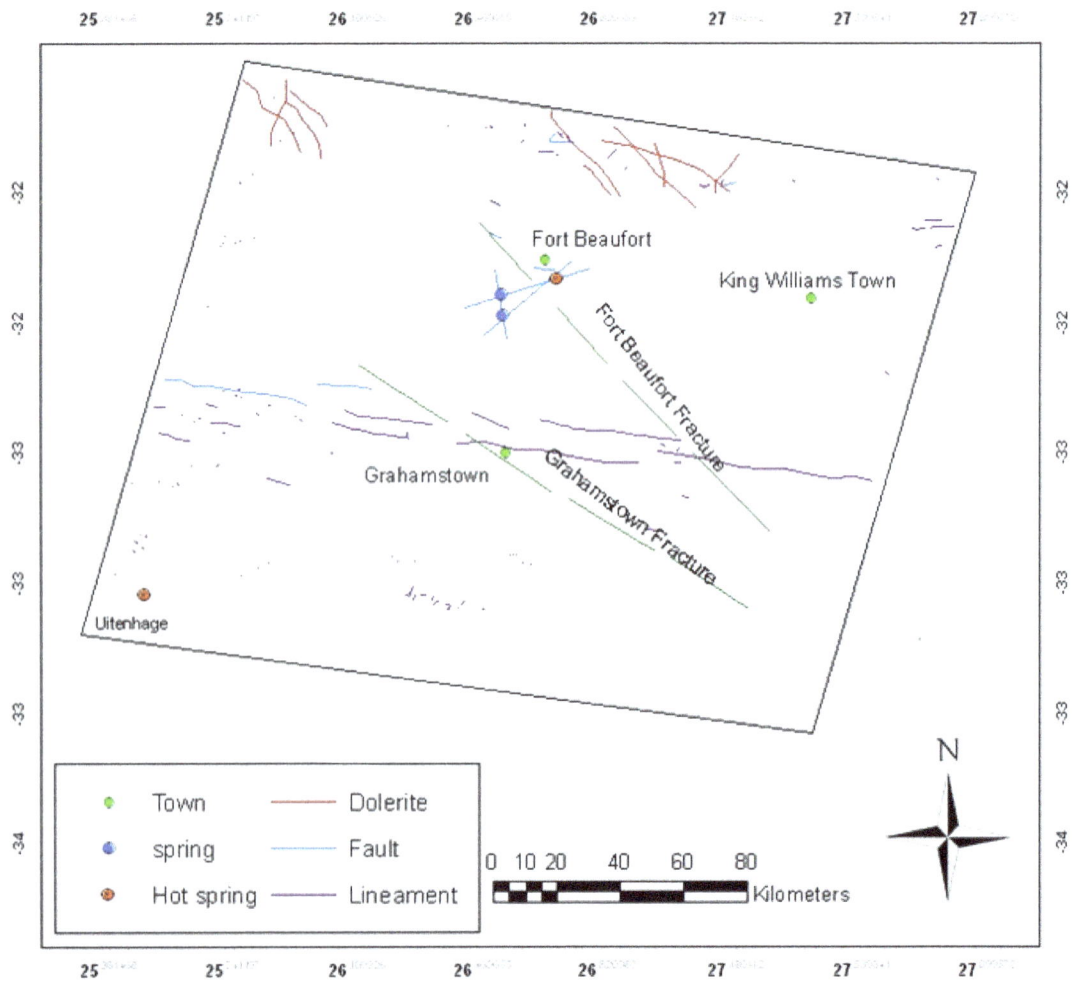

Figure 10. Structures extracted from scene 170/083/.

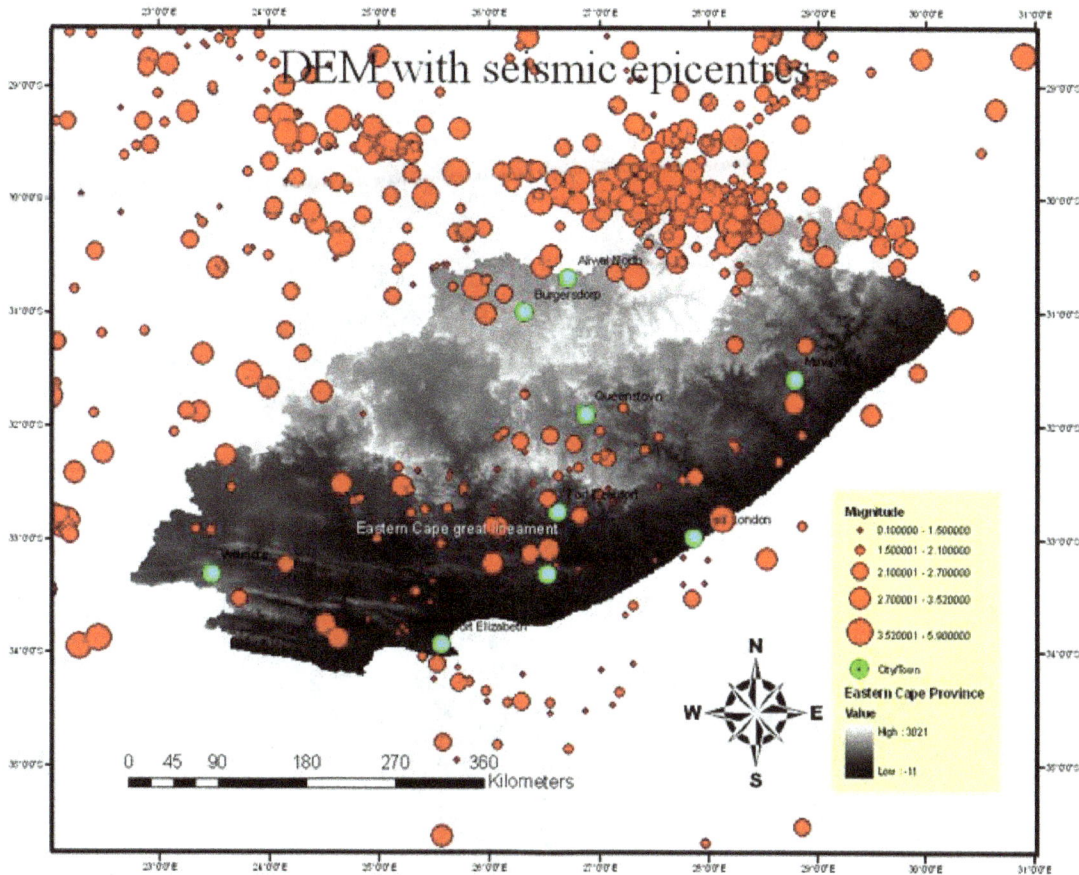

Figure 11. Digital elevation model of the Eastern Cape Province, the great lineament has seismic epicenters within it.

026.32628. Though it generally strikes in an east-west direction, at this point there is a small change, and the recorded strike was N 120°. The space between the two surfaces of the normal faults, which are part of the graben, is almost 200 meters. One surface of the fault has outcrop of quartzites of the Witteberg Group (Figure 12 left), the graben feature can also be seen in Figure 12 (right). This graben forming the major lineament of the Eastern Cape Province has some seismic epicenters; these epicenters can be followed all the way from the border of the Eastern and Western Cape Provinces, they can be seen in Figure 11; this indicate that this lineament is a neotectonic domain, its shape is undoubtedly a good catchment, and these seismic epicenters indicate possible reactivations, which is good for groundwater target. Some extensional fractures were found on a quartzite outcrop, these extensional fractures strike astonishingly the same with the graben itself (N120°), their development is possibly concomitant with extensional movement that is responsible for the graben formation, or there are related to extensional normal faulting that triggers earthquakes in and around Grahamstown, these extensional fractured can be seen

in Figure 13 (left), these fractures are in place filled with quartz veins, either in the quartzite or in some quartzo-phyllites.

Normalized difference vegetation index (NDVI)

The NDVI can be in most of the cases associated in the study and exploration of groundwater exploration, it normally indicates the degree of live green vegetation. Live green vegetation can be the surface expression of presence of considerable amount of water below the surface, which can either be related to fractures and faults, and to a certain extent to catchments in which the water accumulates. The NDVI is calculated from the individual measurements as follows:

$$\text{NDVI} = \frac{(NIR - \text{Re}\,d)}{(NIR + \text{Re}\,d)}$$

Red and NIR stand for the spectral reflectance measurements acquired in the red and near infrared

Figure 12. Left: Fault of the graben with quartzite of the Witteberg Group, Cape Supergroup; Right: The graben morphology.

Figure 13. Left: Quartzite affected with extensional fractures, quartz veins in quartzo-phyllite.

regions, respectively. The NDVI varies itself between -1.0 and +1.0. Healthy vegetation absorbs most of the visible light that hits it, and reflects a large portion of the near infrared light. Unhealthy or sparse vegetation reflects more visible light and less near infrared light. Dark areas have low chlorophyll and light areas have more.

In the above paragraph it was indicated that the eastern neotectonic belt is highlighted as the most neotectonic zone because of high lineament densities, this is only when this is referred to the uplift along the Amatole-Swaziland axis; the NDVI is applied only to this zone in the purpose of groundwater exploration. Four scenes (168/081, 168/082, 169/082, 169/083) were chosen, these scenes can be complemented with one scene in the northern neotectonic belt located in the Senqu seismo-tectonic belt. The NDVI was computed on the band 3 and Near Infrared band using ENVI 4.8

software according to the following band math expression:

$$\frac{(float(B4) - (B3))}{(float(B4) + (B3))}$$

In this equation B4 represents the Near Infrared band and B3 the visible band (Red). Images were then interpreted in terms of colour differentiation, lighter areas being representatives of more chlorophyll, and accordingly potential good catchments, and dark areas being representatives of possible poor catchments or areas not affected by fractures and faults, mainly in the neotectonic zones. Scenes chosen for NDVI are those that were depicted as being more affected by high density lineaments as typical examples, mainly Path/Row: 168/082; and 168/083. All the output maps can

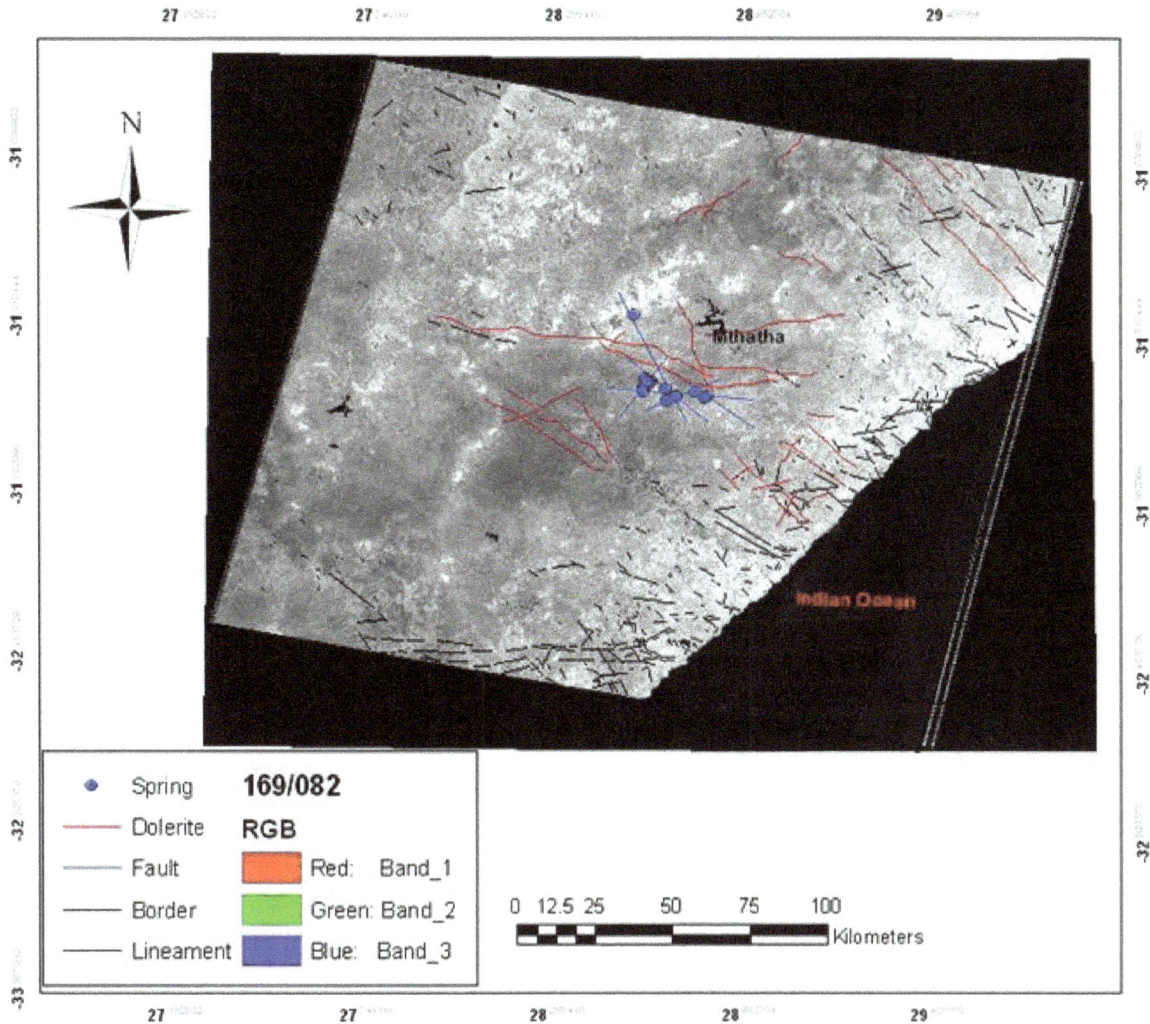

Figure 14. NDVI from the scene 169/082, very potential groundwater resources is highlighted in the east coast.

be seen in Figures 14 and 15.

General trend of lineaments

The lineament trends for all the nine scenes in the three neotectonics belts were deduced in ArcMap using the Rose Plot 4.0 extension. Different notable orientations were categorized using different colours in order to have the major predominant trend that can be considered for future possible groundwater exploration. All rose diagrams for all the three neotectonic belts can be seen in Figure 16.

From the rose plots, it was found that the scenes (169/081, 170/081, 169/083, 170/083, 172/083, 169/082) are characterised by a unifrequency of angle interval between 111 and 140° in the ten examined scenes. Six of the ten scenes show this major trend, which is to be

taken into account during groundwater exploration. Histograms in Figure 17 indicate that five scenes, three in the northern neotectonic belt (170/081, 169/081 and 168/081) and two in the eastern neotectonic zone (169/082 and 169/083) have high lineament density. As mentioned above the east coast would be highly considered morphotectonic induced neotectonic zone due to the positive topography related to the uplift that occurred in the Quaternary than the northern neotectonic belt characterised by the Kokstad-Koffienfontein belt and the presence of several hot springs, it can be said that a morphotectonic induced neotectonics would generate more lineaments than a seismically active zone. Kumanan (2001) has found in India that lineament density maxima zones and lineament intersections were buffered out as possible neotectonic zones. In the same order of idea, no considerable high density of lineament was depicted from the Queenstown area (scene 170/082).

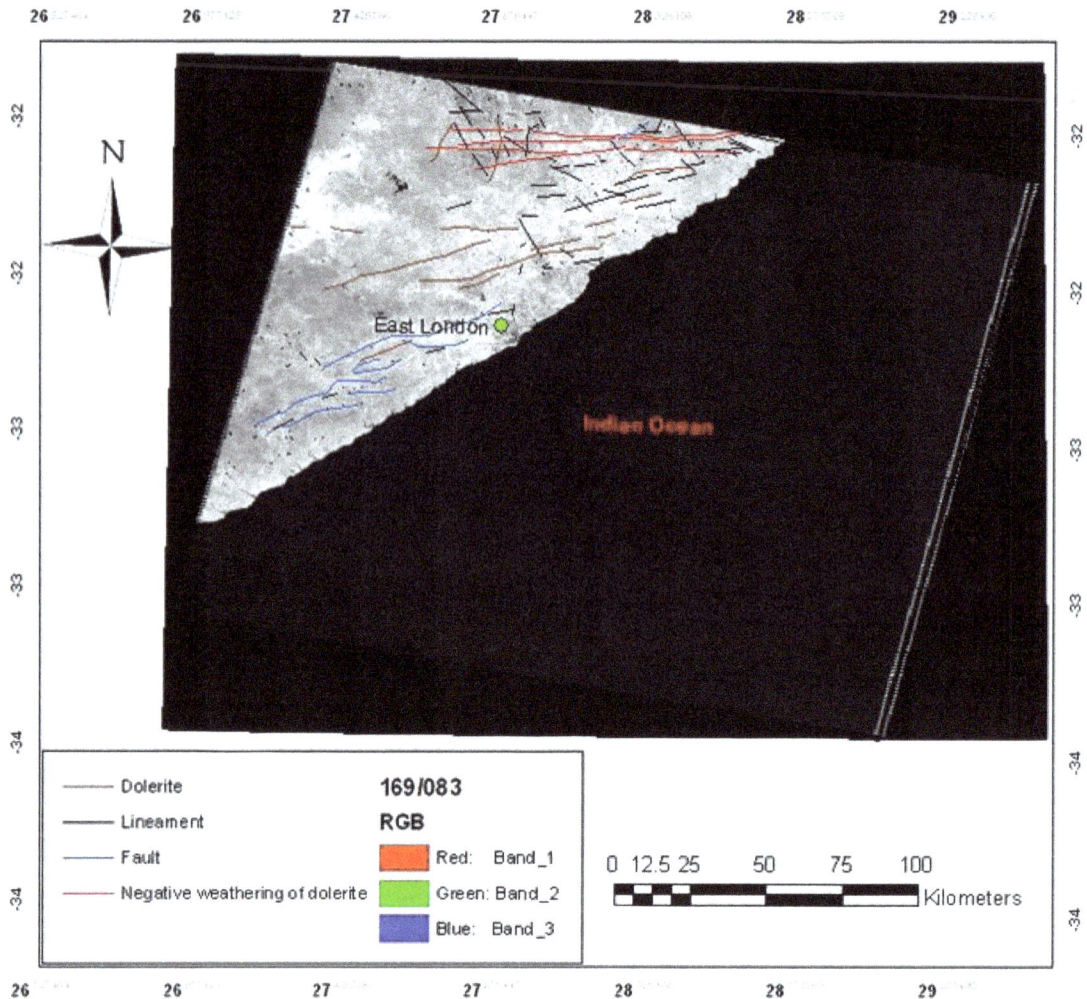

Figure 15. NDVI from the scene 169/083, good potential for groundwater is highlighted.

Geormphologic landform from digital elevation models

Digital Elevation Models can be used as support for hydrological application. With data acquired from the South African National Geo-Spatial Information in relation with the three neotectonic belt were processed using Surfer 10, a small comparative study between these three neotectonic has been done only on the basis of surface topography and vector gradients, the vectors gradients give and indication on the surface water flow. In each neotectonic zone, a grid was picked up randomly, each grid being subdivided in four parts (A, B, C, and D).

In the southern neotectonic belt, many faults such as the Kango Bavianskoof were reactivated in the Quaternary; from DEMs as was mentioned above there is another remarkable lineament, this linear feature is located in the area between south of East London and north of Grahamstown, it is also clearly appearing on the

DEMs in the grid 3327d (Figure 18); this remarkable and striking feature is now known as the Eastern Cape Great Lineament (ECGL). It appears from the image in the grid 3327d that altitude does not exceed 125 m, on the hydrological point of view, if the gradient vectors are considered, the flow direction from the surface topography is mostly oriented east-west (Arrows, Figure 18).

In the eastern neotectonic zone, four images of DEMs from the grid 3228 were also considered in order to see what the surface topography and what should the drainage pattern look like. Only one sample was chosen among the four for illustration (Figure 19). It appears from all these figures that the elevation increases reaching 1020 m above sea level with varied geomorphic surface, this might be a consequence of the uplift that the took place in Cenozoic, or this can be a result of dolerite dykes that have resisted the erosional processes to form these outstanding and conspicuous morphologies at such

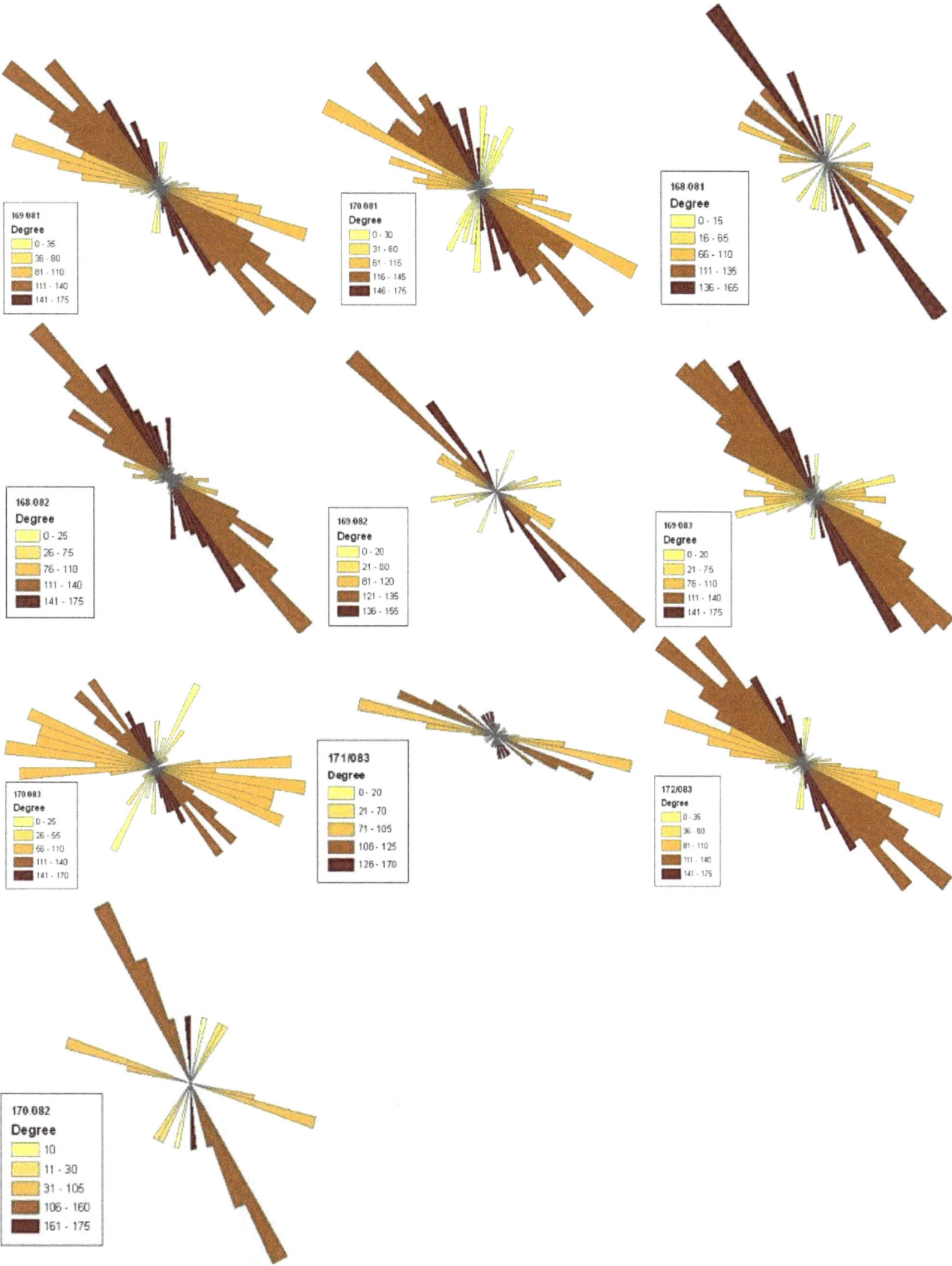

Figure 16. Rose plots derived from the lineament maps in the neotectonic belts.

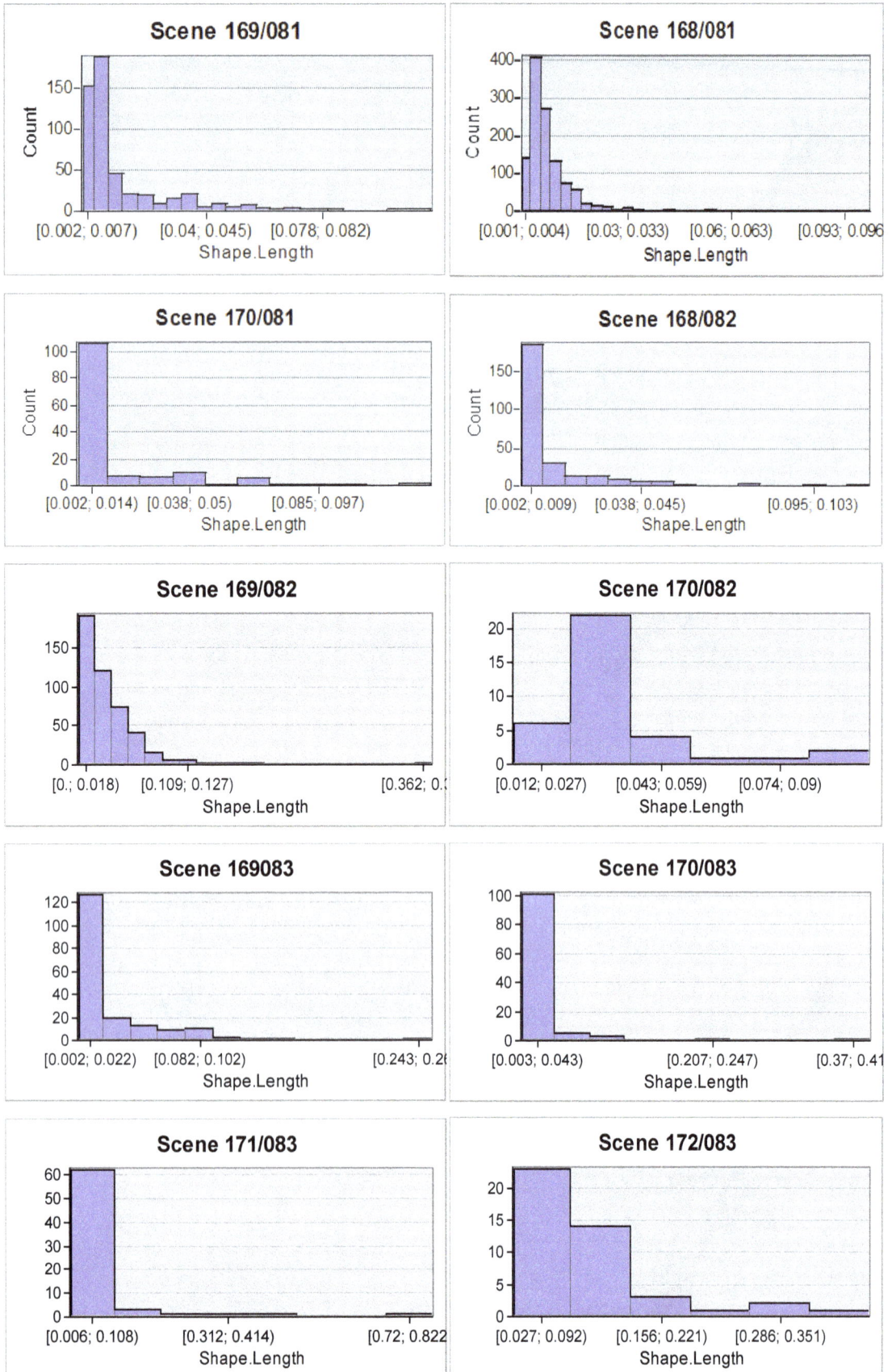

Figure 17. Histograms deduced from lineament maps in the Eastern Cape neotectonic belts.

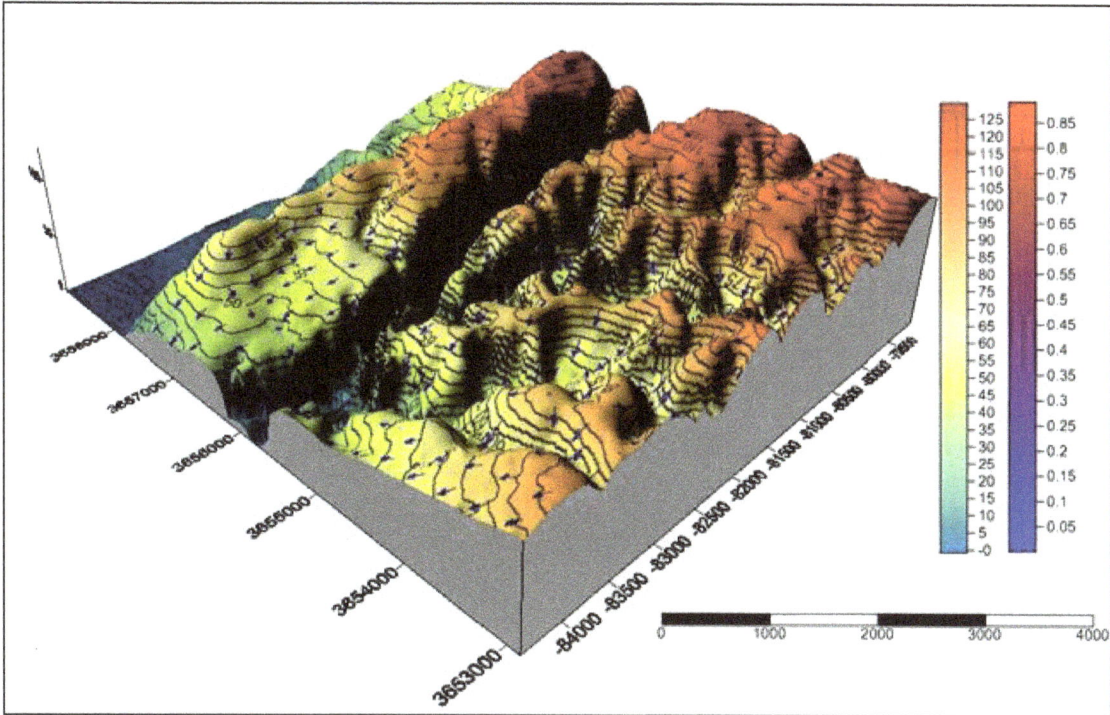

Figure 18. DEM from grid 3327d, note the E-W structure (ECGL).

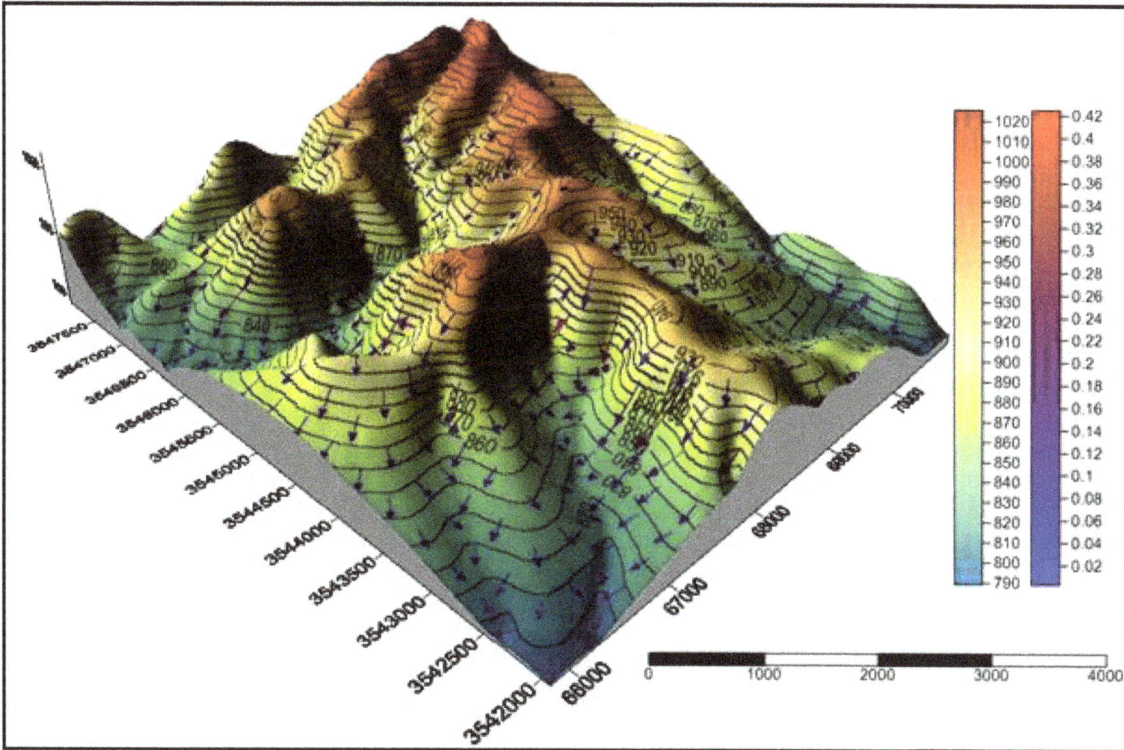

Figure 19. DEM from grid 3228.

an altidue, but the hypothesis of uplift is more and more favoured. It can be noted that in the eastern neotectonic belt, altitudes are not uniforms changing from 190 m, to 500 m, 700 m and 1020 m. The vector gradients indicate that the surface water would flow from west to east in general.

Samples of Digital Elevation Model in the northern neotectonic belt near the country of Lesotho were taken from the grid 3028, the grid 3028 is subdivided in four areas (ABCD), the vector gradients have an orientation mostly in the east-west and north-south direction (Figure 20), this orientation of vector gradients in a quasi east-west trend would be an important factor to consider for future hydrogeological works because it indicates the possible water flow direction. The Digital Elevation Model clearly shows that high altitude that can reach 2000 m characterize the northern neotectonic belt. The legend scales indicate the contour lines and magnitudes of different vector gradients.

DISCUSSION

In the exploration and characterization of potentially high yield aquifers, remote sensing has always played an important part; first because it can be used to find the maximum lineament density, this can be in turn correlated with the strike; which is useful in the exploration of groundwater. Most of the lineaments that were depicted using remote sensing have a predominant NW-SE orientation. The greatest lineament is an important feature to be reckoned with in the Eastern Cape in as far as works related to groundwater is concerned, mostly because of alignement of seismic epicenters related to earthquakes, which can reactivate the structures to allow movement of groundwater. The eastern neotectonic belt has high density of lineament, which makes it another important neotectonic domain.

In continental intraplate configuration like the case of South Africa, seismic epicenters are located on lineaments; these lineaments can be targeted for groundwater exploration. Lineaments can be in form of neotectonic faults, the works of Holford et al. (2011) showed that the southern Australian continental margin has been undergoing mild levels of deformation over the past ~10 Myr, manifested today by high levels of seismicity for a stable intraplate region. However, this deformation is partitioned, with zones of abundant neotectonic faults with evidence for Pliocene–Quaternary displacement.

On the geomorphologic point of view, the modeling of surface topography has indicated that elevations at the western side of the province (for example, grid 3327) do not exceed 450 m. To the contrary the grid 3228 which is more in the east has elevations that can reach 1020 m; though the elevations are not uniform in this grid, the higher surface topography of 1020 m can be related to Cenozoic uplift (Esthuizen, 2008). In the northern neotectonics belt near the country of Lesotho the elevation can reach 2000 m; this higher topography is the source of increasing stress in the lithosphere triggering

earthquakes (Steinberger et al., 2001). The east-west big seismic belt that stretches from the east coast to Koffiefontein is located in this area, and belongs to the Senqu seismotectonic belt, better known as the Kokstad-Koffiefontein seismic belt.

The Sobel operator has been proved to be efficient technique in remote sensing to extract lineaments. From lineament extraction the eastern neotectonics belt can be considered as the second neotectonics domain of the province because of the intensity of lineaments. Kumanan (2001) has found that lineament density maxima zones and lineament intersections were buffered out as possible neotectonics zones. This can be a good highlight from the present project that can help to target productive aquifers. On the other hand what has been called the Eastern Cape great lineament north of the town of Grahamstown, has a geomorphology that is highly favourable to be considered as a major catchment the seismic epicenters present within it has possibly reactivated faults and fractures, and thus contribute to the circulation of groundwater. The Grahamstown fracture is considered as a splay of the structure, which is qualified as a graben.

The neotectonics belt near the country of Lesotho is also known for the occurrence of many hot springs, which are indicative of circulation of groundwater at great depths, the occurrence of these springs indicate that they are connected by a major neotectonics fault that stretches from west (Aliwal North) to east (Polile Tshisa). The flooding that occurred in an abnormally well developed fissure zone, through which lot of water gushed out during the construction of the Orange-Fish tunnel (Olivier, 1975) is a proof that neotectonics is at work in the northern neotectonics belt. (Whittingan pers. com.) noted that the flow of the thermal spring at Badsfontein increased markedly for a period of at least three months after a local earthquake.

Conclusion

Neotectonics is active in the Eastern Cape Province as can be seen from the recent map of seismic epicenters. It is an important tool in targeting potentially high yield aquifers for two reasons: 1) it can reactivate old fractures or faults; 2) it can create new fractures or faults. All these two reasons imply more opportunities for the circulation of groundwater.

Surface topography modeling in this study has highlighted the flow direction of water, it was found that most vectors are oriented in the E-W direction; this can be taken into consideration during groundwater target and exploration. The eastern neotectonics belt is a good target for groundwater, it is characterized by a high density of lineaments, most of them are oriented NW-SE on the other hand the normalized difference vegetation index (NDVI) performed on some satellite images clearly

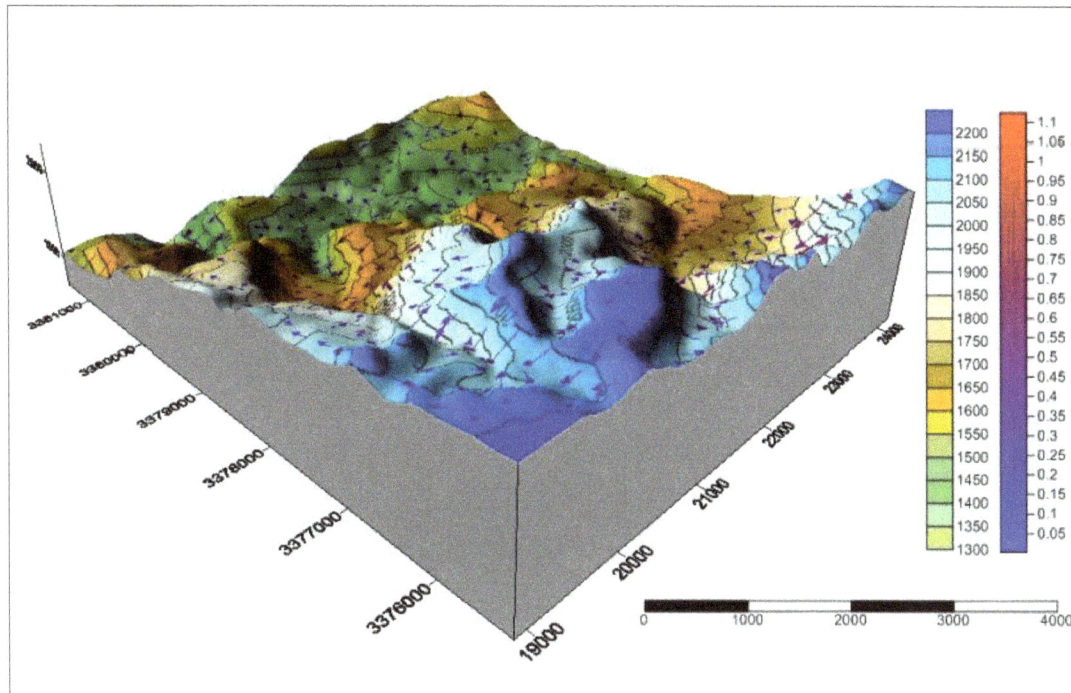

Figure 20. Digital elevation model sample and vector gradients of Grid 3028DB.

indicate that the quality of chlorophyll is higher, this can only be related to a healthy vegetation, which is in turn connected to more groundwater percolation favoured by the occurrence of lineaments.

The hot springs in the Eastern Cape northern neotectonic belt are connected or occur along an E-W neotectonic fault, a regional structure.

This project has given fundemental highlights to characterize the major and potential good yield aquifers by examining the neotectonics, the seismicity, and the structures. This project is important since it has also brought in the environmental impact through examination of seismic risk assessment, which can help in targetting place for dumping nuclear wastes so that the groundwater may not be contaminated.

ACKNOWLEDGEMENTS

This work was made possible by the sponsorship from the National Research Foundation (NRF) and the Govan Mbeki Research and Development Centre (GMRDC). Dr M. A. G. Andreoli is also remembered for his input for the understanding of neotectonics in the Eastern Cape and Southern Africa in general.

REFERENCES

Ali EA, El Khidir SO, Babikir IAA, Abdelrahman EM (2012). Landsat ETM+7 Digital Image Processing Techniques for Lithological and Structural Lineament Enhancement: Case Study Around Abidiya Area, Sudan. Open Rem. Sens. J. 5:83-89.

Andreoli MAG, Doucouré M, Van Bever Donker J, Brandt D, Andersen JB (1996). Neotectonics of Southern Africa. Afr. Geosci. Rev. 3(1):1-16.

Artyushkov EV, Hoffman AW (1986). Neotectonic crustal uplift on the continents and its possible mechanisms, the case of Southern Africa. Surv. Geophys. 19:369-415.

Bejaichund M, Kijko A, Durrheim R (2009). Seismotectonic Models for South Africa: Synthesis of Geoscientific Information, Problems, and the Way Forward. Seismol. Res. Lett. 80:65-33.

Burnett DO (2011). Use of Remote Sensing for Groundwater Mapping in Haiti. Earthzine.

Contes A, Carla A (2011). Lineament mapping for groundwater exploration using remotely sensed imagery in a karst terrain: Rio Tanama and Rio de Arecibo basins in the northern karst of Puerto Rico. Michigan Technol. Univ. 77:1505814.

Dobson KJ, McDonald R, Brown RW, Gallagher KS, Finlay M (2010). Dating the emergence of the Africa Superswell: A window into mantle processes using combined (U-Th)/He and AFT thermochronology. EGU General Assembly, Vienna, Austria, pp. 5167.

Esterhuizen A (2008). Mineral sands deposits of Africa. In: Africa Uncovered: Mineral Resources for the Future. SEG-GSSA 2008 Conference, Johannesburg, South Africa

Elmahdy SI, Mohamed MM (2012). Geological Lineament Detection, Characterization and Association with Groundwater Contamination in Carbonate Rocks of Musandam Peninsular Using Digital Elevation Model (DEM). Open Hydrol. J. 6:45-51.

Gonzalez R, Woods R (1992). Digital Image Processing, Addison-Wesley Publishing Company, New York.

Holford SP, Richard R, Hillis RR, Hand M, Sandiford M (2011). Thermal weakening localizes intraplate deformation along the southern Australian continental margin. Earth Planet. Sci. Lett. p. 8.

Kumanan CJ (2001). Remote sensing revealed morphotectonic anomalies as a tool to neotectonic mapping–experience from South India. In Proceedings of the 22nd Asian Conf. Remote Sensing.

Maouche S, Abtout A, Merabet NE, Aïfa T, Lamali A, Bouyahiaoui B, Bougchiche S, Ayache M (2013). Tectonic and Hydrothermal

Activities in Debagh, Guelma Basin (Algeria). J. Geol. Res. Article ID 409475, 13 p.

Mogaji KA, Aboyeji OS, Omosuyi GO (2011). Mapping of lineaments for groundwater targeting in basement complex area of Ondo State using remote sensed data. Int. J. Water Resour. Environ. Eng. 3(7):150-160.

Mohamed A (2010). Significance of surface lineaments for gas and oil exploration in part of Sabatayn Basin-Yemen, J. Geogr. Geol. 2(1).

Ölgen MK (2004). Determining lineaments and geomorphic features using landsat 5-TM data on the Lower Bakircay Plain, Western Turkey. Aegean Geogr. J. 13:47-57.

Olivier HJ (1975). Geohydrological investigation of the flooding at shaft 2, Orange-Fish tunnel, North-Eastern Cape Province. Trans. Geol. Soc. S. Africa 75:197-224.

Partridge TC, Maud RR (2000). Macro-scale geomorphic evolution of southern Africa. In: T.C.Partridge and R.R. Maud (Editors), The Cenozoic of southern Africa. Oxford University Press, Oxford, pp. 3-18.

Steinberger BM, Schmeling BMH, Marquart G (2001). Large-scale lithospheric stress field induced by global mantle circulation. Earth Planet. Sci. Lett. 186:75-91.

11

Groundwater management using a new coupled model of flow analytical solution and particle swarm optimization

Hamdy A. El-Ghandour[1]* and Ahmed Elsaid[2]

[1]Department of Irrigation and Hydraulics, Faculty of Engineering, Mansoura University, Egypt.
[2]Department of Mathematics and Engineering Physics, Faculty of Engineering, Mansoura University, Egypt.

In this research, a steady state analytical solution is suggested and derived for the groundwater flow equation in a homogeneous unconfined aquifer. The solution can be used for calculating the hydraulic heads in a complex flow field caused by multi injection-pumping wells having different rates in a domain subjected to a uniform recharge. The analytical method, used for flow simulation, is coupled with the Particle Swarm Optimization (PSO) method, used for optimization, to solve the groundwater management problem. A new model called Analytic-PSO, which consists of the two methods, is originally developed. The performance of the proposed model was tested on a popular hypothetical example to maximize the total pumping rate from located well system at steady state condition. The results show the superiority of the proposed model to obtain the maximum pumping rate compared with other methods of previous work. The application was extended to use the Analytic-PSO model for determination of both the optimal location and maximum pumping rate for each well for the pre-specified number of wells in the same hypothetical example. Obtained results of the hypothetical example illustrate the ability of the Analytic-PSO model to solve efficiently the groundwater management problem in the real field aquifers.

Key words: Groundwater management, analytical solution, particle swarm optimization, optimization methods, Fourier series.

INTRODUCTION

Groundwater is considered an important source of freshwater especially in arid semi-arid zones which is used for several life purposes such as drinking, domestic, industrial, and irrigation uses. Indiscriminate exploitation of this source causes environmental hazards including decline of groundwater level and well interference. Consequently, sustainable management strategies have to be developed by decision makers to optimally utilize the groundwater resources. Groundwater management problems are typically solved by researchers using the simulation-optimization approach. In the simulation-optimization approach, a coupled optimization and groundwater flow model is used to obtain the optimum strategy.

During the past two decades, several computer codes have been established to deal with groundwater management problems by linking groundwater flow and optimization models (McKinney and Lin, 1994; Wang and Zheng, 1998; Wu et al., 1999; Wu and Zhu, 2006; Zhu et al., 2006; Ayvas, 2009; and Gaur et al., 2011a). These codes differ in the used numerical model to simulate the groundwater flow system, the type of groundwater management problems and the approaches used to solve these management problems (Gaur et al., 2011b). In most of the previous groundwater management studies, the flow models were based on the finite difference method (FDM) or finite element method (FEM). These two methods, which are used to predict the hydraulic heads for the whole flow domain, have several limitations such as domain discretization error, selection of appropriate boundary conditions, numerical stability, and

*Corresponding author. E-mail: Eng_hamd@yahoo.com.

approximate location of well over the cell. Analytic element method (AEM) is considered one of the analytical methods used to simulate the groundwater flow. AEM can give an exact analytic solution for groundwater flow problems and is capable of simulating streams, lakes, and complex boundary conditions (Strack, 1989).

Optimization techniques are categorized into two types. The first one is deterministic optimization technique including Linear Programming (LP), Non-Linear Programming (NLP), and Dynamic Programming (DP). The second type is the stochastic optimization including Genetic Algorithm (GA), Particle Swarm Optimization (PSO), Shuffled Complex Evolution developed at University of Arizona (SCE-UA), Simulating Annealing (SA),...etc. Groundwater management problems are usually nonlinear and non-convex mathematical programming problems (McKinney and Lin, 1994). For such problems, using deterministic optimization techniques may result in some unexpected situations. These techniques usually require good initial solutions to produce an optimal solution. Also, they rely on local gradients of the objective function to determine the search direction, and thus, may converge to local optimal solutions (Ayvas, 2009). Therefore, use of the stochastic optimization techniques is usually preferred due to their ability of finding solutions without requiring gradients and initial solutions.

There are several studies dealing with the solution of groundwater management problems using stochastic optimization methods. One of the first applications was performed by McKinney and Lin (1994). In that study, the GA-based groundwater simulation optimization models were developed to solve three groundwater management problems. They found that genetic algorithms could effectively and efficiently be used to obtain globally (or, at least near globally) optimal solutions to these groundwater management problems. Wang and Zheng (1998) compared the performance of GA and SA for maximization of pumping and minimization of the cost.

Their results showed that both methods yield nearly identical and better solutions than various other programming methods. Wu et al. (1999) developed a GA based SA penalty function approach (GASAPF) to solve a groundwater management model. Their results showed that GASAPF model can effectively solve the groundwater management model. Wu and Zhu (2006) applied (SCE-UA) to solve groundwater management models. Using the developed solution algorithm, two management models were developed for an unconfined aquifer: linear model of maximum pumping and nonlinear model of minimum pumping cost. In a later study, Zhu et al. (2006) compared the performances of SCE-UA and GA methods in the solution of a management model for deep groundwater resources of the Yangtze Delta, which is a multi-aquifer system with large area and complicated geology conditions. Their results showed that the SCE-UA is more effective than GA in the solution of the management model. Ayvas (2009) tested Harmony

Search (HS) algorithm on three separate groundwater management problems. Their results showed that the HS yields nearly same or better solutions than the previous solution methods and may be used to solve management problems in groundwater modeling. Gaur et al. (2011) developed two models for the solution of groundwater management problem. The first one consists of a linkage between analytical element and particle swarm optimization methods (AEM-PSO) whereas the second model consists of finite difference and particle swarm optimization methods (FDM-PSO). The comparative analysis was performed between AEM and FDM, and the abilities of the AEM method to solve groundwater management problems were investigated. Also Gaur et al. (2011) applied (AEM-PSO) model to the Dore river basin, France, to solve two groundwater hydraulic management problems. They examined the effect of piping length in the total developed cost for new wells. They also used the (AEM-PSO) model to determine optimal locations, discharges and optimum number of wells.

There are two main objectives of this study; the first one is to derive a new analytical solution for the groundwater flow equation to predict the hydraulic heads in the unconfined aquifer, and the other objective is to develop a groundwater resources management model that combines a new analytical and particle swarm optimization methods (Analytic-PSO). The proposed management model is tested on the most popular hypothetical example to obtain the maximum pumping rate and a comparison is carried out with the corresponding ones given by other previous studies. Also (Analytic-PSO) model is used to determine both the optimum locations and discharges of wells for the pre-specified number of wells.

PROBLEM FORMULATION

In the design of ground water management systems there are usually two sets of variables: decision variables and state variables. In the considered model, the decision variables include the well locations and pumping rates. These are the variables that can be specified, managed, or controlled by the designer. The purpose of the design process is to identify the best combination of these decision variables. On the other hand, the state variable is the hydraulic head, which is the dependent variable in the groundwater flow equation (Wang and Zheng 1998). The governing equation describing the three dimensional movement of ground water is as follows (Bear, 1979):

$$\frac{\partial}{\partial x}\left(K_{xx}\frac{\partial h}{\partial x}\right)+\frac{\partial}{\partial y}\left(K_{yy}\frac{\partial h}{\partial y}\right)+\frac{\partial}{\partial z}\left(K_{zz}\frac{\partial h}{\partial z}\right)-f=S_{s}\frac{\partial h}{\partial t} \quad (1)$$

in which, K_{xx}, K_{yy}, K_{zz}: the principal components of hydraulic conductivity aligned along the x, y, and z

coordinate axes, respectively; h: the hydraulic head; f: a flux term that incorporates pumping, recharge, or other sources or sinks; S_s: the specific storage; and t: the time.

In this study, two separate groundwater management problems are solved. The objective function in the two problem is to maximize the total pumping rate from an aquifer. For the first management problem, it is assumed that the numbers and the locations of the wells are known whereas in the second management problem the number of wells is only known. Consequently, decision variables are only the pumping rates in the first management problem and both the pumping rates and locations of the wells in the second management problem. The two management problems are subjected to constraints on lower and upper bounds of pumping rates. Also, there is an additional constraint on hydraulic heads at well locations such that they must be greater than a specified lower bound. The management model can mathematically be stated as follows (Ayvas, 2009):

$$Obj. = \max \left[\sum_{i=1}^{N_W} Q_i - P(h) \right] \tag{2}$$

Subjected to:

$$h_i \geq h_{i,\,min}, \ i = 1, 2, 3, ..., N_W \tag{3}$$

$$Q_{i\,min} \leq Q_i \leq Q_{i\,max}, \ i = 1, 2, 3, ..., N_W \tag{4}$$

$$P(h) = \begin{cases} h_{i\,min} - h_i & if \quad h_i < h_{i\,min} \\ 0 & if \quad h_i \geq h_{i\,min} \end{cases} \quad i = 1, 2, 3, ..., N_W \tag{5}$$

in which, Q_i: the pumping rate of well i; N_W: the number of wells; $P(h)$: penalty term; $h_{i,\,min}$: the minimum hydraulic head value at well i; and $Q_{i\,min}$ and $Q_{i\,max}$: the minimum and maximum bounds of the pumping rates at well i, respectively.

ANALYTICAL SOLUTION

There are several previous studies that suggested different analytical solutions for the groundwater flow equation in confined and unconfined aquifers (Grubb, 1993; Shan, 1999; Kim and Ann, 2001; Yeo and Lee, 2003; Ma et al., 2009). Lee and Yeo (2003) suggested an analytical steady state solution using double Fourier transformation to deal with arbitrarily located multi-injection/pumping wells in anisotropic homogeneous confined aquifer. This methodology is adopted in the present study with the following modifications:

1. The analytical solution is carried out for the groundwater flow equation that describes unconfined aquifer.

2. The effect of uniform recharge on the studied domain is taken into consideration.
3. Different types of boundary conditions (BCs) are considered in the studied domain (two Dirichlet BC and two Neumann BCs).
4. The suitable number of Fourier coefficients is determined.
5. Application of the sigma-approximation technique, given by Jerri (1998), to reduce the effect of Gibbs phenomenon.

For the unconfined aquifer shown in Figure 1, the following assumptions are taken into consideration for the derivation of analytical solution: (1) the Dupuit's hydraulic assumption is employed to vertically integrate the flow equation, reducing it from three dimensional geometry to two dimensional, (2) the aquifer specific storage is ignored such that the governing equation becomes time independent, (3) the wells fully penetrate the aquifer thickness, (4) the impervious bed of the aquifer is considered horizontal, (5) hydraulic conductivity is assumed constant throughout the studied domain, and (6) two Dirichlet boundary conditions are assumed having the same value.

According to the previous assumptions, Equation 1 can be written as follows:

$$\frac{\partial}{\partial x}\left(Kh\frac{\partial h}{\partial x}\right) + \frac{\partial}{\partial y}\left(Kh\frac{\partial h}{\partial y}\right) = f(x, y) \tag{6}$$

on the rectangular domain $(x, y) \in \Omega = [0, L] \times [0, H]$ subject to the boundary conditions:

$$h(x, 0) = h(x, H) = h_o \tag{7}$$

$$\frac{\partial h}{\partial x}(0, y) = \frac{\partial h}{\partial x}(L, y) = 0 \tag{8}$$

in which, L: length of the studied domain in the direction of x-axis, H: length of the studied domain in the direction of y-axis, and h_o: boundary condition constant head.

Equation 6 can be linearized by the substitution $\phi = h^2/2$ and the equation then takes the form:

$$K\frac{\partial^2 \phi}{\partial x^2} + K\frac{\partial^2 \phi}{\partial y^2} = f(x, y) \tag{9}$$

With the boundary conditions:

$$\phi(x, 0) = \phi(x, H) = \frac{h_o^2}{2} \quad \text{(Dirichlet BCs)} \tag{10}$$

$$\frac{\partial \phi}{\partial x}(0, y) = \frac{\partial \phi}{\partial x}(L, y) = 0 \quad \text{(Neumann BCs)} \tag{11}$$

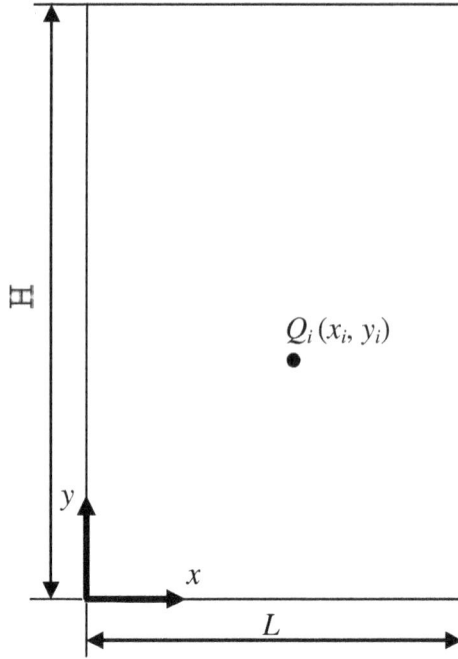

Figure 1. Definition sketch of the studied domain.

The analytical solution of Equation 9 can be written as a double Fourier series of the form:

$$\phi(x, y) = c + \left[\frac{\tilde{a}_0}{2} + \sum_{n=1}^{\infty} \tilde{a}_n \cos\frac{n\pi x}{L} + \tilde{b}_n \sin\frac{n\pi x}{L} \right]$$
$$\left[\frac{\tilde{c}_0}{2} + \sum_{m=1}^{\infty} \tilde{c}_m \cos\frac{m\pi y}{H} + \tilde{d}_m \sin\frac{m\pi y}{H} \right] \quad (12)$$

where the constant c is obtained from the boundary conditions to be:

$$c = \frac{h_o^2}{2} \quad (13)$$

From condition given in Equation 11.:

$$\left[\sum_{n=1}^{\infty} \frac{n\pi}{L} \tilde{b}_n \right]\left[\frac{\tilde{c}_0}{2} + \sum_{m=1}^{\infty} \tilde{c}_m \cos\frac{m\pi y}{H} + \tilde{d}_m \sin\frac{m\pi y}{H} \right] = 0 \quad (14)$$

and

$$\left[\sum_{n=1}^{\infty} \frac{n\pi}{L}(-1)^n \tilde{b}_n \right]\left[\frac{\tilde{c}_0}{2} + \sum_{m=1}^{\infty} \tilde{c}_m \cos\frac{m\pi y}{H} + \tilde{d}_m \sin\frac{m\pi y}{H} \right] = 0 \quad (15)$$

which implies that $\tilde{b}_n = 0$. From condition given in Equation 10:

$$\left[\frac{\tilde{a}_0}{2} + \sum_{n=1}^{\infty} \tilde{a}_n \cos\frac{n\pi x}{L} \right]\left[\frac{\tilde{c}_0}{2} + \sum_{m=1}^{\infty} \tilde{c}_m \right] = 0 \quad (16)$$

and

$$\left[\frac{\tilde{a}_0}{2} + \sum_{n=1}^{\infty} \tilde{a}_n \cos\frac{n\pi x}{L} \right]\left[\frac{\tilde{c}_0}{2} + \sum_{m=1}^{\infty} (-1)^m \tilde{c}_m \right] = 0 \quad (17)$$

which implies that $\tilde{c}_0 = \tilde{c}_m = 0$. According to these conditions, the analytical solution of Equation 9 can be written as double cosine-sine Fourier series in the form:

$$\phi(x, y) = \frac{h_0^2}{2} + \sum_{m=1}^{\infty}\left[\frac{c_{0,m}}{2} + \sum_{n=1}^{\infty} c_{n,m} \cos\frac{n\pi x}{L} \right]\sin\frac{m\pi y}{H} \quad (18)$$

To evaluate the Fourier coefficients $c_{n,m}$, the solution given in Equation 18 is substituted in Equation 9 to obtain the following form:

$$\sum_{m=1}^{\infty}\left[\frac{b_{0,m}}{2} + \sum_{n=1}^{\infty} b_{n,m} \cos\frac{n\pi x}{L} \right]\sin\frac{m\pi y}{H} = f(x, y) \quad (19)$$

where:

$$b_{n,m} = -K\pi^2\left(\frac{n^2 H^2 + m^2 L^2}{H^2 L^2} \right) c_{n,m} \quad (20)$$

in which, $b_{n,m}$ represents the coefficient of the double cosine-sine Fourier series of the function $f(x, y)$ and can be obtained by the canonical form:

$$b_{n,m} = \frac{4}{HL}\int_0^L \int_0^H f(x, y)\cos\frac{n\pi x}{L}\sin\frac{m\pi y}{H}\, dy\, dx \quad (21)$$

hence, $c_{n,m}$ is given by:

$$c_{n,m} = \frac{-4HL}{K\pi^2}\int_0^L \int_0^H \frac{f(x, y)}{n^2 H^2 + m^2 L^2}\cos\frac{n\pi x}{L}\sin\frac{m\pi y}{H}\, dy\, dx \quad (22)$$

Consider the case where:

$$f(x, y) = W - \sum_{i=1}^{N_W} Q_i \delta(x - x_i)\delta(y - y_i) \quad (23)$$

in which, W: the uniform rainfall or uniform evaporation (W takes minus sign in case of uniform evaporation), Q_i: the injection or pumping rate of the ith well (Q_i takes minus sign in case of injection), δ: the Dirac delta function,

N_W: number of wells. Then, from Equation 22 and for integer $m \geq 1$:

$$c_{0,m} = \frac{-4H}{K\pi^2 m^2 L}\left\{[(-1)^m - 1]\frac{WLH}{m\pi} + \sum_{i=1}^{N_W} Q_i \sin\frac{m\pi y_i}{H}\right\} \quad (24)$$

$$c_{n,m} = \frac{-4HL}{K\pi^2(n^2 H^2 + m^2 L^2)}\sum_{i=1}^{N_W} Q_i \cos\frac{n\pi x_i}{L}\sin\frac{m\pi y_i}{H}, \quad (25)$$
$$n \neq 0$$

In actual computations, a truncated Fourier series is used. To achieve a certain degree of accuracy in computations, the number of Fourier coefficients is chosen such that it ensures that $|c_{n,m}| < \varepsilon$ for a sufficiently small positive constant ε. From Equation 22:

$$|c_{n,m}| = \left|\frac{4HL}{K\pi^2(n^2 H^2 + m^2 L^2)}\sum_{i=1}^{N_W} Q_i \cos\frac{n\pi x_i}{L}\sin\frac{m\pi y_i}{H}\right|$$
$$\leq \frac{4HLN_W}{K\pi^2(n^2 H^2 + m^2 L^2)}\max_{1\leq i\leq N_W}|Q_i| \quad (26)$$

and if equal number of Fourier coefficients is set for both variables x and y and denote it by M, then the following inequality holds:

$$M > \sqrt{\frac{4HLN_W}{K\pi^2\varepsilon(H^2 + L^2)}\max_{1\leq i\leq N_W}|Q_i|} \quad (27)$$

Finally, since the function $f(x, y)$ considered in the proposed model contains the Dirac delta function, Equation 18, the obtained Fourier series is affected by the Gibbs phenomenon. This phenomenon describes that the partial sum of the Fourier series has large oscillations near the jump, which might increase the maximum of the partial sum above that of the function itself. Several methods are available to reduce the Gibbs effect such as Fejer summation (Carslaw, 1921) and sigma approximation (Jerri, 1998). The sigma-approximation technique is used and the truncated series solution then takes the form:

$$\phi(x, y) = c + \sum_{m=1}^{M}\frac{\sin\left(\frac{m\pi}{M+1}\right)}{\left(\frac{m\pi}{M+1}\right)}\left[\frac{c_{0,m}}{2} + \sum_{n=1}^{N} c_{n,m}\frac{\sin\left(\frac{n\pi}{N+1}\right)}{\left(\frac{n\pi}{N+1}\right)}\cos\frac{n\pi x}{L}\right]\sin\frac{m\pi y}{H} \quad (28)$$

The hydraulic head can be computed using the following equation:

$$h(x, y) = \sqrt{2\phi(x, y)} \quad (29)$$

PARTICLE SWARM OPTIMIZATION

PSO was developed by Kennedy and Eberhart (1995). The PSO is inspired by the social behavior of a flock of migrating birds trying to reach an unknown destination. In PSO, each solution is a 'bird' in the flock and is referred to as a 'particle'. The birds in the population only evolve their social behavior and accordingly their movement towards a destination (Shi and Eberhart, 1998).

Physically, this mimics a flock of birds that communicate together as they fly. Each bird looks in a specific direction, and then when communicating together, they identify the bird that is in the best location. Accordingly, each bird speeds towards the best bird using a velocity that depends on its current position. Each bird, then, investigates the search space from its new local position, and the process is repeated until the flock reaches a desired destination. It is important to note that the process involves both social interaction and intelligence so that birds learn from their own experience (local search) and also from the experience of others around them (global search).

The process is initialized with a group of random particles (solutions). The ith particle is represented by its position as a point in a S-dimensional space, where S is the number of variables. Throughout the process, each particle i monitors three values: its current position (X_i); the best position it reached in previous cycles (P_i); its flying velocity (V_i). These three values are represented as follows (Elbeltagi et al. 2005):

Current position $X_i = (x_{i1}, x_{i2},..., x_{iS})$

Best previous position $P_i = (p_{i1}, p_{i2},..., p_{iS})$

Flying velocity $V_i = (v_{i1}, v_{i2},..., v_{iS})$ (30)

In each time interval (cycle), the position (P_g) of the best particle (g) is calculated as the best fitness of all particles. Accordingly, each particle updates its velocity V_i to catch up with the best particle g, as follows (Shi and Eberhart 1998):

$$New \ V_i = \omega \times current \ V_i + c_1 \times RN_1 \times (P_i - X_i) + c_2 \times RN_2 \times (P_g - X_i) \quad (31)$$

using the new velocity V_i, the particles update their position as follows:

$$New \ position \ X_i = current \ position \ X_i + New \ V_i; V_{max} \geq V_i \geq -V_{max} \quad (32)$$

in which, c_1 and c_2: two positive constants named learning factors ($c_1 = c_2 = 2$); RN_1 and RN_2: two random functions in the range [0, 1]; V_{max}: upper limit on the maximum change of particle velocity (Kennedy and Eberhart, 1995); and ω: an inertia weight employed as an improvement proposed by Shi and Eberhart (1998) to

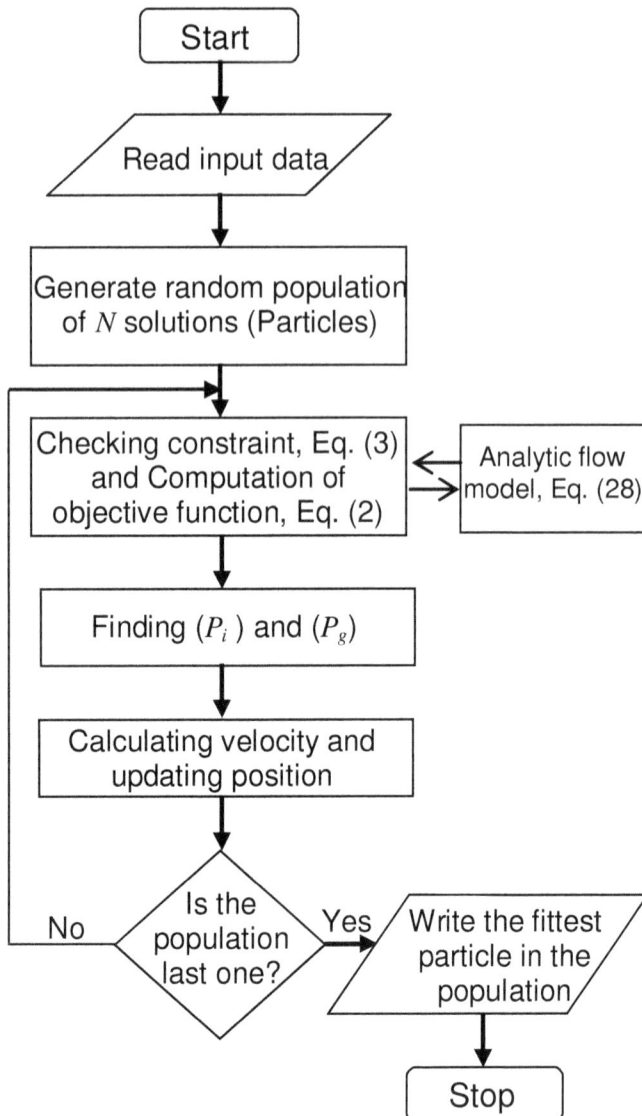

Figure 2. Flow chart for coupled Analytic-PSO model.

the new position of a particle is calculated using Equation 32, the particle then flies towards it (Shi and Eberhart 1998). The main parameters used in the PSO technique are the population size (number of birds), number of cycles, the maximum change of a particle velocity V_{max}, and the inertia weight ω.

SIMULATION-OPTIMIZATION MODEL

After developing the simulation and optimization models, mentioned in the two previous sections, both models are coupled to solve groundwater management problems. Figure 2 shows the flow chart of the Analytic–PSO model. The coupled Analytic–PSO model is particularly developed to apply the principles of simulation–optimization approach, where the optimization model repeatedly calls the simulation model to find the optimum solution of the problem. The optimization model calls simulation model to predict the state variables (hydraulic heads at well locations). The values of those state variables are used to check the constraints and then penalty value is considered if constraint violations occurred. The whole solution procedure is successively repeated to generate new solution (well discharges in the first application and set of co-ordinates and discharge of the wells in the second application, which excessively increases the computational burden) until the global (or near global) solution is obtained.

NUMERICAL APPLICATION

To investigate the performance of applying Analytic-PSO model to solve groundwater management problems, two problems are used as examples. The first one concerns with determining the maximum pumping form pre-specified well system, whereas the second problem includes determination of both coordinates and maximum pumping from a pre-specified number of wells. The two applications are performed in unconfined aquifer system given by McKinney and Lin (1994). Figure 3 shows the plan view and cross-sectional elevation of the studied aquifer. As can be seen from this figure, the aquifer having dimensions of 4500 × 10000 m². The boundary conditions include the Dirichlet at the north (river) and south (swamp) sides (h_0 = 20 m); and the no-flow at the east and west sides (mountains). The aquifer is composed from sand and gravel and it is assumed that porous medium is homogeneous and isotropic. The hydraulic conductivity and the areal recharge rate (K and W) are 50 m/day and 0.001 m/day, respectively. There are 10 pumping wells having locations listed in Table 1.

Groundwater management problem (1)

The first problem deals with the maximization of total

control the impact of the previous history of velocities on the current velocity. The operator ω plays the role of balancing the global search and the local search; and was proposed to decrease linearly with time from a value of 1.4 to 0.5 (Shi and Eberhart, 1998). As such, global search starts with a large weight and then decreases with time to favor local search over global search (Eberhart and Shi, 1998).

It is noted that the second term in Equation 31 represents *cognition*, or the private thinking of the particle when comparing its current position to its own best. The third term in Equation 31, on the other hand, represents the *social* collaboration among the particles, which compares a particle's current position to that of the best particle (Kennedy 1997). Also, to control the change of particles' velocities, upper and lower bounds for velocity change is limited to a user-specified value of V_{max}. Once

Pumping from an unconfined aquifer using a pre-specified system of wells, McKinney and Lin (1994). This typical problem was solved by various researchers (McKinney and Lin, 1994; Wang and Zheng, 1998; Wu et al., 1999; Wu and Zhu, 2006; Ayvas, 2009; Gaur et al., 2011) using different optimization methods such as LP, GA, SA, SCF-UA, HS, and PSO. The purpose of this section is to perform a comparison between results of Analytic-PSO model and the corresponding ones given by other previous researchers. The objective function for this problem was taken as Equation 2. Two constraints were considered, (1) hydraulic head should be above the aquifer bottom ($h_{i,min}$ = 0), and (2) discharge range of the pumping wells should be within the limits of 0–7000 m³/day to prevent aquifer dewatering. The suitable number of Fourier coefficients is determined, in the worst case, by taking $\max |Q_i|$ = 7000 m³/day, N_W = 10, H = 10000 m, L = 4500 m, K = 50 m/day, π = (22/7), and M = 60, (Equation 27.) in the analytic flow model.

The sensitivity analysis is carried out to determine the PSO solution parameters as follows: number of particles = 40, number of cycles = 300, and V_{max} = 200.

The factor α is also set as a linear function decreasing with the increase of number of generations where, at any generation i,

$$\alpha = 0.4 + 0.8 \times \text{(number of generations} - i) / \text{(number of generations} - 1) \qquad (33)$$

such that α = 1.2 and 0.4 at the first and last generation, respectively.

After applying the model to the problem, it is found that the values of most of $|c_{n,m}|$ are of order 10^{-6}, and the maximum value in the last 60 term does not exceed 0.007.

The relationship between optimal total pumping rates and number of cycles is shown in Figure 4. From this figure, it can be shown that the Analytic-PSO model can achieve a fairly good solution for pumping rate (59000 m³/day) after only 71 cycles whereas, it converges to the optimal value of 59463.76 m³/day at the last cycle. Table 2 presented a comparison between the maximum discharge by 10 pumping wells in the present study and those given by other previous ones. As can be seen from the table, the Analytic-PSO model gives a higher value for total pumping rate (59463.76 m³/day) in comparison with other models. This result closely agrees with the PSO 2 solution (59425 m³/day) given by Gaur et al. (2011). Results of other previous solution methods are found to be 59300 m³/day in LP, 58000 m³/day and 59000 m³/day in GA, 59078 m³/day in GASAPF, 59266 m³/day in SCE-UA, 59400 m3/day in SA, and 59279 m³/day in HS. The hydraulic heads at wells and hydraulic head contours corresponding to the obtained total pumping rate are shown in Table 3 and Figure 5, respectively.

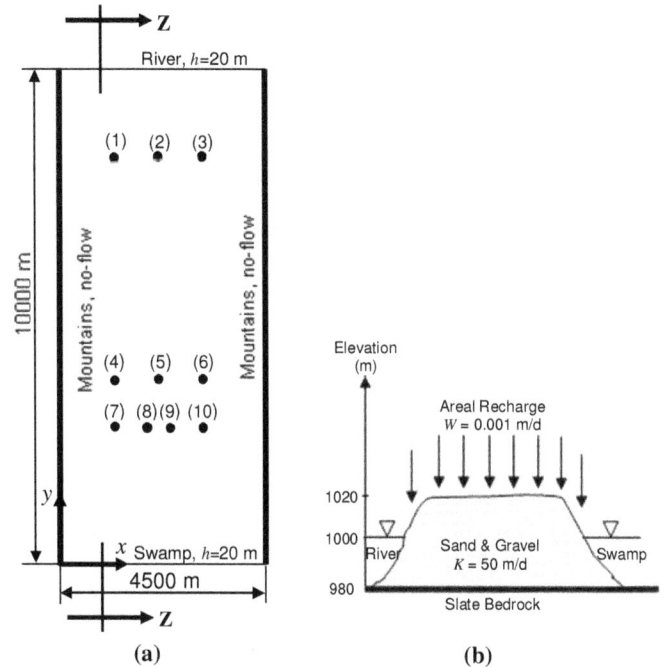

Figure 3. Unconfined aquifer model under consideration: (a) Plan view and (b) Sectional elevation **Z-Z** (After McKinney and Lin1994).

Table 1. Coordinates of pumping wells, shown in Figure 3 (McKinney and Lin, 1994).

Well number	x-coordinate (m)	y-coordinate (m)
1	1250	8500
2	2250	8500
3	3250	8500
4	1250	5000
5	2250	5000
6	3250	5000
7	1250	3500
8	2000	3500
9	2500	3500
10	3250	3500

The hydraulic heads at wells shown in Table 3 indicate that results from the proposed model satisfied constrains of the optimization model. Consequently, the Analytic-PSO model can effectively and efficiently be used to solve groundwater management problems.

Groundwater management problem (2)

The objective of the second problem is also maximizing the total pumping rate from the unconfined aquifer shown in Figure 3 as well as determining the best coordinates for the ten wells. The used objective function and hydraulic constrains are the same ones as in the problem

Table 2. Maximum discharge by 10 pumping wells using different optimization techniques (units: m^3/day).

Well number	Optimizations techniques									
	(LP) McKinney and Lin (1994)	(GA) McKinney and Lin (1994)	(GA) Wang and Zheng (1998)	(SA) Wang and Zheng (1998)	GASAPF Wu et al. (1999)	(SCE-UA) Wu and Zhu (2006)	(HS) Ayvaz (2009)	(PSO 1) Gaur (2011)	(PSO 2) Gaur (2011)	Present study
1	7000	7000	7000	7000	7000	7000	7000	7000	7000	7000.00
2	7000	7000	7000	7000	7000	7000	7000	7000	7000	7000.00
3	7000	7000	7000	7000	7000	7000	7000	7000	7000	7000.00
4	6000	7000	5000	6200	6056	5987	5904	6300	6315	5960.43
5	4500	2000	5000	6700	4290	4477	4590	4600	4600	4503.70
6	6000	6000	6000	6200	6056	5986	5904	6150	6150	5949.65
7	6800	6000	6650	6650	6774	6814	6821	6500	6600	6729.79
8	4100	7000	7000	4000	4064	4094	4121	4100	4055	4282.12
9	4100	4000	4000	4000	4064	4094	4120	4100	4100	4230.97
10	6800	7000	7000	6650	6774	6814	6820	6600	6605	6807.09
Total pumping	59300	58000	59000	59400	59078	59266	59279	59350	59425	59463.76

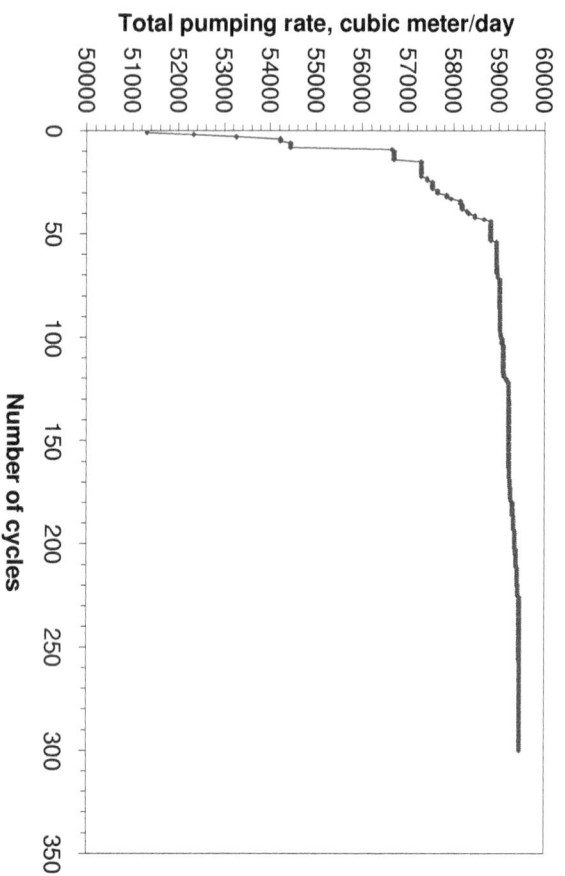

Figure 4. Relationship between optimal total pumping rates and number of cycles.

Total pumping rate, cubic meter/day
(Y-axis: 50000, 51000, 52000, 53000, 54000, 55000, 56000, 57000, 58000, 59000, 60000)

Number of cycles
(X-axis: 0, 50, 100, 150, 200, 250, 300, 350)

Table 3. Hydraulic head at wells corresponding to the obtained total pumping in problem 1.

Well number	Head (m)
1	12.18
2	11.40
3	12.18
4	0.12
5	0.33
6	0.03
7	1.22
8	0.34
9	0.68
10	0.17

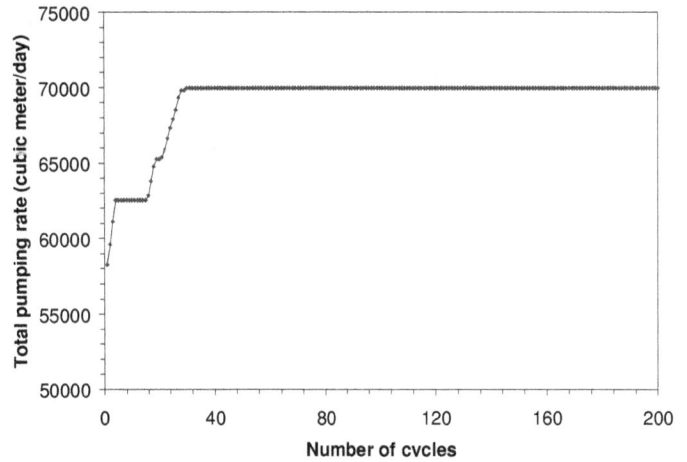

Figure 6. Relationship between optimal total pumping rates and number of cycles.

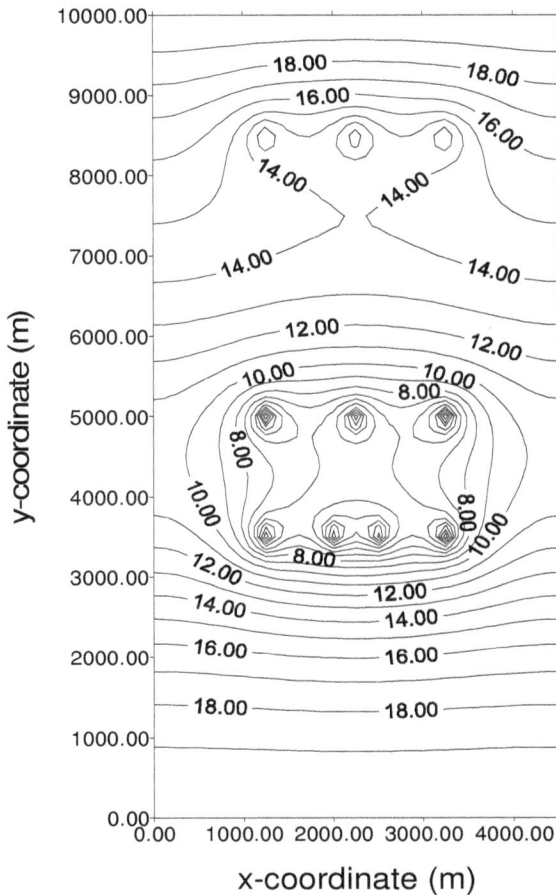

Figure 5. Contour heads corresponding to the obtained total pumping in problem 1.

1. The sensitivity analysis is performed to determine the Analytic-PSO model parameters as follows: $M = 60$, number of particles = 30, number of cycles = 200, $V_{max} = 350$ and the factor a is considered as previously mentioned.

After applying the model it is found that each well can pump its maximum bound of the pumping rate (7000 m^3/day) without violation of the minimum hydraulic head value. Figure 6 demonstrates the relationship between optimal total pumping rates and number of cycles. As seen from this figure, the Analytic-PSO model can obtain an optimum solution for pumping rate (70000 m^3/day) after only 30 cycles performing 900 simulations. Table 4 lists the obtained coordinates of wells and the hydraulic heads at well locations. Figure 7 shows the groundwater head contours generated by the Analytic-PSO model for the maximum pumping values.

Conclusions

A steady state analytical solution is suggested and derived for the groundwater flow equation in a homogeneous unconfined aquifer. The analytical solution is based on double Fourier series and is suitable to deal with any number of pumping or injection wells or combination of them. In addition, uniform rainfall or evaporation can be taken into consideration. The new analytical method is linked with the Particle Swarm Optimization method to solve the groundwater management problems. A new model called Analytic-PSO, which consists of the two methods, is originally developed. The Analytic-PSO model is verified on the most popular hypothetical example to obtain the maximum pumping rate and a comparison was carried out with the corresponding ones given by other previous studies. In addition, the model is used to determine both the optimum locations and discharges of wells for the pre-specified number of wells. The results showed that the Analytic-PSO model can effectively and efficiently be used to solve real groundwater management problems.

ACKNOWLEDGEMENT

The authors would like to thank Prof. Emad Elbeltagi Professor of Construction Projects Management,

Table 4. The obtained results for problem 2

Well no.	x-coordinate (m)	y-coordinate (m)	Head (m)
1	1538.81	553.1	16.15
2	4053.67	3398.12	4.18
3	885.40	2145.87	13.23
4	4238.68	8515.75	13.75
5	4154.02	1205.85	12.22
6	3339.77	3548.14	6.50
7	1369.39	6814.13	15.50
8	3442.24	9710.71	16.58
9	2565.95	4191.12	10.44
10	3521.94	9908.74	18.36

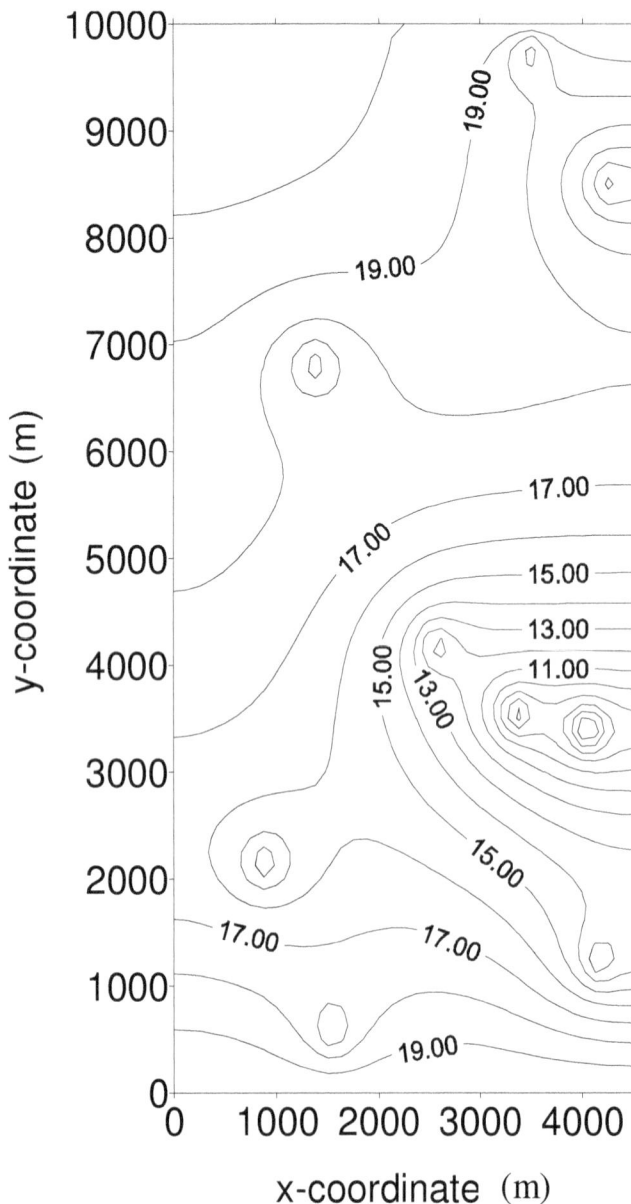

Figure 7. Contour heads corresponding to the obtained total pumping in problem 2.

Department of Structural Engineering, Mansoura University Mansoura, Egypt, for his enormous help and fruitfu advice.

REFERENCES

Ayvas MT (2009). "Application of harmony search algorithm to the solution of groundwater management models". Adv. Water Resour 32:916-924.

Bear J (1979). "Hydraulics of Groundwater" McGraw-Hill Inc, p. 569.

Carslaw HS (1921). "Introduction to the theory of Fourier's series and integrals" 2nd ed. New York: Dover Publications Inc.

Eberhart R, Shi Y (1998). "Comparison between genetic algorithms and particle swarm optimization" Proceedings of the 7th Annu. Conf. Evol Prog. Berlin, pp. 611-618.

Elbeltagi E, Hegazy T, Grierson D (2005). "Comparison among Five Evolutionary-Based Optimization Algorithms." J. Adv. Eng. Inf. Elsevier Sci. 19(1):43-53.

Gaur S, Chahar BR, Graillot D (2011a). "Analytic element method and particle swarm optimization based simulation–optimization model for groundwater management". J. Hydrol. 402:217-227.

Gaur S, Mimoum D, Graillot D (2011b). "Advantages of the analytic element method for the solution of groundwater managemen problems". Hydrol. Proc. 25:3426-3436.

Grubb S (1993). "Analytical model for estimation of steady-state capture zones of pumping wells in confined and unconfined aquifers". Ground Water 31(1):27-32.

Jerri AJ (1998). "The Gibbs phenomenon in Fourier analysis, splines and wavelet approximations". Boston: Kluwer Academic.

Kennedy J (1997). "The particle swarm: Social adaptation of knowledge". Proceedings of the IEEE International Conference on Evolutionary Computation, Piscataway, pp. 303-308.

Kennedy J, Eberhart R (1995). "Particle swarm optimization". Proceedings of the IEEE International Conference on Neura Networks (Perth, Australia), 1942-1948.

Kim DJ, Ann MJ (2001). "Analytical solutions of water table variation in a horizontal unconfined aquifer: Constant recharge and bounded by parallel streams". Hydrol. Proc. 15:2691-2699.

Ma XY, Li SG, Zhu WS (2009). "A new method in ground water flow modeling". J. Hydrodyn. 21(2):245-254.

McKinney DC, Lin MD (1994). "Genetic algorithm solution of groundwater management models". Water Resour. Res. 30:1897-1906.

Shan C (1999). "An analytical solution for the capture zone of two arbitrarily located wells". J. Hydrol. 222:123-128.

Shi Y, Eberhart R (1998). "A modified particle swarm optimizer" Proceedings of the IEEE International Conference on Evolutionary Computation, Piscataway, pp. 69-73.

Strack ODL (1989). "Groundwater Mechanics" Prentice-Hall: Englewood Cliffs, NJ.

Yeo W, Lee KK (2003). "Analytical solution for arbitrarily located multiwells in an anisotropic homogeneous confined aquifer". Water Resour. Res. 39(5):TNN 3-1-TNN 3-5.

Wang M, Zheng C (1998). "Ground water management optimization using genetic algorithms and simulated annealing: Formulation and comparison". J. Am. Water Works Assoc. 34:519-530.

Wu J, Zhu X (2006). "Using the shuffled complex evolution global optimization method to solve groundwater management models". Lect. Note Comp. Sci. 3841:986-995.

Wu J, Zhu X, Liu J (1999). "Using genetic algorithm based simulated annealing penalty function to solve groundwater management model". Sci. China (Series E) 42(5):521-529.

Zhu X, Wu J, Wu J (2006). "Application of SCE-UA to optimize the management model of groundwater resources in deep aquifers of the Yangtze Delta". Proceedings of the First International Multi-Symposiums on Computer and Computational Sciences (IMSCCS'06).

12

Study of major and trace elements in groundwater of Birsinghpur Area, Satna District Madhya Pradesh, India

R.N. Tiwari[1] , Shankar Mishra[2] and Prabhat Pandey[3]

[1]Department of Geology, Government P.G. Science College Rewa, Madhya Pradesh, India.
[2]Department of Chemistry, P.G. College Semariya, Rewa, Madhya Pradesh, India.
[3]Department of Physics, Government P.G. Science College Rewa, Madhya Pradesh India.

The present paper deals with major and trace elements geochemistry from the groundwater of Birsinghpur area, Satna district, Madhya Pradesh, India. Geologically, the area comprises sandstone and shale formations of the Proterozoic Upper Vindhyan Supergroup. The study reveals that the water is hard to very hard (The classification as hard to very hard is possibly based on incorrect analytical data); the elevated hardness is attributed to the calcareous nature of the aquifer. The concentrations of cations are characterized by Ca >Mg> Na> K. Elevated concentrations of calcium (possibly incorrect analytical data) in some localities are related to the aquifer lithology. Concentrations of magnesium, sodium and potassium are generally within the permissible limits. The concentration of anions are characterized by $HCO_3 > SO_4 > Cl > NO_3 > F$. Bicarbonate and sulphate concentrations exceed the permissible limits in a few samples; elevated concentrations appear to be related to the aquifer lithology. The fluoride concentration exceeds the desirable limit in some areas; the elevated fluoride values are attributed to the application of chemical fertilizers in agriculture and to the occurrence of fluoride bearing minerals in shale formation. Six trace elements (Fe, Cu, Ni, Pb, Mn, Cd) were also analysed. The iron concentrations ranged from 0.27 to 2.98 mg/L, exceeding the permissible limit in drinking water in 4% of the samples mainly in lateritic aquifers Nickel, and cadmium concentrations are well within the permissible limits. The manganese concentrations ranged from 0.20 to 0.282 mg/L, 30% of the samples exceeding the desirable limit: the elevated manganese concentrations are associated with iron ores (?) as well as lateritic mining. The elevated concentrations of trace elements are combined effects of geogenic sources as well as of mining activities and excessive use of chemical fertilizers. It is recommended to control anthropogenic activities adequately in order to minimise the pollution problems.

Key words: Major and trace elements, groundwater, Birsinghpur, Satna, Madhya Pradesh, India.

INTRODUCTION

Water is one of the most vital resources for the sustenance of human, plants and other living beings. It is required in all aspects of life and health for producing food, agricultural activity and energy generation. Groundwater is rarely treated and presumed to be naturally protected, it is considered to be free from impurities, which are associated with surface water, because it comes from deeper parts of the earth. In India, almost 80% of the rural population depends on untreated groundwater for potable water supplies. Due to rapid growth of population, urbanisation, industrialization, water resources of our country are now getting stressed with declining per capita availability and deteriorating quality. The groundwater used for drinking should be free from

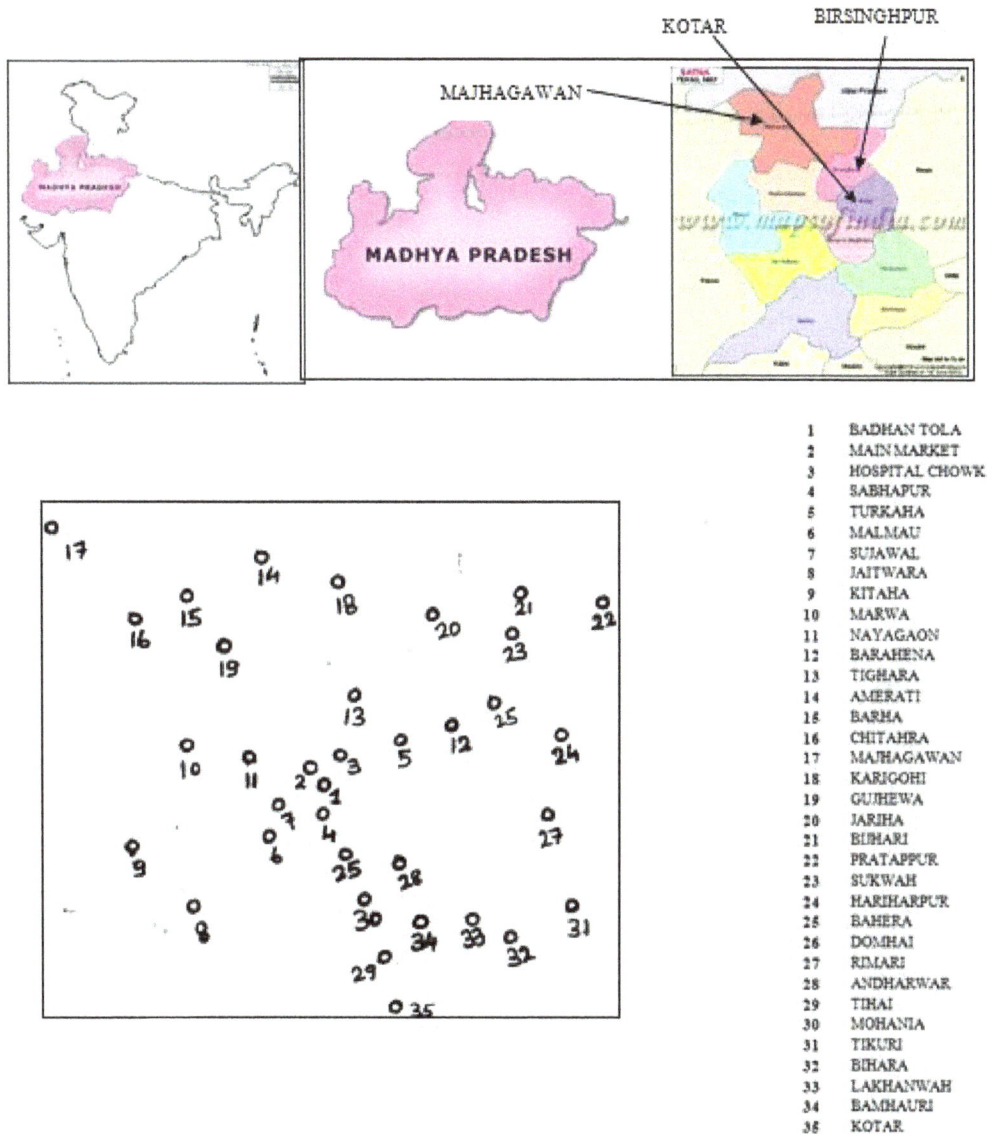

Figure 1. Location Map of the Study Area Showing Ground Water Sampling points

any toxic elements, living and non living organisms; and excessive amounts of minerals that may be hazardous to health. Assessment of groundwater quality requires determination of ion concentrations which decide the suitability for drinking, agricultural and industrial uses (Tiwari, 2011). Some heavy metals are very essential in human health, but they may cause various health problems, if present in higher concentrations. The contamination of groundwater by heavy metals and pesticides may cause problems due to their generally non biodegradable nature (Eugenia et al., 1996). The study of major and heavy metals contaminations in groundwater from different parts of world was carried out by various workers (Henphem et al., 1983; Applin and Zhao, 1989; Barzilay, 1999; Sharma et al., 2000; Dixit et al., 2003; Sudhakhar and Mamta, 2004; Subba Rao, 2006; Chiman and Jiu, 2006; Demirel, 2007; Khan et al., 2010;

Madhulakshmi et al., 2012; Nag and Ghosh, 2013). The prime objective of this study is to analyse the major and trace elements present in groundwater in the Birsinghpur area and their comparison with WHO (1993) and ISI (1991) standards for drinking purposes. Besides, the prime cause of elevated concentrations of dissolved constituents and remedial measures is also discussed by the authors.

Study area

The present study is carried out at Birsinghpur area of Satna district, Madhya Pradesh which is geographically bounded by 24°40' to 24°50' N latitude and 80°45' to 81 E longitude covering an area of above 900 km^2 (Figure 1). The area is historically and mythologically well known and famous for lord Gabinath (Shiva) temple and is rich in natural resources namely: bauxite, laterite, flagstone and Ramraj (a type of clay used in paint). Accordingly, intensive mining activities have been carried by various agencies. The waste materials of mining area are dumped on the spot causing pollution problems. The region receives 900 mm average rainfall from the south west monsoon during the months of June to September and the climate is sub-humid. The temperature varies from 42°C (summer season) to 4°C (winter season).

Geologically, the area comprises formations of the Proterozoic Vindhyan Supergroup. The main rock types are Kamour sandstone and shales of Upper Vindhyan. Bauxite and laterite are associated with the Rewa Group of Vindhyan Supergroup. The general geological succession of the area is given in Table 1. Hydrogeologically, the study area belongs to the Precambrian hard rock province. The groundwater occurs in semi-confined to confined conditions. The water level fluctuations are generally 3 to 4 m.

MATERIALS AND METHODS

A total number of 35 groundwater samples were collected in 1 L polythene bottles from different borewells and handpumps from the study area in the month of May 2012. The pH, EC and TDS have been determined with portable kits. Total hardness as $CaCO_3$ was analysed by EDTA. The major cations and anions were analysed per standard methods.

For the determination of heavy metals, the collected water samples were immediately acidified with HNO_3 to bring the pH below 2 to avoid the precipitation of the metals. The samples were concentrated and subjected to nitric acid digestion. Selected trace elements such as Fe, Pb, Mn, Cu, Ni and Cd were analysed per standard method.

RESULTS AND DISCUSSION

Major elements

The pH of the groundwater samples ranges from 6.8 to 8.5 indicating weakly acidic to slightly alkaline conditions (Tables 2 and 3). The electrical conductivity (EC) which is a measure of water's capacity to conduct electric current varies from 244 to 816 µS/cm. Total hardness of groundwater of the study area varies from 166 to 885 mg/L, 70% of the samples are categorized hard to very hard (Possibly incorrect analytical data) (Sawyer and McCarty, 1967). The hardness of the water is caused by the presence of alkaline earth such as calcium and magnesium. The hard water is marginally suitable for drinking, cooking, cleaning and laundry jobs and may needs proper treatment (Categorization as hard to very hard is possibly based on incorretc analytical data).

The cation composition of the groundwater shows an order of Ca > Mg>Na > K. The calcium of groundwater varies from 39.7 to 215 mg/L (possibly incorrect analytical data); about 60% of samples exceed the desirable limit of

75 mg/L (BIS, 1991). The continuous high intake o calcium may cause stone problem which is also noticed in the area. The sources of calcium are limestone anc calcareous sandstone present in the area. The concentration of magnesium varies from 8.4 to 70.4 mg/L (possibly incorrect analytical data). The mica rich shale and micaceous sandstone are the main source o magnesium. The sodium and potassium are within permissible limits and their source rocks are K and Na feldspar (orthoclase and microcline) associated with shale and sandstone. The anion composition of the groundwater is characterized by $HCO_3 > SO_4 > Cl > NO_3 > F$. The bicarbonate concentrations ranges from 53 tc 305 mg/L; about 10% of samples exceed the desirable limit of 200 mg/L. Bicarbonate is usually the primary anion in groundwater and is derived from the carbor dioxide released by the organic decomposition in the soi (Todd, 1985).

$$CO_3 + H_2O \rightarrow HCO_3 + H^+$$

The sulphate concentrations varies from 60 to 305 mg/L. About 40% of samples exceed the desirable limit of 200 mg/L. The source of sulphate in the groundwater are gypsiferous bands associated with shale formation at higher depth. Elevated sulphate concentration may cause laxative effect. Being a hydrous sulfate of calcium, gypsum gets dissolved by circulating groundwater, thus imparting the water a permanent sulphate hardness. Also, gypsum is reduced by anerobic organisms in the deeper parts of the aquifer (Tiwari, 2011). The resulting reduction, leads to the formation of hydrogen sulphide gas (H_2s) which is easily soluble in groundwater and gives it an unpleasant odour. This peculiar odour is also noticed in a few localities. The concentrations of chloride and nitrate are within the permissible limits. The chemical fertilizers used by farmers are the main source of nitrate whereas industrial effluents may be the source of chloride. The concentration of fluoride ranges from 0.37 to 1.53 mg/L, exceeding the maximum desirable limit (1 mg/L, BIS, 1991) in a few localities. The sources of fluorides are chemical fertilizers as well as micaceous shale.

(b) Trace Elements (Tables 2 and 3) Iron

Iron is an essential element in human body (Moore, 1973) and is found in groundwater all over the world; higher concentrations of iron cause bad taste, discoloration, staining, turbidity, esthetic and operational problem in water supply systems (Dart, 1974; Vigneshwaran and Vishwanathan, 1995). Deficiency of iron results in hyochromic macrobiotic anemia; one of the world's common health problems.

The limit of concentration of iron in drinking water ranges between 0.3 (desirable limit) to 1.0 mg/L (permissible limit). In the study area, a minimum value

Table 1. General geological succession of the area.

Supergroup		Group	Formation	Thickness (m)
		Recent	Alluvium	4
		Pleistocene	Laterite, Bauxite	45
		Rewa	Upper Rewa sandstone	120
			Jhiri shale	18
Vindhyan (Proterozoic)		Kaimur	Kaimur Sandstone	180

Age	Supergroup			
Recent Pleistocene			Alluvium	4
			Laterite, Bauxite	45
Proterozoic	Vindhyan	Rewa	Upper Rewa sandstone	120
			Jhiri shale	18
		Kaimur	Kaimur Sandstone	180

(0.27 mg/L) of iron is observed in Hariharpur whereas the maximum value (2.98 mg/L) has been observed in Lakhanmah area. In about 45% of the samples, the maximum permissible limit of iron in water is exceeded; the elevated iron concentrations are derived from laterite and ferruginous sandstone (Tiwari and Dubey, 2012). Besides natural sources, the corrosive nature of casing pipe used for water supply might contribute to the elevated iron concentration.

Copper

Copper is an essential element, concentrated in several enzymes, and its presence in trace concentrations is essential for the formation of hemoglobin. An over dose of copper may lead to neurological complication, hypertension, liver and kidney dysfunctions (Krishna and Govil, 2004; Khan et al., 2010). Ingestion of copper causes infant death, short lived vomitting diarrhea etc (Barzilay, 1999). In the present study its value ranged from nil to 0.134 mg/L which is within the permissible limit as suggested by WHO (1993) and BIS (1991).

Nickel

Nickel is present in a number of enzymes in plants and microorganism. In the human body, nickel influences iron adsorption, metabolism and may be an essential component of the haemopoiitic process. Acute exposure of nickel in the human body is associated with a variety of chemical symptoms and signs such as nausea, vomiting, headache, giddiness etc. (Barzilay, 1999).

The BIS (1991) has recommended 0.02 mg/L as maximum permissible concentration in drinking water. In the study area, Ni concentrations range from nil to 0.025 mg/L. The primary source of nickel in drinking water is leaching from metals in contact with drinking water such

as pipes and fittings. However, it may also be present in some ground waters as a consequence of dissolution from nickel ore bearing rocks. The geogenic source appears to be responsible mainly for the nickel concentrations in groundwater of the study area (Tiwari and Dubey, 2012).

Lead

Lead occurs geologically in association with sulphide minerals and may be present in generally elevated concentration in areas with ores and coal (Reimanne and Decarital, 1998). Lead is toxic to the central and peripheral nervous system causing neurological and behavior effects. The consumption of lead in higher quantity may cause hearing loss, blood pressure and hypertension and eventually it may be prove to be fatal.

In the present study, lead concentration ranged from 0.007 mg/l to 0.065 mg/l. It is observed that all groundwater samples except sample no. 8 (Jaitauara) and sample no. 21 (Bijhari) have lead values within the permissible limit (BIS, 1991). The comparatively higher concentrations in a few localities are attributed to the chemistry of aquifer.

Manganese

Manganese is one of the most abundant elements in the earth's crust, it usually occurs together with iron and is widely distributed in soil, sedimentary rocks and water. The most abundant compounds of manganese are sulphide, oxide, carbonate and silicate. In the present study, the concentration of manganese ranged from 0.020 to 0.282 mg/L. About 30% of samples exceed the desirable limit however they do not exceed the permissible limit (0.3 mg/L). Higher concentrations are related to the geology of the area (Tiwari and Dubey, 2012)

Table 2. Analytical Results of Groundwater Samples of the Study Area.

S. No.	Sampling locality	pH	EC	TDS	TH	Cl	F	SO4	NO3	HCO3	Pb	Fe	Mn	Cu	Ni	Cd	Ca	Mg	Na	K
1	Badhan tola	7.4	584	365	448	28.6	0.85	31.4	10.45	140	0.010	0.56	0.055	0.085	0.012	0.0032	160.3	62.5	12.2	2.4
2	Main market	7.9	536	335	568	32.4	0.87	34.1	8.35	110	0.024	0.66	0.035	0.058	0.013	0.0036	147.2	50.6	22.1	3.2
3	Hospital Chowk	7.9	525	328	582	33.1	0.72	20.2	7.88	60	0.019	0.73	0.042	0.053	0.012	0.0021	175.3	35.5	15.5	3.4
4	Sabhapur	7.9	597	373	567	33.6	0.94	39.7	7.40	142	0.021	0.91	0.062	0.070	0.015	0.0038	158.3	41.7	4.8	0.8
5	Turkaha	7.4	430	268	320	26.8	0.68	29.4	8.76	145	0.012	0.53	0.042	0.047	ND	0.0024	80.2	30.0	5.3	0.8
6	Malmau	7.2	244	152	310	21.3	0.99	30.9	6.95	135	0.013	0.64	0.048	0.042	0.015	0.0028	59.2	43.0	6.7	1.2
7	Sujawal	7.6	413	258	282	25.8	0.78	29.2	6.90	117	0.015	0.28	0.049	0.028	ND	ND	75.8	23.0	15.0	2.2
8	Jaitwara	7.4	432	270	349	33.7	0.92	30.5	10.27	68	0.065	1.63	0.127	0.042	0.025	0.0049	94.6	28.6	19.1	12.2
9	Kitaha	7.7	389	243	253	26.8	0.88	35.4	8.36	103	0.018	0.76	0.070	0.041	0.016	0.0030	65.1	22.6	17.0	3.5
10	Marwa	8.5	364	227	256	36.6	1.21	33.3	7.40	105	0.015	0.63	0.049	0.045	0.017	0.0022	56.5	28.7	18.0	4.0
11	Nayagaon	7.6	365	228	235	24.9	1.26	33.2	6.48	107	0.007	0.47	0.024	0.040	0.018	ND	46.4	29.8	16.0	2.8
12	Barahena	7.5	468	292	462	24.2	0.57	33.7	6.10	82	0.009	0.29	0.045	0.046	0.014	0.0054	100.1	53.5	25.2	3.3
13	Tighara	7.0	349	218	252	27.8	0.72	25.2	6.88	62	0.016	1.35	0.083	0.049	ND	0.0026	58.4	26.8	15.2	4.1
14	Amerati	7.0	301	188	196	27.9	0.75	16.2	3.56	80	0.032	0.53	0.213	0.069	0.019	ND	39.7	24.1	18.3	1.3
15	Barha	6.8	421	263	328	12.0	0.98	30.5	4.80	103	0.049	2.10	0.182	0.079	0.013	0.0027	104.8	16.5	22.3	0.63
16	Chitahra	7.5	557	348	540	27.2	1.06	34.5	7.18	109	0.034	1.35	0.079	0.042	0.017	0.0029	147.3	42.8	18.5	1.2
17	Majhagawan	7.7	657	410	703	33.4	1.52	41.1	8.26	54	0.036	1.42	0.143	0.048	0.019	0.0031	179.2	64.0	35.3	7.0
18	Karigohi	7.4	360	225	255	24.3	0.69	27.7	7.34	79	0.032	1.32	0.050	0.134	0.014	0.0030	48.6	32.9	22.0	3.5
19	Gujhewa	7.0	340	212	195	15.8	0.89	26.4	3.65	89	0.025	1.56	0.078	0.044	0.015	0.0026	65.0	8.4	32.3	5.8
20	Jariha	6.8	416	260	341	25.7	0.75	13.8	4.45	94	0.020	2.72	0.246	0.070	0.011	0.0029	60.5	47.6	42.2	3.5
21	Bijhari	6.9	490	306	395	19.5	0.37	17.9	1.98	202	0.052	2.47	0.230	0.054	0.018	ND	64.2	58.7	14.8	3.5
22	Pratappur	7.0	736	460	502	29.7	0.87	28.9	7.10	305	0.013	1.53	0.185	0.087	0.018	0.0027	139.8	38.1	22.3	4.5
23	Sukwah	6.8	816	510	693	33.4	0.86	31.7	4.92	109	0.011	0.89	0.208	0.082	0.016	0.0048	165.4	70.2	30.1	8.1
24	Hariharpur	7.5	456	285	514	27.1	1.30	30.0	7.05	185	0.010	0.27	0.042	0.046	0.015	0.0020	140.0	41.2	18.3	2.1
25	Bahera	7.2	536	335	202	30.2	0.62	24.6	5.30	106	0.017	0.79	0.034	ND	0.013	0.0019	55.2	16.4	15.3	3.1
26	Domhai	7.3	423	264	260	25.2	0.79	38.4	5.18	107	0.009	0.30	0.020	ND	0.010	0.0040	69.0	21.8	22.3	2.3
27	Rimari	7.9	496	310	403	20.3	0.82	45.2	7.58	103	0.024	0.78	0.021	0.035	0.015	ND	110.4	32.2	27.0	2.5
28	Andharwar	7.8	430	268	376	18.7	0.94	35.4	5.86	109	0.018	0.76	0.031	0.040	0.016	ND	98.5	32.4	32.00	2.4
29	Tihai	6.7	600	375	490	22.4	1.17	39.2	5.05	203	0.012	1.28	0.084	0.072	0.019	0.0023	127.3	43.5	12.3	1.2
30	Mohania	6.8	488	305	621	25.5	0.96	47.2	4.47	53	0.021	1.53	0.089	0.067	0.017	0.0028	118.1	55.2	18.0	1.3
31	Tikuri	7.5	548	342	764	37.4	1.53	37.8	2.40	55	0.017	1.26	0.070	0.092	0.014	ND	149.3	62.6	15.3	1.5
32	Bihara	7.3	666	416	652	33.1	0.80	39.8	8.59	62	0.021	1.38	0.089	0.048	0.017	0.0021	210.0	59.5	14.2	4.1
33	Lakhanwah	6.9	676	422	710	29.3	0.91	49.5	6.10	164	0.016	2.98	0.148	0.075	0.015	0.0026	162.8	61.4	15.2	1.6
34	Bamhauri	7.0	648	405	822	25.4	0.97	45.2	6.90	159	0.013	2.35	0.282	0.046	0.023	0.0018	190.5	58.1	10.3	1.2
35	Kotar	7.4	772	482	885	32.7	1.20	50.8	9.80	110	0.025	1.78	0.102	0.102	0.022	0.0028	215.8	70.4	15.4	5.3

Unit: EC μS/cm, pH unitless, other are in ppm.

Table 3. Comparison of the quality parameters of groundwater of the study area with WHO and ISI for drinking purpose.

S. No.	Water quality parameter	WHO (1993)		BIS (1991)		Concentration in study area
		Max. desirable	Max. permissible	Max. desirable	Max. permissible	
1.	pH	7.0	8.5	6.5	8.5	6.8-8.5
2.	TH	300	500	300	600	166-885
3.	TDS	500	1500	500	1000	152-510
4.	Ca	75	200	75	200	39.7-215
5.	Mg	30	150	30	100	8.4-70.4
6.	Na	-	200	-	200	4.8-42.2
7.	K	-	12	-	-	0.8-12.2
8.	Cl	200	600	250	1000	12-37.4
9.	SO$_4$	200	400	150	400	60-305
10.	HCO$_3$	-	-	200	600	53-305
11.	F	1	1.5	1	1.5	0.37-1.53
12.	Lead	-	0.05	0.05	No relaxation	0.007-0.06
13.	Iron	0.3	1.0	0.3	1.0	0.28-2.48
14.	Manganese	0.1	--	0.1	0.3	0.20-0.282
15.	Copper	0.05	1.0	0.05	1.5	Nil-0.134
16.	Nickel	--	0.02	..	0.02	Nil-0.025
17.	Cadmium	...	0.005	0.01	No relaxation	Nil – 0.005

Except pH all values are in ppm.

Manganese is regarded as one of the least toxic elements but its excess amount in the human body may cause growth retardation, fever; fatigue and eye blindness, and may affect reproduction.

Cadmium

Cadmium is a cumulative environmental pollutant and its exposure to the body results damage of the kidney, and causes renal dysfunction, arteriosclerosis, cancer etc. (Goel, 1997; Robards and Worsfold, 1991). In the present study, the concentration of cadmium ranged from nil to 0.005 mg/L which is well within the permissible limit as recommended by BIS (1991) and WHO (1993) respectively. The concentration of cadmium in water samples of the study area may be attributed to the runoff from the agricultural sector where pesticides as well as cadmium phosphatic fertilizer are being used.

Conclusions

The analysed groundwater samples of the study area indicate a slightly alkaline nature of groundwater. The water is hard to very hard (the categorization as hard to very hard is possibly based on incorrect analytical data) due to the presence of calcium and magnesium in the aquifers. The hard water may require chemical treatment before use for drinking. The calcium is high in few localities due to the calcareous nature of the sandstone. The other cations are generally within the permissible limit and their sources are the shale and sandstone

aquifers. The concentration of sulphate is elevated due to the presence of gypsum bands associated with shale formations. The gypsum easily dissolves into the groundwater, attributing to the permanent hardness. The associated intake of sulphate in higher concentrations may cause laxative effect. Chloride and nitrate concentrations are within the permissible limits. The fluoride concentration exceeds the maximum desirable limit in a few localities (BIS > 1 mg/L) due to the fluorine rich mica and phosphate fertilizers.

The iron is generally elevated in most localities, originating from laterite bearing geologic formation. The continuous higher intake of iron may cause toxic effects to the human health. At greater depth, with reducing conditions, the solubility of Fe-bearing minerals in water increases leading to an enrichment of dissolved iron in groundwater (Applin and Zhao, 1989; White et al., 1991). The concentrations of copper and nickel are within the permissible limit and their sources are geologic as reported by Tiwari and Dubey (2012). The concentration of manganese is elevated in a few localities due to its affinity with iron. Higher concentrations of manganese may cause metabolic disorder. The study reveals that iron and manganese are exceeding permissible limits and their main sources are combined effects of geogenic and anthropogenic sources. Proper monitoring is needed to avoid anthropogenic contamination.

REFERENCES

Applin KR, Zhao N (1989). The kinetics of Fe (II) oxidation and well screen encrustation. Groundwater 27:168-174.

Barzilay JI (1999). The water we drink: Water quality and its effect on health. New Brunswick New Jursey, Rutgers University Press, p. 152.

BIS (1991). Bureau of Indian Standard Specification for Drinking Water ISI : 10500.

Chiman L, Jiu JS (2006). Heavy metal and trace element distribution in groundwater in natural stops and highly urbanised spaces in Mid level area Howngkong. Water Res. 40(4):753-76.

Dart FJ (1974). The hazard of iron. Ottawa Water and Pollution Control, Canada.

Demirel Z (2007). Monitoring of heavy metal pollution of groundwater in a phreatic aquifer in Mersin-Turkey. Environ. Monit. Assess. 132(1-3):15-23.

Dixit RC, Verma SR, Nitna Ware V, Thackere NP (2003). Heavy metal contamination in surface and groundwater supply of an urban city. Indian J. Environ. Health 45(2):107-112.

Eugenia GG, Vicente A, Rafael B (1996). Heavy metals incidence in the application of inorganic fertilizers and pesticides to rice forming soils. Environ. Poll. 92:19-25.

Goel PK (1997). Water pollution causes effect and control. New Age International Publishers, p. 269.

Henphem DD, Wixson BG, Gale NL, Clevenger TE (1983). Dispersal of heavy metals into the environment as a result of mining activities. Proc. Int. Conf. Heavy Metals Environ. Heidelberg, pp. 917-924.

Khan MQMA, Umar R, Latch H (2010). Study of trace elements in groundwater of Uttar Pradesh, India. Sci. Res. Essays 5(20):3175-3182.

Krishna AK, Govil PK (2004). Heavy metal contamination of soil around Pal. industrial area, Rajasthan, India. Environ. Geol. 47:38-44.

Madhulakshmi L, Rauma A, Kannan N (2012). Seasonal distribution of some heavy metal concentrations in groundwater of Virudhunagar district, Tamilnadu, South India. J. Environ. Agric. Food Chem. 11(2):32-37.

Moore CV (1973). Iron: Modern nutrition in health and disease, Philadelphia. Lea and Fiibeger, p. 297.

Nag SK, Ghosh P (2013). Variation in groundwater levels and water quality in Chhatna Block, Bankura district, West Bengal- A GIS approach. J. Geol. Soc. India. 81:262-280.

Reimanne C, Decaritat P (1998). Chemical elements in the environment. Springer Verlag, p. 398.

Robards M, Worsfold P (1991). Cadmium toxicology and analysis - A review: Analyst 116:549-560.

Sawyer CN, McCarty PL (1967). Chemistry for salinity engineers (2nd edn.). McGraw Hill, New York, p. 518.

Sharma BC, Mishra AK, Bhattarcharyya KG (2000). Metal in drinking water in a predominantly rural area. Indian J. Environ. Prot 21(4):315-322.

Subba RN (2006). Seasonal variation of ground water quality in a part of Guntur district, Andhra Pradesh, India. Environ. Geol. 49:413-429.

Sudhakar MR, Mamatha P (2004). Water quality in sustainable water management. Curr. Sci. 87(7):942-947.

Tiwari RN (2011). Geochemical studies of Groundwater in Semariya Teheil, Rewa District Madhya Pradesh, India. Proc. Int. Groundwater Conf. Madurai, India, pp. 679-678.

Tiwari RN, Dubey DP (2012). A study of Bauxite deposit of Tikar Plateau Rewa district, M.P. Gond. Geol. Soc. sp. 13:111-118.

Todd DK (1985). Groundwater Hydrology. New York, John Wiley and Sons Ltd. Songapore, 535 p.

Vigneswaran S, Viswanathan C (1995). Water treatment process Simple options, New York : CRC 11 p.

White AF, Benson SM, Yee AW, Woo LHA, Flexser S (1991). Groundwater contamination at the Kesterson reservoir, California - Geochemical parameters influencing selenium mobility. Water Resour. Res. 27:1085-1098.

WHO (1993) Guidelines for drinking water. Vol. 2, Recommendations. World Health Organization Geneva.

13

Surface water quality and risk assessment in the vicinity of Sylhet City

Abdur Rahman[1]*, M. A. Zafor[2] and Mursheda Rahman[1]

[1]Department of Civil Engineering, Stamford University Bangladesh, Dhaka, Bangladesh.
[2]Department of Civil Engineering, Leading University, Sylhet, Bangladesh.

This paper is aimed to determine the surface water quality in an around Sylhet City. We used surface water for our keen purpose. Surface water as well as ground water has been contaminated by humans. Nowadays, water pollution is a major global problem which requires ongoing evaluation and revision of water resource policy at all levels. In addition to the acute problems of water pollution in developing countries, developed countries continue to struggle with pollution problems as well. Discharges from various contaminated sources are a very common thing in city life which deteriorates the water quality of any main water source and necessary treatment is required to make water potable. Sylhet, one of the six divisional cities of Bangladesh is suffering from shortage of domestic water supply and adequate sewerage lines, drainage and solid waste disposal problems. This paper will focus on evaluation of the existing condition of surface water in Sylhet City and investigating some physical and chemical quality of water throughout the year 2008 to 2009. The selected parameters for assessing the water quality are pH, total dissolved solid (TDS), biochemical oxygen demand (BOD), dissolved oxygen (DO), ammonia nitrogen, nitrate and turbidity in the vicinity of Sylhet City.

Key words: Ammonia, biochemical oxygen demand (BOD), coliform, dissolved oxygen (DO), treatment, pH.

INTRODUCTION

Water quality assessment is the process of overall evaluation of the physical, chemical and biological nature of the water. The need to verify whether the observed water quality is suitable for intended uses is the main reason for the assessment of the quality of aquatic environment. Water is one of the major components of the life environment. The growing global awareness in the maintenance of a "clean world", public and private agencies have come to realize the importance of surface water to a national economy. Knowledge of water quality thus plays a significant role in the development of water quality control and management (Lohani and Todino, 1984). Sylhet City, one of the rapidly developing urban areas is located in the northeast region of Bangladesh and situated at 28.85° latitude and 98.80° longitude. The

region is in the hilly portion of the country. The city occupies a total area of 26.5 km^2 with a population of around 0.5 million (Sylhet City Corporation, 2005). A total of nine natural drainage channels (locally called chara) are responsible for draining storm water from city area to the Surma and Khushiara River. The main study area Originated from Lakkatura tea graden, it passes over Baluchar, Shibgang, Sobhani Ghat, Chalibondar, Chararpar. At Masimpur the chara falls in the Surma and Khushiara River. As the quality of waste water is not satisfactory, problems like pollution to Surma and Khushiara River and the streams, deterioration of the environment and health sanitation have become serious. The water quality criteria are developed on the basis of pollutants upon a specific use of water. The criteria, therefore, are defined as the acceptable levels of concentration of pollutants for a particular use and describe the water quality requirements for protecting aquatic life. Therefore, the present study has been undertaken to assess water quality parameters which will

*Corresponding author. E-mail: ceeabdur09@gamil.com or ceeabdur09@yahoo.com.

help to evaluate the potential risks, as well as to suggest measures for the prevention of the water pollution.

MATERIALS AND METHODS

The steps that have been adopted to attain the objectives of the study are as follows: Primarily recent stream water quality data, photos and information of surface water in and around Sylhet City has been procured from field observation and analyzed to assess the present water quality scenario. For this study, field observation was needed to know about the existing physical and environmental condition of the study area. During the study of the water quality of surface water in and around Sylhet City, a laboratory test program was under taken to monitor the different parameters which were the main causes of water pollution.

Sample collection

The most important task of water quality analysis is sampling. In order to get the information about the water quality of differents Chara and Surma, Khushiara river grab sampling procedure is applied. Water samples were collected in plastic containers with stopper from surface and from two feet below the top of water surface from different sampling points at second week in every month over the year. Plastic containers of capacity greater than 2000 cm^3 were used for sampling, and 2000 cm^3 of each sample in each location was collected for the study. Water samples were collected from seven locations namely Baluchar, Shibgonj, Uposhahar, Chalibondar, Masimpur, Surma River and Khushiara River.

Water quality testing

The water collected from different points in and around Sylhet City was tested monthly throughout the year. The sample was tested for the parameters like pH, ammonia, dissolved oxygen, BOD$_5$, and fecal coliform. For fecal coliform testing, membrane filter technique was used. The potable pH meter HI 8014 by HANNA Instruments was used to test pH. For turbidity testing, microprocessor turbidity meter HI 93703 by HANNA Instruments was used. Iron and manganese were tested using HACH UV Spectrophotometer DR/4000U. Suspended solids, dissolved solids, dissolved oxygen were tested by standard Methods developed by The American Public Health Association (APHA), American Water Works Association (AWWA), Water Pollution Control Federation WPCF (1998).

Field visit and questionnaire survey

All the sampling points were visited by walking and by Rickshaw and photograph taken with camera. Questionnaire surveys were distributed to key informants and interviews were done by group discussions with the people living along the Chara.

RESULTS AND DISCUSSION

pH

The standard pH value of surface water is 6.5 to 8.5 (WHO), in that respect, pH at all the points are within the

range. The highest value of Goalichara is 7.18 and Surma River is 7.2. The following figure shows the pH data of different points. The highest pH was found at Baluchar and the lowest at Uposhahor. From Figure 1 we see that Goalichara downstream and Surma downstream pH is decreasing. pH decrease at downstream due to acidic waste which are produced from different sources. For these reason we get pH lower at downstream than upstream.

Fecal coliform

Average fecal coliform at Baluchar is 15.5/100 ml, at Shibgonj 26.03/100 ml, at Uposhahor 30.5/100 ml, at Chalibonbar 36.25/100 ml, at Masimpur 48.91/100 ml, at Surma u/s 29.66/100 ml, and at Surma d/s 36.66/100 ml. Highest number of fecal are present in Masimpur at Goalichara d/s and lowest are present at Baluchar at Goalichara u/s. Highest fecal coliform at Goalichara due to domestic sewage discharged to the channel directly polluting water continuously. Figure 2 shows the fecal coliform at seven points. Open defecation, sanitary sewage and domestic sullage water are the main causes of fecal coliform at this point. Fecal coliform at goalichara D/S are higher than Goalichara U/S, when the channel flows through Sylhet city. Sanitary wastes are disposed into the channel which carry high amount of fecal coliform. Fecal coliform to Surma River increases while Goalichara channel falling into it should be zero. At every sampling location, fecal coliform exceeded the limit of Environment Conservation Rules (ECR) (1997). Because of urbanization impact fecal coliform of Goalichara d/s is greater than Goalichara u/s.

Dissolved oxygen

The main reason DO levels might fall is the presence of organic waste. Organic waste comes from something living or that was once living. It comes from raw or poorly treated sewage; runoff from farms and animal feedlots; and natural sources like decaying aquatic plants and animals and fallen leaves in water. Warmer water holds less oxygen than cold water. The time of yearly and many other factors also affect the amount of DO in water. DO levels can also fall to any human activity that heats the water. Average dissolved oxygen at Baluchar is 6.24 mg/L, at Shibgonj 5.41 mg/L, at Uposhahor 5.71 mg/L, at Chalibonbar 5.61 mg/L, at Masimpur 5.28 mg/L, at Surma u/s 6.88 mg/L, at Surma d/s 6.22 mg/L. The normal limit of DO in rivers is 6.0 mg/L. Dissolved oxygen is highest at Baluchar of Goalichara u/s and lowest at Masimpur of Goalichara d/s. Highest dissolved oxygen is reduced because oxygen is consumed by bacteria to degrade waste. Figure 3 shows the dissolved oxygen at seven points. DO at Goalichara u/s is high but

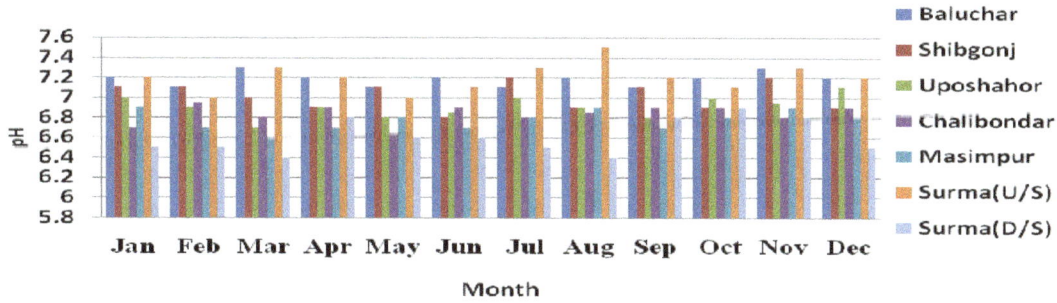

Figure 1. Variation of pH at different location.

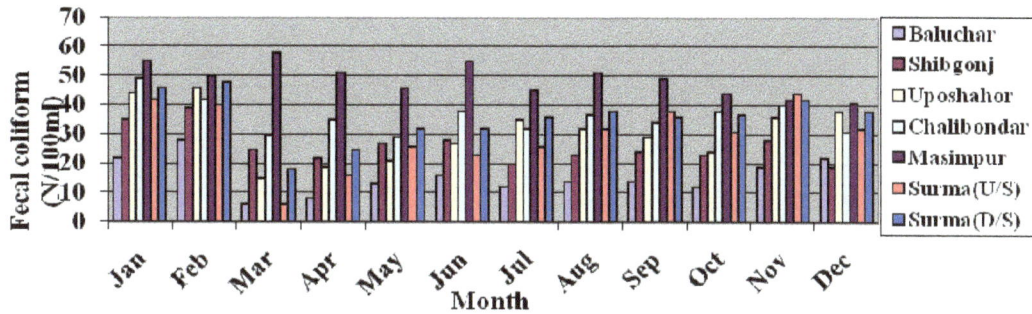

Figure 2. Variation of fecal coliform at different location.

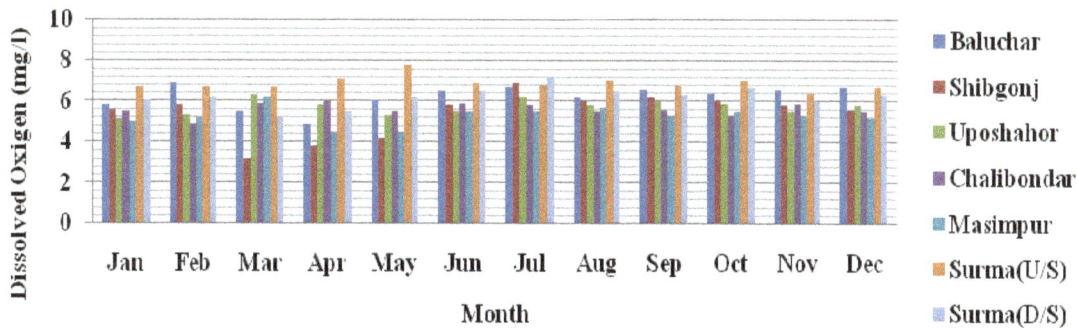

Figure 3. Variation of Dissolved Oxigen at different location.

it gradually decreases when Goalichara channel flows through the Sylhet City. Flowing to the downstream, at Masimpur it becomes low because of high amount of waste at Goalichara are being degraded with the passage of time at different locations.

Biochemical oxygen demand

The water of Surma River is not suitable for drinking purpose. Average BOD at Baluchar is 27.33 mg/L, at Shibgonj 33.83 mg/L, at Uposhahor 34.83 mg/L, at Chalibonbar 42.42 mg/L, at Masimpur 44.33 mg/L, at Surma u/s 39.5 mg/L, at Surma d/s 47.17 mg/L. BOD at Goalichara channel and Surma river is higher than 2 mg/L permissible value of Environment Conservation

Rules (ECR) (1997) due to high concentration of sewage discharged directly to the water. Figure 4 shows the BOD at seven points. BOD at Goalichara u/s is comparatively low but it becomes high at Goalichara d/s may result from domestic or industrial waste water discharge. BOD at Surma river is also high because of various pollution sources are connected with the river (Table 1). Because of urbanization impact BOD of Goalichara d/s is much greater than Goalichara u/s.

Ammonia

The water of Surma River is not suitable for drinking purpose. Average ammonia at Baluchar is 0.512 mg/L, at Shibgonj 1.025 mg/L, at Uposhahor 1.652 mg/L, at

Table 1. Test results of different pollution parameters.

Place	Parameter of various months											
	January	February	March	April	May	June	July	August	September	October	November	December
BOD5 (mg/L)												
Baluchar	33	38	10	16	16	25	32	29	32	32	35	35
Shibgonj	38	44	15	24	24	32	35	38	34	38	40	44
Uposhahar	44	41	21	32	38	35	31	39	38	33	33	33
Chalibondar	48	51	24	38	46	40	42	42	42	45	47	44
Masimpur	55	42	32	46	35	50	43	48	49	41	50	41
Surma(U/S)	51	53	25	35	40	38	35	42	36	42	38	39
Surma(D/S)	65	60	42	45	45	41	43	48	46	46	42	41
DO (mg/L)												
Baluchar	5.8	6.9	5.5	4.9	6.1	6.5	6.7	6.2	6.6	6.4	6.6	6.7
Shibgonj	5.6	5.8	3.2	3.8	4.2	5.8	6.9	6.0	6.2	6.1	5.8	5.6
Uposhahar	5.1	5.3	6.3	5.8	5.3	5.5	6.2	5.8	5.6	5.9	5.5	6.8
Chalibondar	5.5	4.9	5.8	6.0	5.5	5.9	5.8	5.5	5.3	5.3	5.9	5.8
Masimpur	5.0	5.2	6.2	4.5	4.5	5.5	5.5	5.7	6.8	5.5	5.3	5.2
Surma(U/S)	6.7	6.7	6.7	7.1	7.8	6.9	6.8	7.0	6.3	7.0	6.4	6.7
Surma(D/S	6.1	6.2	5.2	5.5	6.2	6.5	7.2	6.5	6.3	6.7	6.0	6.3
NH3 (mg/L)												
Baluchar	0.42	.88	.45	.3	.35	.38	.28	.25	.29	.25	.45	.8
Shibgonj	1.1	1.6	.85	.55	.45	.45	.38	.55	.41	3.39	.55	1.2
Uposhahar	1.6	1.9	2.03	1.08	1.2	.89	.83	1.2	1.1	1.5	1.5	1.5
Chalibondar	1.85	2.5	3.08	2.03	2.2	1.87	2.1	1.8	1.75	1.65	1.75	2.3
Masimpur	2.4	2.95	2.51	2.1	2.5	1.98	1.75	1.9	1.65	1.45	1.85	2.54
Surma(U/S)	2.85	3.5	1.85	1.92	1.85	2.5	2.2	2.62	2.2	2.1	2.25	2.86
Surma(D/S)	3.11	3.85	4.05	4.35	3.5	3.65	3.25	3.85	3.25	2.75	3.75	3.33
pH												
Baluchar	7.2	7.1	7.3	7.2	7.1	7.2	7.1	7.2	7.1	7.2	7.3	7.2
Shibgonj	7.1	7.1	7.0	6.9	7.1	6.8	7.2	6.9	7.1	6.9	7.2	6.9
Uposhahar	7.0	6.9	6.7	6.9	6.8	6.85	7.0	6.9	6.8	7.0	6.95	7.1
Chalibondar	6.7	6.95	6.8	6.9	6.65	6.9	6.8	6.85	6.9	6.9	6.8	6.9
Masimpur	6.9	6.7	6.6	6.7	6.8	6.7	6.8	6.9	6.7	6.8	6.9	6.8
Surma(U/S)	7.2	7.0	7.3	7.2	7.0	7.1	7.3	7.5	7.2	7.1	7.3	7.2
Surma(D/S)	6.5	6.5	6.4	6.8	6.6	6.6	6.5	6.4	6.8	6.9	6.8	6.5
Turbidity (NTU)												
Baluchar	6.2	6.4	6.4	6.5	6.8	8.1	8.9	7.3	7.1	6.6	6.5	6.3
Shibgonj	5.4	5.5	5.3	5.4	6.8	6	6.4	6.5	6.5	6.8	6.1	5.4
Uposhahar	4.5	5.1	5.1	5.2	5.8	5.9	5.8	6	6.1	5.4	4.9	4.4
Chalibondar	5.4	5.4	5.7	5.8	5.5	6.3	6.4	6.5	6.2	6.1	5.8	5.4
Masimpur	5.5	5.7	5.8	5.9	5.8	6.3	6.6	6.8	7	6.4	6.1	5
Surma(U/S)	7.5	7.9	8.3	8.5	6	8.8	8.5	8.6	8.9	9	8	7.5
Surma(D/S)	8.5	8.4	8.3	8.5	8.6	8.8	8.5	8.6	8.9	9	8	7.5

Chalibonbar 2.365 mg/L, at Masimpur 2.49 mg/L, at Surma u/s 2.53 mg/L, at Surma d/s 3.84 mg/L. At every point except Goalichara u/s, ammonia contents are higher than expected. Following Figure 5 graphically shows the ammonia content at seven sampling points. Ammonia at Goalichara d/s is higher than Goalichara u/s due to animal waste, open defecation, chemica (particularly chemical fertilizer) and domestic waste water

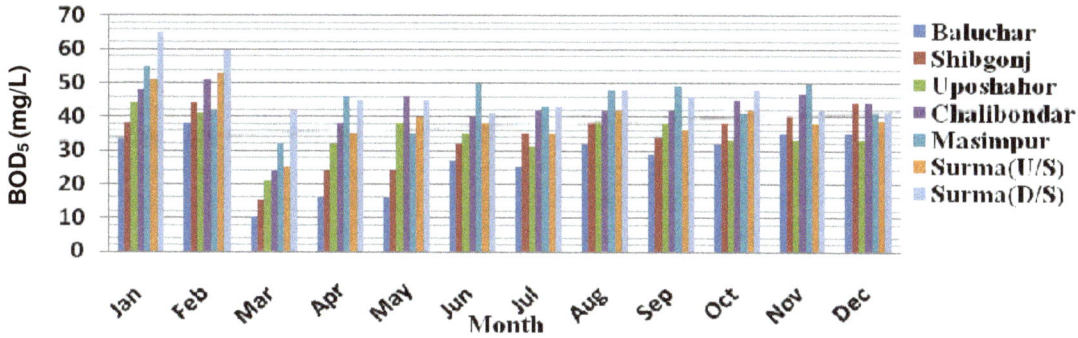

Figure 4. Variation of BOD$_5$ at different Location.

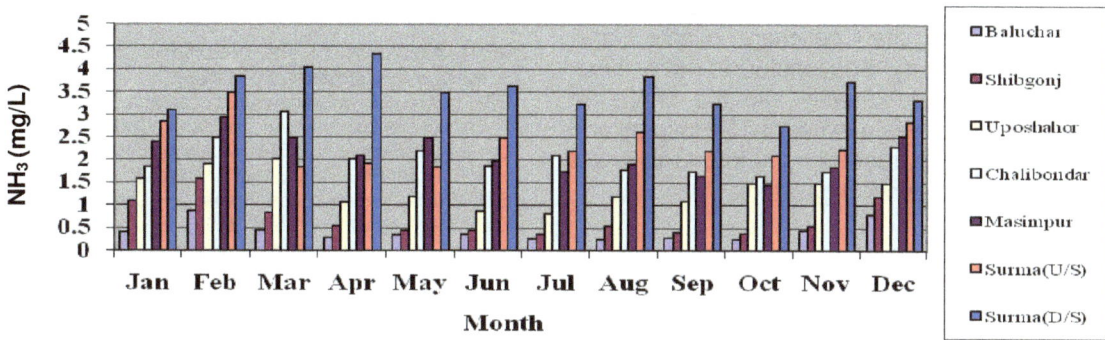

Figure 5. Variation of NH$_3$ at different location.

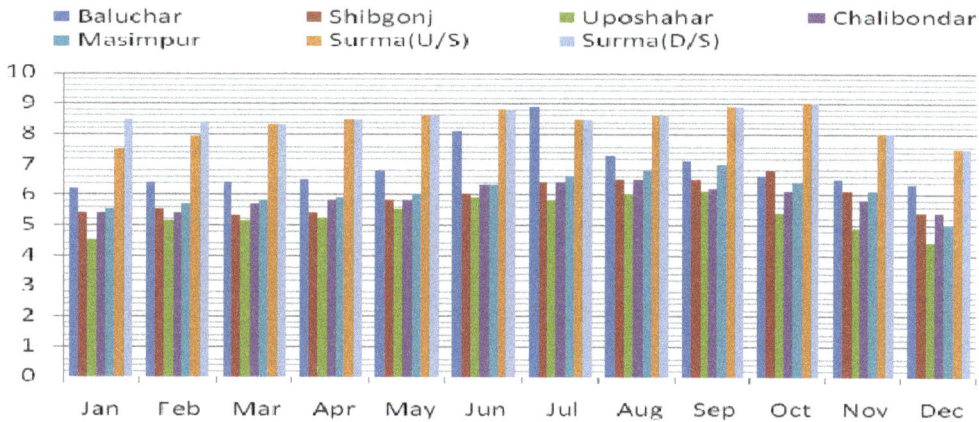

Figure 6. Variation of Turbidity (NTU) at different location.

discharged from household and hospital waste water discharge. At rainy season, concentration of ammonia decreases than the dry season. Desired value of ammonia according to Environment Conservation Rules ECR (1997) is 0.5 mg/L. At Goalichara u/s ammonia value is within the desired value but Goalichara d/s this value are higher than desired value because of urbanization impact.

Turbidity

Turbidity in and around Sylhet City ranges from 4 to 10 variation of turbidity (NTU) (Figure 6). Turbidity varies with season and usually remains high in flood period. Turbidity data procured by Institute of water modeling (IWM) from Department of environment DOE show significant variation between that of 2009 and 2011.

Encroachment

Encroachment of Goalichara is the main cause of reducing the width of the chara. Encroachment has occurred in the urban catchment of the chara. Local peoples are also reducing the channel width by filling the side of the channel with soil to meet the accommodation needs or for other purposes. In Baluchar, Shibgang, Uposhahor, Chalibandar and Masimpur encroachment have been found at the time of survey work.

Sewage and waste water disposal

As the sewerage system has not been established in the city, on site septic tank has been used for sewage treatment. However along the Goalichara, middle class, low income and temporary householders have been found to discharge sewage directly into the chara. In Baluchar and Uposhahor, waste water of TB Hospital and other private clinic is discharged directly into Goalichara deteriorating the quality of receiving waters.

Open defecation

Lack of proper maintenance and awareness people are not using the channel properly in many region of the channel; people are using channel as a place of defecation. They think it is less harmful and they are also habituated to this action. Children's open defecation in the Chara is a common scenario along the long way of Chara throughout the city. Lack of proper knowledge open defecation practices is very common to the children that can be easily reduced if parents of children are well conscious. Poor people of Chalibandar and Masimpur are evenly habituated with the open defecation in the channel.

Conclusion

The main waste water quality parameters such as DO, BOD_5, fecal coliform, ammonia and turbidity etc. have greatly exceeded the acceptable level of a good water source for water supply. The obtained values for pH ranges from 7.18 to 6.6, for total dissolved solids ranges from 162.75 to 328.75 mg/L, for dissolved oxygen from 6.24 to 5.28 mg/L, ammonia from 0.155 to 0.3333 mg/L, fecal coliform from 15.5 to 48.91N/100 ml and turbidity from 4 to 10 NTU. The results from data analysis show that, the water is certainly unfit for drinking purposes without any form of treatment.

REFERENCES

APHA, AWWA, WPCF (1998). Standard Methods for the examination of water and wastewater. 19th Edition.

ECR (1997). Environmental Conservation Rules (ECR), Department of Environment, Government of Bangladesh.

Lohani BN, Todino N (1984). Water quality index for Chao Praya River J. Environ. Eng. ASCE 110(6):1163.

SCC (2005). Annual report, Sylhet City Corporation, Bangladesh.

Zafor MA, Chowdhury MMH (2009). "Environmental condition and water quality of Goalichara in the vicinity of Sylhet city." B.Sc. Engineering Thesis, Department of Civil and Engineering, SUST, Sylhet.

14

Artificial neural network simulation of ground water levels in uplands of a coastal tropical riparian

Shivakumar. J. Nyamathi[1], Bharath. R[1] and A. V. Hegde[2]

Civil Engineering Department, UVCE, Bangalore University, Bangalore, India.
Applied Mechanics Department, NITK, Surathkal, Mangalore, India.

Wetlands play an important role in the ecological balance of the coastal region. Understanding groundwater level behaviour in uplands is important for the management and the development of coastal tropical riparian wetland. Artificial Neural Networks has proved to be robust techniques in modeling and prediction of hydrological processes. This paper presents the application of ANNs to model groundwater levels in uplands around a wetland environment. Weekly hydro meteorological observations have been used as an input to model groundwater fluctuation observed in sevel open wells in the region. A comparison of different training algorithms has also been carried out. The results obtained show that the use of Artificial Neural Networks in modeling the groundwater levels was successful. With Root Mean Square Error values in the range of 0.09 to 0.16, the study also reasserts that the same training algorithm need not provide the best results for different conditions.

Key words: Artificial neural networks, wetlands, stepwise linear regression.

INTRODUCTION

The Ramsar convention (Ramsar, 1971) defines wetlands as areas of soil covered by a shallow layer of seasonal of permanent, flowing or static, salt or freshwater. Wetlands can be natural or artificial and include areas of marine water. Riparian wetlands are the wetlands along lakes, rivers and streams. Riparian wetlands are very productive ecosystems providing vital habitat and hence their conservation is very important. Wetlands form a major ecological part of a watershed. It helps regulate the water levels within the watershed, helps in eutrophication of lakes, reduces flood and storm damages, and provides an important habitat for flora and fauna. Wetland management is an integral part of watershed management. These wetlands were not given their due importance till recently. These wetlands have been drained and converted to farmlands, filled up for housing and infrastructure thus reducing their area and their purpose. Anthropogenic activities continue to affect

the working of the wetland hydrology. The 'uplands' of a wetland is that region which is adjoining to the wetlands which are at a slightly higher altitude. Groundwater discharge to the wetlands usually occurs near the edge where the plains meet regional uplands. The wetland's surface water is dependent upon the groundwater levels of uplands. Drilling wells in upland to supply water for development or agriculture will reduce the ground water level and decrease the depth of water and the hydroperiod (the length of time the surface is inundated) in the nearby wetland. This decrease can cause changes in the structure and composition of the wetland community. The usual result following drainage of a wetland is a replacement of the plant and animal life adapted for deeper water and a longer hydroperiod with those species adapted for shallower water and/or shorter hydroperiods. Small wetlands are the most vulnerable to changes in water levels and small wetlands with short hydroperiods can easily be completely eliminated.

Although conceptual and physically based models are main tools for depicting hydrological variables and understanding the physical processes involved in the dynamics of groundwater levels, they do have practical limitations. When data is not sufficient and getting accurate responses is more important than conceiving the actual physics, empirical models remain a good alternative method, and can provide useful results without a costly calibration time. Artificial Neural Network (ANN) models are such 'black box' models with particular properties which are greatly suited to dynamic nonlinear system modeling (Coulibaly et al., 2001a).

The ANN technology is an alternate computational approach based on theories of the massive interconnection and parallel processing architecture of biological systems. The main theme of ANN research focuses on modeling of the brain as a parallel computational device for various computational tasks that were performed poorly by traditional serial computers. ANNs have a number of interconnected processing elements (nodes) that usually operate in parallel and are configured in regular architectures. The collective behavior of ANN, like a human brain, demonstrates the ability to learn, recall, and generalize from training patterns or data (Balkhair, 2002). Artificial neural networks (ANNs) have proved to be a robust tool that can be applied to simulation and prediction of non linear hydrological processes. Literature reviews reveal that the ANNs have been successfully used in modeling hydrological processes (ASCE, 2000a, b; Maier and Dandy, 1998, 2000; Aytek et al., 2008; Dawson and Wilby, 1998; Ahmed and Simonovic, 2005; Peters et al., 2006). In the groundwater modeling, ANNs have been used in prediction of water levels (Coulibaly et al., 2000; Daliakopoulos et al., 2005; Nayak et al., 2006) and aquifer parameter estimations (Lin and Chen, 2006).

The present study aims at modeling the water table fluctuations in the uplands of a coastal riparian wetland which are the main source of water to the wetlands, on a weekly basis, determining the ability of Artificial Neural Networks in simulating the water level fluctuations using only hydro-meteorological inputs and to evaluate the performance of different network types and training algorithms in modeling the hydrological process. A feed forward neural network is designed to observe water level fluctuations in eight wells in the study area and its performance evaluated. It is observed that the neural networks have successfully modeled the fluctuations with low RMSE values.

Study area and data description

Coastal Karnataka forms a part of the Malabar Coast in the south-west of India with a long coastline running 290 km, indented with promontories, headlands, picturesque estuaries, encompassing tidal wetlands essaying complex mangroves and long linear beaches. The study area is the humid tropical 'Padre Wetland' ($13°00'0''N$ to $13°01'4''N$ and $74°47'35''E$ to $74°48'35''E$), near to National Institute of Technology Karnataka (NITK) Surathkal, Mangalore city in Karnataka State of India (Figure 1). It is a coastal flood plain wetland of 1.5 km^2 has altitude range of +0.0 m to +4 m with respect to MSL in the lowlands of the Nandini sub-watershed, which is about 17 km^2 in area and at an elevation ranging from +0.0 to +68 m above mean sea level in the coastal region, around 21 km north of Mangalore. It is just upstream of a vented dam, which was constructed in the year 1965 to control salt-water intrusion.

There are mainly two types of soils in the study area: coastal alluvium and laterite soils. The coastal alluvium exists along with silt and clay, which is evident from laboratory tests. The clay deposits differ from each other that depend upon geological processes such as sea level changes, erosion of superimposed load and desiccation.

The hydraulic conductivity in the uplands ranges from $2 × 10^{-2}$ to $5 × 10^{-3}$ cm/s in lateratic formations and in the order of 10^{-4} cm/s in clayey formations. Since then, the humid tropical wetland complex has been degrading rapidly by conversion of Padre Wetland into agricultural, horticultural, residential and for other purposes by anthropogenic activities (Nyamathi, 2008).

Daily data of rainfall, evaporation, maximum temperature and average temperature were obtained for the study from the meteorological station at NITK, Surathkal. The daily data was transformed to weekly data to be given as input. This was done as the remaining input parameters namely: stream levels and well observations were taken on a weekly basis. Data of two years (May 2004 to May 2006) were used for the study. Antecedent data of two weeks were considered as inputs also.

ARTIFICIAL NEURAL NETWORKS

An ANN is a massively parallel distributed information processing system that has certain performance characteristics resembling biological neural networks of the human brain (Haykin, 1994). ANNs are arranged into three basic layers—*input, hidden* and *output*. The input nodes in this representation perform no computation but are used to distribute inputs into the network. This kind of network is called a feed forward network as information passes one way through the network from the input layer, through the hidden layer and finally to the output layer. Recurrent networks, such as Hopfield nets allow feedback between layers. Figure 2 provides an overview of ANN topology.

The following equation sums up the calculations that undergo in each neuron represented by Figure 3.

$$Sj = \sum_{i=1}^{n} w_{ij} x_j + b_{oj} \qquad (1)$$

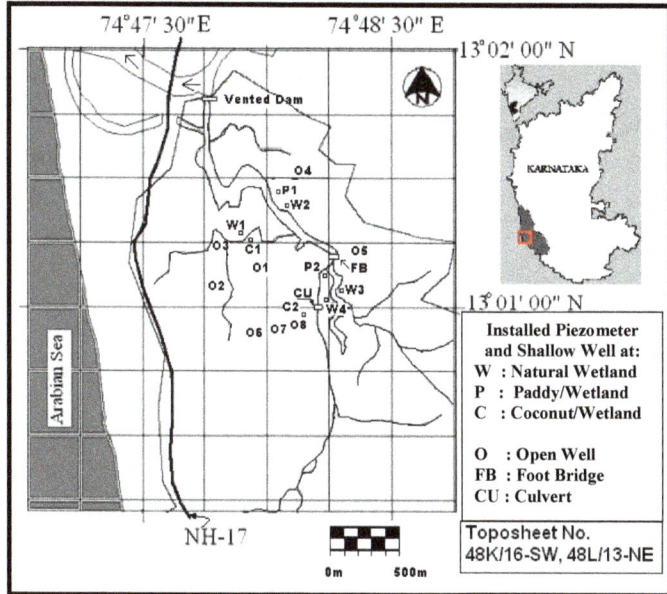

Figure 1. Location map of Padre Wetland.

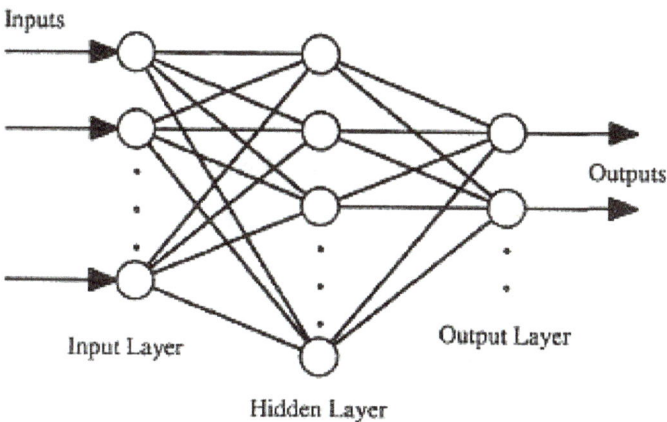

Figure 2. A basic overview of artificial neural network topology.

Where, w_{ij} is the weight which represents the strength or amplitude of a connection between two nodes. The final relationship between the input and the output of a network depends is detremined by the weights assigned to each neuron. The weights are initialized randomly and are updated by iterations using optimization techniques. , x_j the input at node j, and b_{oj} is called the bias term which is a constant. For the present study, the back – propogation algorithm was applied. An activation function is applied to the value S_j, to provide the final output from the neuron. This activation function can be linear, nonlinear, discrete, or some other continuous distribution functions. However, in order to use the back-propagation algorithm to train a network, this function must have the

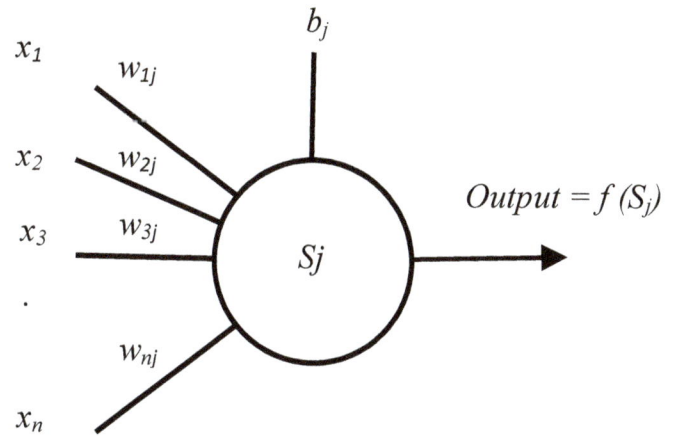

Figure 3. A single neuron.

property of being everywhere differentiable. The sigmoid function which is a smooth nonlinear activation function satisfies this criterion of having a positive derivative everywhere. It is the function generally used in most feed forward neural network applications and is represented by Equation 2. The readers are encouraged to refer to ASCE (2000a) for a detailed description of the backpropagation algorithm and the mathematics of weight updating and error backpropogation.

$$f(S_j) = \frac{1}{1 + e^{-S_j}} \tag{2}$$

METHODOLOGY

ANN model development

An optimal architecture may be considered the one that yields the best performance in terms of error minimization while retaining a simple and compact structure. No unified theory exists for determination of such an optimal ANN. A feed forward network with a single hidden layer can approximate any continuous function (ASCE, 2000a). For the present study, a static three layer feed forward back propagation neural network was used. The tan-sigmoidal transfer function was used for both the hidden layer and the output layer. The selection of network architecture and network inputs is discussed below.

Selection of input parameters

Not all of the potential input variables will be equally informative since some may be correlated, noisy or have no significant relationship with the output variable being modeled (Maier and Dandy, 2000). Any dependent variables (output parameters) may be a function of one or more independent variables (input parameters). The weekly rainfall, stream levels, average temperature, maximum temperature and evaporation along with their antecedent values of two weeks have been taken as input. To determine the optimum number of input parameters, Stepwise Linear Regression (SLR) was carried out between all the inputs and the output (Figure 4). The partial regression coefficient was

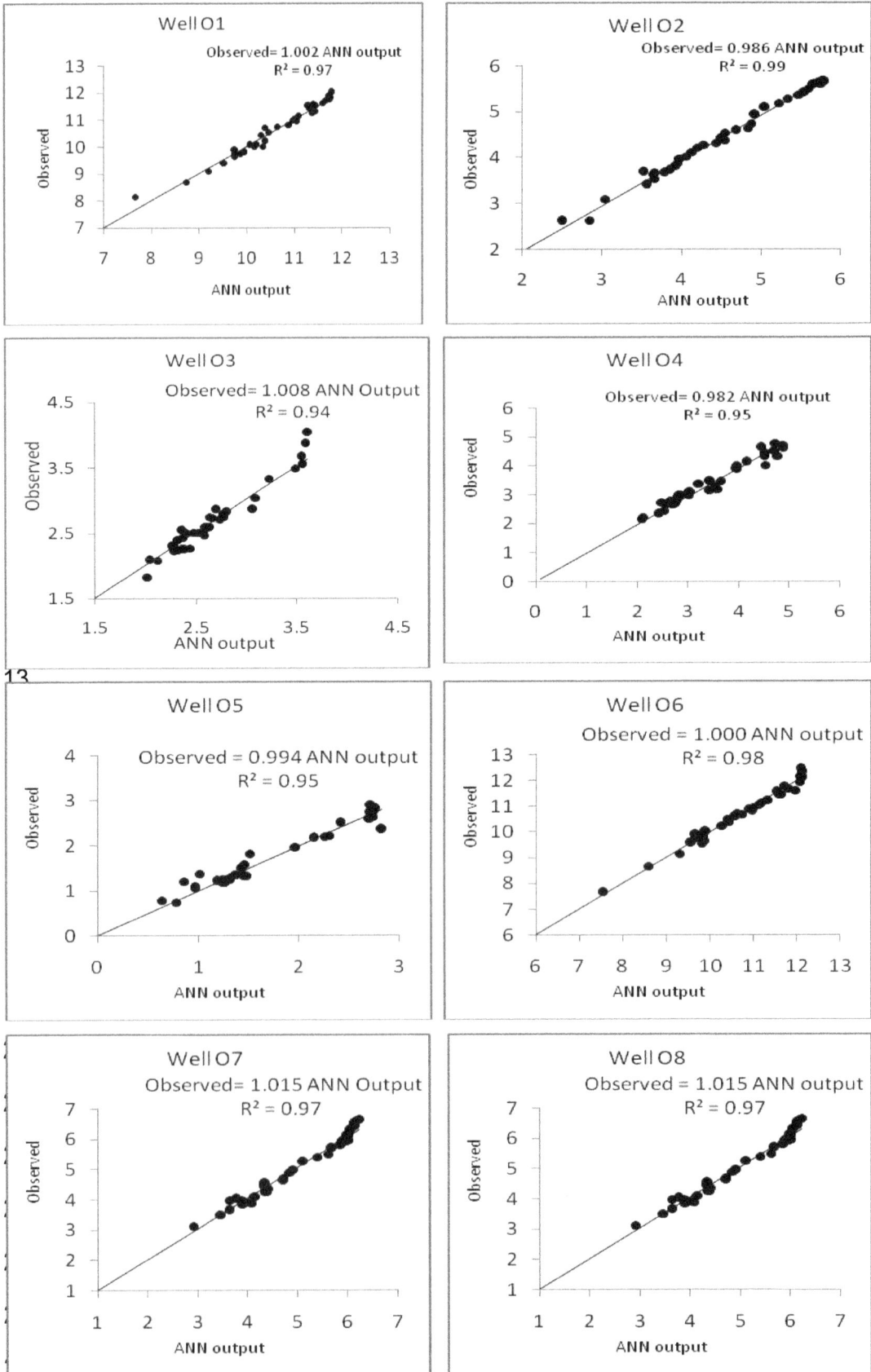

Figure 4. Regression plots between observed and obtained water levels for wells O1 to O7. An R^2 value close to 1 is considered a good fit between model output and observed data.

Table 1. Input variables considered for simulation studies. The variables considered are Rainfall (P), stream level (Str. Lvl.), Evaporation (E), Mean temperature (T_{mean}), Max. Temperature (T_{max}), water level at influencing wells (Lvl@Ox).

S/no.	Variable	Time steps at well O1	Time steps at well O2	Time steps at well O3	Time steps at well O4	Time steps at well O5	Time steps at well O6	Time steps at well O7	Time steps at well O8
1.	P	t, t−1	T, t−1, t−2	t, t−1, t−2	t, t−1, t−2	t, t−1, t−2	t, t−1	t	t, t−1
2.	Str. lvl	t−1, t−2	-	t−1, t−2	-	t−1, t−2	-	t, t−2	t
3.	E	t, t−1	t, t−1	t, t−1, t−2	t, t−1, t−2	t, t−1, t−2	t−1, t−2	t, t−1, t−2	t
4.	Tmean	t, t−1, t−2	t, t−1, t−2	t, t−1, t−2	t, t−1, t−2	t, t−1, t−2	t, t−1, t−2	t, t−1, t−2	−, −2
5.	Tmax	t−1, t−2	t, t−1, t−2	t, t−1, t−2	t, t−1	t, t−1, t−2	t−1, t−2	t−1, t−2	t−1, t−2
6.	Lvl@Ox	O2: t−1, t−2; O1: t−1	O2: t−1, t−2	O3: t−1, t−2	O4: t−1, t−2	O5: t−1, t−2	O7: t−1, t−2; O6: t−1	O7: t−1, t2	O7: t, t−1; O8: t−1, t−2

determined and the parameters which did not show any effect on the output were omitted. This procedure was carried out for each of the wells and the input neurons were determined. The major disadvantage with such an analytical method of input selection is that the nonlinear influence of the input parameters with the output cannot be determined and hence is should be used when there is no clear understanding of the physical relationships of the hydrological process. However in the present study, even with the input parameters that have only linear dependence the networks have generalized very well. SLR was carried out for each of the wells and input nodes determined. Table 1 provides information on the input parameters after identification of most influencing parameters using SLR.

Standardization of input parameters

Due to the nature of the sigmoidal function used in the backpropagation algorithm, it is prudent to standardize all input values before passing into the neural network. Without this standardization, large values input into an ANN would require extremely small weighting values which can cause a lot of problems (Dawson and Wilby, 1998). For the present study, the built-in function of the neural network toolbox of MATLAB 7.5 "mapminmax" was used to limit the inputs to the range of (−1, 1).

Network training

The network was trained using four training algorithms namely, the Levenberg-Marquardt algorithm (LM), the resilient backpropagation (RP), the scaled conjugate gradient (SCG) algorithm and the Broyden Fletcher Goldfarb Shanno (BFGS) quasi Newton algorithm. Each training algorithm uses the gradient of the performance function to determine how to adjust the weights to minimize the performance function. A detailed description of these training algorithms can be found in Demuth et al (2007). All the selected training algorithms have a speed faster than the conventional gradient descent algorithm.

Providing the complete data for training will result in redundant information as there will not be any new data to be provided as input to test the performance of the network. Hence the data was divided into two sets, one for training and the other for testing. The training set comprised of the first 15 months that had great variations as the data contained two monsoon seasons in it and the testing set comprised of 9 months which had relatively less variations.

Determination of hidden neurons

The size of the hidden layer influences the output significantly. Though many empirical relationships to determine the number of hidden neurons have been suggested (Maier and Dandy, 2000), since network architecture is always problem dependent, it is widely accepted that the neurons are best decided by trial and error. For the present study, the network was trained with only 2 hidden neurons and increased by a step size of one neuron until there was a reduction in the performance of the network or no increase in efficiency with neuron increase was observed in the network output.

Criteria for evaluation

The R^2 statistic and the root mean square error were the criteria used to evaluate the goodness of fit of the network. While the R^2 statistic gives the overall performance of the network, the RMSE provides the global goodness of fit.

RESULTS AND DISCUSSION

All network programming was carried out using the neural network toolbox available in Matlab 7.5. The number of input nodes varied from well to well based on the variable selection using SLR. Time lags up to 2 weeks were considered for the inputs. The number of hidden neurons was determined by trial and error. Three error measures namely, Root mean square error (RMSE) and the coefficient of determination (R^2) were used to determine the best fitting training algorithm. A low value of RMSE and high values of R^2 were considered to finalize a combination as the best fitting combination of architecture-training algorithm. The best fit combination of each of the network–algorithm combination has been shown in Table 1.

It is observed from Table 1 that the selected training algorithm is not consistent and varies from well to well. This also reconfirms the general notion that

Table 2. Optimum network architecture considered for each training algorithm.

Algorithm	LM		RP		SCG		BFGS	
Well no.	Architecture							
	R^2				RMSE (m)			
01	14 - 5 - 1		14 - 6 - 1		14 - 3 - 1		14 - 6 - 1	
	0.96	0.19	0.97	0.19	0.96	0.19	**0.97**	**0.16**
02	13 - 3 – 1		13 – 4 – 1		13 – 5 - 1		13 - 8 - 1	
	0.99	0.10	0.98	0.11	**0.99**	**0.09**	0.99	0.11
03	17 – 3 - 1		17 – 4 - 1		17 - 3 - 1		17 – 5 - 1	
	0.89	0.14	0.87	0.15	**0.94**	**0.12**	0.91	0.13
04	13 – 4 - 1		13 - 7 - 1		13 - 6 - 1		13 – 5 - 1	
	0.94	0.20	0.92	0.21	0.92	0.23	**0.95**	**0.18**
05	17 – 4 - 1		17 - 4 - 1		14 – 7 - 1		14 - 3 - 1	
	0.95	**0.15**	0.89	0.23	0.95	0.15	0.94	0.17
06	12 – 5 - 1		12 – 7 - 1		12 – 4 - 1		12 – 3 - 1	
	0.97	0.18	0.97	0.19	**0.98**	**0.18**	0.98	0.19
07	13 - 4 - 1		13 – 4 - 1		13 - 3 - 1		13 – 6 - 1	
	0.97	0.18	0.96	0.20	**0.97**	**0.17**	0.97	0.18
08	11 – 4 - 1		11 - 6 - 1		11 - 5 - 1		11 – 6 - 1	
	0.97	0.16	0.97	0.20	0.97	0.18	**0.97**	**0.17**

the ANN methodology is problem specific and its generalization is not advisable. The values obtained in Table 1 shows that the network has simulated the water level fluctuations satisfactorily with RMSE values ranging from 0.09 to 0.19 m for different wells. The algorithm that has provided fair results for more wells is the scaled conjugate gradient algorithm, followed by the BFGS quasi Newton algorithm (Table 2). Figure represents the regression plots between observed water levels and model outputs. An R^2 value greater than 90% is observed consistently thus showing the accuracy of ANN in simulating the water levels.

Conclusions

This study was carried out to determine the performance of ANNs in modeling water level fluctuations on a weekly basis. Results obtained have shown that the network has been able to model the process with low RSME values. The different network algorithms that have simulated the water levels indicate that that the same algorithm need not provide the same result for different conditions. It is

also observed that a clear understanding of the fielc conditions is also required for the decision of inpu parameters. Also it can be concluded that the network performs well even with antecedent conditions of only twc weeks considered for input along with the inpu parameters at current values. However this study provides an insight into the application of ANN in wetlanc management and further understanding of the physica processes will increase the performance of the network.

REFERENCES

Ahmed S, Simonovic SP (2005). An artificial neural network model for generating hydrographs from hydrometeorological parameters. J Hydrol. 315:236-251.

ASCE Task Committee (2000a). Artificial neural network in Hydrology I Preliminary concepts. J. Hydrol. Eng. 5(2):115-137.

Aytek AG, Yuce MI, Aksoy H (2008). An explicit neural network formulation for evapotranspiration. Hydrol. Sci. J. 53(4):893-904.

Bhalkhair KS (2002). Aquifer parameter determination for large diameter wells using Neural network approach. J. Hydrol. 265:118 128.

Coulibaly P, Anctil F, Aravena R, Bobe´e B (2001a). ANN modeling o water table fluctuations. Water Resour. Res. 37(4):885-896.

Coulibaly P, Anctil F, Bobee B (2000). Daily reservoir inflow forecasting using ANN. J. Hydrol. 230:244-257.

Daliakopoulos NI, Coulibaly P, Tsanis IK (2005). Groundwater level forecasting using artificial neural networks. J. Hydrol. 309:229-240.

Dawson CW, Wilby RL (1998). An artificial neural network approach to rainfall runoff modeling. Hyd. Sci. J. 43(1):47-66.

Demuth H, Beale M, Hagan M (2007). Neural network toolbox User's Guide, Mathworks Inc.

Haykin S (1999). Neural Networks, A Comprehensive Foundation. second edition. Prentice-Hall, Englewood Cliffs, NJ.

Lin GW, Chen GR (2006). An improved neural network approach to the determination of aquifer parameters. J. Hydrol. 316:281–289.

Maier HR, Dandy GC (1998). Understanding the behavior and optimizing the performance of back propagation neural network: An empirical study. Environ. Mod. Soft. 13:179-191.

Maier HR, Dandy GC (2000). Neural networks for prediction and forecasting of water resources variables: A review of modeling issues and applications. Environ. Mod. Soft. 15:101-124.

Nayak PC, Rao SYR, Sudheer KP (2006). Groundwater level forecasting ina a shallow aquifer using artificial neural networks. Water Res. Manage. 20:77-90.

Nyamathi SJ (2008). Characterizing Hydrological Responses of Coastal Humid Tropical Wetland. Unpublished Ph.D thesis, NITK Surathkal, Karnataka, India.

Peters R, Schmitz G, Cullmann J (2006). Flood routing modeling using ANN. Adv. Geol. 9:131-136.

Ramsar Convention (1971). www.ramsar.org/ ris/ key_ris_types.htm, viewed on 28/04/09.

15

Assessment of groundwater quality for irrigation of Bhaskar Rao Kunta watershed, Nalgonda District, India

K. Srinivasa Reddy

Central Research Institute for Dryland Agriculture (CRIDA), Santoshnagar, Hyderabad-500 059, India.

Semi-arid region of Bhaskar Rao Kunta watershed groundwater quality was evaluated for suitability irrigation; in this situation twenty groundwater samples were collected at identical locations from deeper bore wells. The American Public Health Association (APHA) standard methods were followed and the concentrations of physicochemical parameters of pH, electrical conductivity (EC), total hardness (TH), Ca^{2+}, Mg^{2+}, Na^+, K^+, HCO_3^-, SO_4^{2-}, Cl^- and NO_3^- analyzed. The results of the concentrations were interpreted and measured with different irrigation indexes like EC, sodium percent (SP), sodium adsorption ratio (SAR), residual sodium carbonate (RSC), permeability index (PI) and Kelly's ratio (KR). The interpreted results were indicated that the groundwater quality stands on EC values; 20 and 80% of the samples fall under medium to high salinity category in pre and post-monsoon seasons and stands on sodium percent (SP) values, 25 and 75% of the samples fall under excellent to good category in both seasons. The remaining indexes SAR, RSC and KR values stands on 100% of the samples and fall under the excellent and excellent to good category in both seasons. Hence, the indexes results were concluded that the quality of groundwater in general was suitable for irrigation.

Key words: Water quality, irrigation, quality indexes.

INTRODUCTION

The ground water quality study reveals that the water is suitable for drinking, agricultural and industrial purposes. Particular in arid and semi-arid areas their natural ground water resources were used as poor quality of water for irrigation. In this context irrigation water quality is important for successful crop production. The poor quality of the irrigation water may affect crop yields and soil physical conditions (Talukder et al., 1998). For example, the average yield of wheat decreased by 24% (Datta et al., 2000), rice decreased by 39% (Bai, 1988), vegetables decreased by 30% (Chang et al., 2001), and corn decreased by 21% (Lindhjem, 2007) over normal yield when poor quality water was used. The major irrigation water is judged by four important measures of salinity hazard, sodium hazard, toxicity hazard and residual sodium carbonate hazard (Michael, 1978).

In India unfortunately, salinity hazards is extensive

irrigation regions problem. In addition, different crops require different irrigation water qualities. Therefore testing the irrigation water is prior to contribute to effective management and utilization of the groundwater resources by clarifying relations among many hydrogeological considerations. In the present study, the physiochemical quality of groundwater has assessed and dissimilar index methods which were used like EC, SP, SAR, RSC, PI and KR with reference to their suitability for irrigation.

Study area

Semi-arid region of Bhaskar Rao Kunta watershed geographically lies between northern latitudes from 16° 42' 25" to 16° 37' 58" and eastern longitudes from 79° 28'

15" to 79° 32' 30" of the Krishna lower basin. The watershed elevation ranges between 80 to 140 m above the mean sea level (MSL), slightly undulating terrain with slight to moderate slopes (2 to 3%) and annual normal rain fall ~ 737 mm. The average maximum and minimum temperature was 40 and 28°C respectively. The drainage system was showing dendrite to sub-dendrite pattern, governed by regional slope, homogenous lithology and relief, exhibited by 146 streams and were curved. It contributed to the flow of mostly dry season except for seasonal run-off, which could be either due to structural or topographic control (Figure 2). The study area was 40.25 km^2 out of which 71% of the land was under cultivated, in this 32% of the area was under bore well irrigation, the remaining 39% was under tank and canal irrigation, the major agricultural crops were chilli, paddy, sun flower and cotton.

Geology

Geologically the study area consisted of the Kurnool group of Palnadu sub basin and partially covered by Srisailam succession of Kadapa super group (Figure 1). Srisailam sub basin was exposed with Quartzites rocks. The Quartzites rocks were inter-bedded with thin siltstone units and were usually thick bedded, dense and fine to medium grained. Palnadu sub-basin was exposed with Calcareous (chemical precipitates) sedimentary rocks like quartzites, shales and flaggy-massive limestones with covered red and red sandy soils. General sequence of sub-surface strata was encountered on the top soil, weathered/semi weathered, and shale/quartzite.

MATERIALS AND METHODS

Collection of water samples

Twenty groundwater samples were collected from deeper bore wells (average depth 60 m bgl) in pre and post monsoon seasons at identical same locations in June and December 2009 year, according to prerequisite for the analysis. Locations of sampling points were determined using a Global Positioning System (GPS) (Figure 2). Samples were collected after one hours of pumping and the screen interval of the well represents the average sample depth. The samples were collected in 1000 mL plastic bottles and field filtration was carried out through filter papers to remove suspended solids. They were then carefully sealed, labeled and taken for analyses.

Analytical procedure

Collected samples were analyzed in the laboratory to measure the concentration of the quality parameters using American Public Health Association standard methods (APHA, 1995). pH, EC, TH, Ca^{+2}, Mg^{+2}, Na^+, K^+, CO^{-2}_3, HCO^-_3, NO_3, SO^{-2}_4, and Cl^- were the major ions in groundwater of the study area. Calcium and Magnesium concentrations were determined by Ethylenediaminetetraacetic acid (EDTA) titration using Eriochrome black-T as indicator. Sodium and potassium concentrations were

determined by using a flame photometer. Chloride concentration was measured by silver nitrate titration. Carbonate and bicarbonate concentrations were measured by acid-base titration. Sulphate and nitrate concentrations were measured by using colorimetric-spectrophotometer. The accuracy of the analysis for major ions was cross checked from the ionic balance was within ±7% for all the samples, ions were converted from milligram per litre to milliequivalent per litre. Correlation of geochemical data has been attempted as presented in Tables 1, 2 and 3. The concentrations were interpreted and calculated with irrigation indexes using the following formula of SP, SAR, RSC, PI and KR as follows:

Sodium percentage

This was calculated employing the equation (Todd, 1995) as:

$$Na\% = \frac{(Na^+ + K^+) \div (Ca^{2+} + Mg^{2+} + Na^+ + K^+)}{100}$$

(Concentrations are in meq/L).

Sodium absorption ration

This was calculated employing the equation (Raghunath, 1987) as:

$$SAR = \frac{Na^+}{\sqrt{\frac{(Ca^{2+} + Mg^{2+})}{2}}}$$

(Concentrations are in meq/L)

Residual sodium carbonate

This was calculated employing the equation (Eaton, 1950) as:

$$RSC = [(CO_3^{2-} + HCO_3^-) - (Ca^{2+} + Mg^{2+})]$$

(Concentrations are in meq/L)

Permeability index

This was calculated employing the equation (Domenico, 1990) as:

$$PI = \frac{[(Na^+ + HCO_3^-) \div (Ca^{2+} + Mg^{2+} + Na^+)]}{100}$$

(Concentrations are in meq/L)

Kelly's ratio

This was calculated employing the equation (Kelly, 1963) as:

$$KR = \frac{Na^+}{(Ca^{2+} + Mg^{2+})}$$

(Concentrations are in meq/L)

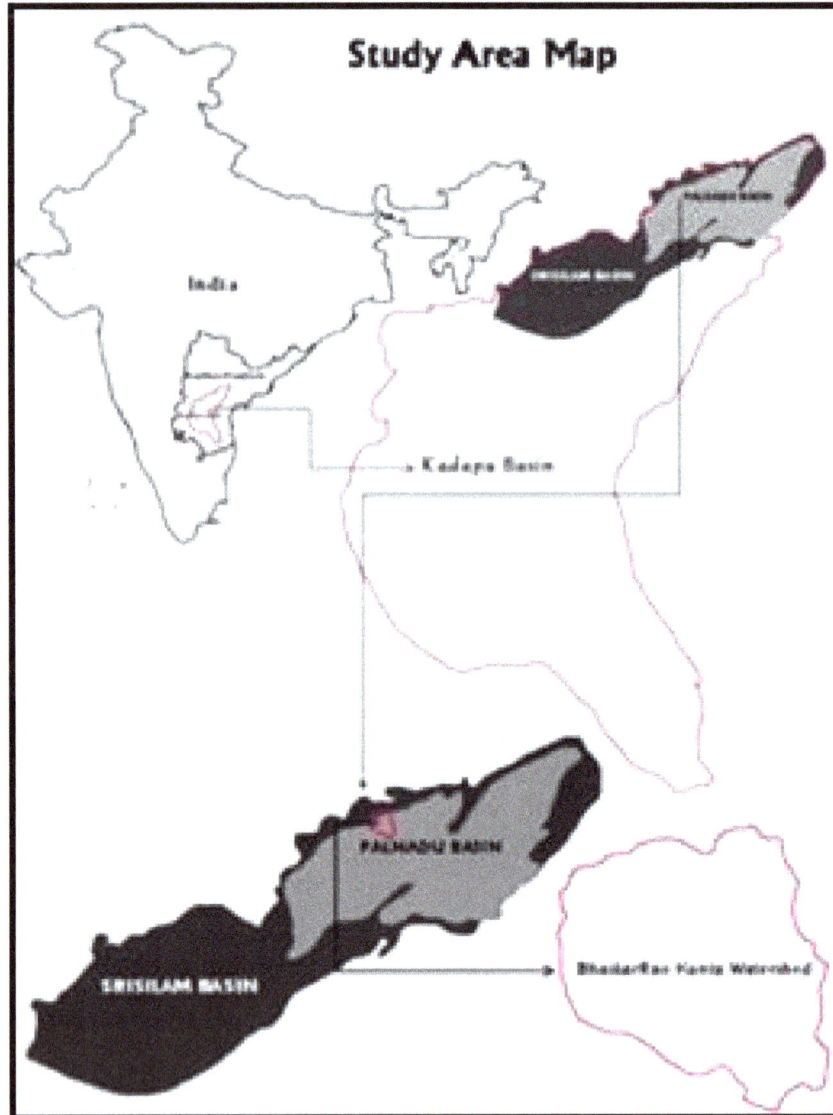

Figure 1. Map showing location of the study area.

RESULTS AND DISCUSSION

The results are obtained from hydrogeochemical analysis of water samples of study area as presented in Table 1. The calculated parameter results were presented in Table 2 and summary statistics of different indexes of groundwater quality are presented in Table 3.

Electrical conductivity

It was a measurement of all soluble salts in samples, the most significant water quality standard on crop productivity which was the water salinity hazard. The primary effect of high EC water on crop productivity was the failure of the plant to compete with ions in the soil solution for water. The higher the Ec, the lesser the water available to plants, even though the soil may show wet because plants can only transpire "pure" water; useable plant water in the soil solution decreases significantly as Ec increases. The amount of water transpired through a crop was directly related to yield; therefore, irrigation water with high EC reduces yield potential. In the study area, the classification for EC is given (Handa, 1969) in Table 4. It indicated that overall the water quality was medium to high EC category.

Sodium percentage

Sodium hazard was an important factor in irrigation water quality. The use of high percentage sodium of water for

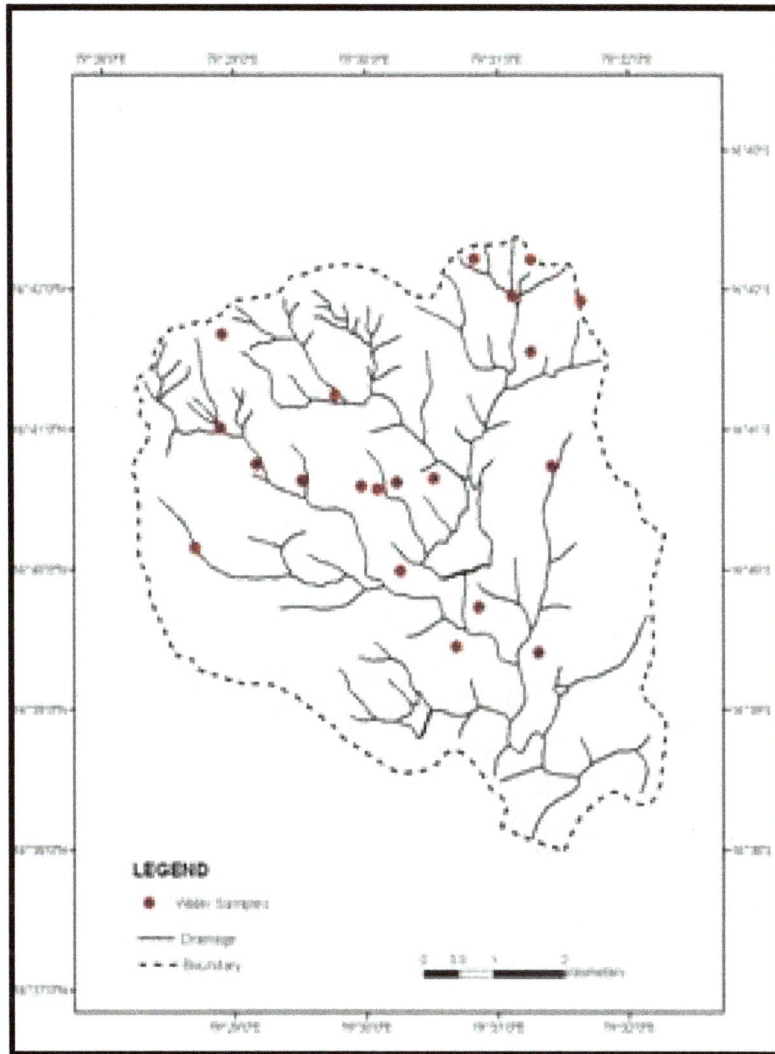

Figure 2. Groundwater Samples Locations Map.

irrigation was stunts, the plant growth and sodium reacts with soil to reduce its permeability (Joshi et al., 2009). The finer the soil texture and the greater the organic matter content, the greater the impact of sodium on water infiltration and aeration. The classification for SP was given (Wilcox, 1955) in Table 5. It is indicating the overall ground water quality of the samples which are falling under excellent to good category in pre and post-monsoon seasons.

Sodium adsorption ratio

It was a significant parameter for the determination of suitability of irrigation water; excess sodium in water produces the undesirable effects of changing soil properties and reducing soil permeability (Biswas et al., 2002). The measure to which irrigation water tend to

penetrate into cation-exchange reactions in soil can be indicated by the sodium adsorption ratio, sodium replacing adsorbed calcium and magnesium was a hazard as it causes damage to the soil structure, it becomes compact and impervious. In the study area all the groundwater samples have SAR values within the excellent class and acceptable for irrigation. The classification for SAR as is given (Richards, 1954) in Table 6.

Residual sodium carbonate

The concentration of bicarbonate and carbonate also influences the suitability of water for irrigation purpose. One of the empirical approaches was based on the assumption that calcium and magnesium precipitate as carbonate, considering this hypothesis (Eaton, 1950)

Table 1. Results of hydrogeochemical analysis of groundwater samples.

S.No	Location		Pre-monsoon										Post-monsoon									
	Longitude	Latitude	pH	Ca^{+2}	Mg^{+2}	Na^+	K^+	HCO_3^-	SO_4^-	Cl^-	NO_3^-	TH	pH	Ca^{+2}	Mg^{+2}	Na^{+2}	K^+	HCO_3^-	SO_4^-	Cl^-	NO_3^-	TH
1	79.48184	16.69462	8.3	96	72	72	0.8	199	23	57	25	407	7.9	103	78	75	0.8	206	27	61	29	418
2	79.52091	16.69261	8.5	79	83	120	1.0	231	39	37	26	446	8.4	82	89	122	1.0	241	41	39	28	450
3	79.51853	16.69924	8.1	88	87	80	1.0	219	33	95	25	495	8.1	92	89	81	0.9	226	34	97	29	500
4	79.50391	16.67703	7.8	127	88	110	1.0	192	28	115	26	423	7.6	132	89	109	0.9	195	29	120	26	425
5	79.50153	16.67625	8.1	97	65	103	0.8	184	31	50	23	422	8.0	101	68	106	0.7	189	32	51	24	425
6	79.49611	16.68746	7.7	140	94	81	1.0	237	23	110	40	523	7.4	144	97	86	1.1	244	25	113	41	525
7	79.49947	16.67665	8.0	116	100	70	0.8	217	19	60	29	396	7.9	118	103	74	0.8	220	19	61	30	400
8	79.49188	16.67732	8.1	118	72	81	0.8	241	31	95	30	446	7.9	122	73	82	0.7	247	32	96	32	450
9	79.52731	16.69863	7.9	108	53	89	0.7	165	23	93	29	323	7.8	112	53	91	0.6	168	26	96	31	325
10	79.51375	16.70373	8.5	86	91	81	0.6	195	27	47	21	397	8.1	89	92	82	0.5	198	29	48	22	400
11	79.52091	16.70356	8.1	91	67	75	1.0	221	43	38	35	370	7.9	92	69	74	1.0	229	46	39	39	375
12	79.47841	16.66934	8.1	95	102	102	0.9	179	25	30	40	573	7.9	96	107	106	0.9	183	26	31	41	575
13	79.48162	16.68361	8.4	73	56	63	0.7	177	19	35	38	349	8.2	72	58	66	0.7	183	19	36	39	350
14	79.48616	16.67927	8.8	130	76	71	0.8	196	39	73	24	447	8.1	131	76	72	0.7	198	41	76	26	450
15	79.52356	16.67901	8.5	85	69	106	1.0	260	36	46	17	347	8.3	88	71	110	0.9	262	38	47	18	350
16	79.51143	16.65753	7.9	114	70	80	1.0	244	37	102	49	396	7.9	116	73	83	0.9	250	39	109	51	400
17	79.51427	16.66217	8.5	151	91	77	1.0	236	35	54	20	520	7.8	152	93	78	1.1	241	37	59	21	525
18	79.50864	16.67751	8.3	124	95	80	0.9	191	24	97	31	542	7.9	128	96	81	0.7	189	25	99	32	550
19	79.50442	16.66661	8.7	97	97	91	1.0	211	27	98	41	497	8.2	101	98	93	1.0	214	29	106	43	500
20	79.52186	16.65682	9.0	119	88	91	1.5	237	38	83	34	536	8.0	127	94	97	1.0	243	42	87	40	550
Mean			8.3	107	81	86	0.9	212	30	71	30	443	8.0	110	83	88	0.8	216	32	74	32	447
Minimum			7.7	73	53	63	0.6	165	19	30	17	323	7.4	72	53	66	0.5	168	19	31	18	325
Maximum			9.0	151	102	120	1.5	260	43	115	49	573	8.4	152	107	122	1.1	262	46	120	51	575

Units are expressed in EC (µS/cm); except all are in mg/L.

proposed by the concept of residual sodium carbonate (RSC) for the measurement of high carbonate waters. The classification for RSC is given (Richards, 1954) in Table 7. In the present study area, RSC values are within the falling safe category in pre and post monsoon seasons respectively, hence, all water samples are considered safe for irrigation.

Permeability index

The soil permeability is affected by long term use of irrigation water. A criterion for assessing the suitability of water for irrigation was based on PI water and can be classified as class I, Class II and Class III orders. Class I and Class II water was categorized as good for irrigation with 75% or

more maximum permeability. Class III water was unsuitable with 25% of maximum permeability (Doneen, 1964; Raghunath, 1987). In the present study area the minimum and maximum permeability is 38 and 61%, 37 and 60% in pre and post monsoon seasons respectively in Table 3; hence, the groundwater quality was suitable for irrigation.

Table 2. Calculated parameters indexes for irrigation quality.

S.No	Location		Pre-monsoon						Post-monsoon					
	Longitude	Latitude	EC (µS/om)	SP (%)	SAR	RSC (meq/L)	PI (%)	KR (meq/L)	EC (µS/cm)	SP (%)	SAR	RSC (meq/L)	PI (%)	KR (meq/L)
1	79.48184	16.69462	859	23	1.35	-7.45	46	0.29	862	22	1.36	-8.18	45	0.28
2	79.52091	16.69261	795	33	2.25	-6.99	56	0.48	798	32	2.22	-7.47	55	0.46
3	79.51853	16.69924	840	23	1.45	-7.96	47	0.30	855	23	1.44	-8.22	47	0.30
4	79.50391	16.67703	1300	26	1.84	-10.44	43	0.35	1304	26	1.80	-10.71	43	0.34
5	79.50153	16.67625	751	31	1.98	-7.18	51	0.44	758	30	2.00	-7.54	51	0.43
6	79.49611	16.68746	1349	19	1.30	-10.84	41	0.24	1355	20	1.36	-11.17	41	0.25
7	79.49947	16.67665	873	18	1.15	-10.45	39	0.22	876	18	1.20	-10.76	39	0.22
8	79.49188	16.67732	965	23	1.45	-7.86	49	0.30	967	23	1.45	-8.05	49	0.29
9	79.52731	16.69863	689	29	1.75	-7.04	48	0.40	698	29	1.77	-7.20	48	0.40
10	79.51375	16.70373	781	23	1.45	-8.58	44	0.30	782	23	1.46	-8.76	44	0.30
11	79.52091	16.70356	829	25	1.46	-6.43	52	0.32	830	24	1.42	-6.52	52	0.31
12	79.47841	16.66934	661	25	1.73	-10.20	42	0.34	662	25	1.77	-10.60	42	0.34
13	79.48162	16.68361	650	25	1.35	-5.34	51	0.33	650	26	1.40	-5.37	52	0.34
14	79.48616	16.67927	852	20	1.22	-9.52	40	0.24	849	20	1.24	-9.54	40	0.24
15	79.52356	16.67901	1001	32	2.07	-5.66	61	0.46	1008	32	2.12	-5.93	60	0.47
16	79.51143	16.65753	1421	23	1.45	-7.45	50	0.30	1428	24	1.49	-7.70	50	0.31
17	79.51427	16.66217	1453	18	1.22	-11.15	39	0.22	1460	18	1.23	-11.29	39	0.22
18	79.50864	16.67751	1027	20	1.32	-10.87	38	0.25	1031	20	1.32	-11.19	37	0.25
19	79.50442	16.66661	901	24	1.56	-9.36	44	0.31	910	24	1.58	-9.61	44	0.31
20	79.52186	16.65682	1497	23	1.54	-9.29	46	0.30	1526	23	1.59	-10.10	45	0.30

Table 3. Summary statistics of different indexes of groundwater quality.

Parameter	Pre-monsoon				Post-monsoon			
	Mean	Minimum	Maximum	SD	Mean	Minimum	Maximum	SD
EC (µS/cm)	975	650	1497	285	980	650	1526	285
SP (%)	24	18	33	4	24	18	32	4
SAR	1.56	1.15	2.25	0.31	1.57	1.20	2.22	0.30
RSC (meq/L)	-8.56	-11.15	-5.34	1.84	-8.83	-11.29	-5.37	1.89
PI (%)	46	38	61	6	46	37	60	6
KR (meq/L)	0.32	0.22	0.48	0.08	0.32	0.22	0.47	0.07

Table 4. Irrigation water quality based on Ec values.

EC (µS/cm)	Class	Samples falling in dissimilar seasons			
		Pre-monsoon		Post-monsoon	
		%	No. of samples and Samples No's	%	No. of samples and Sample's No's
0-250	Low	Nil		Nil	
251-750	Medium	20	4 (5,9,12,13)	20	4 (5,9,12,13)
751-2250	High	80	16 (except 5, 9,12,13)	80	16 (except 5, 9,12,13)
2251-6000	Very high	Nil		Nil	

Kelly's ratio

Based on Kelly's ratios (Kelly, 1963) ground water was classified for irrigation, Kelly's ratio was more than 1 indicating an excess level of sodium in water; therefore the water Kelly's ratio of less than 1 was suitable for irrigation. In the study KR values fall within the safe category in pre and post monsoon seasons in Table 8;

Table 5. Classification of water based on SP values.

SP (%)	Class	Samples falling in dissimilar seasons				
		Pre-monsoon		Post-monsoon		
		%	No. of samples and Samples No.	%	No. of samples and Samples No.	
< 20	Excellent	25	5 (6,7,14,17,18)	25	5 (6,7,14,17,18)	
20-40	Good	75	15 (except 6,7,14,17,18)	75	15 (except 6,7,14,17,18)	
40-60	Permissible	Nil		Nil		
60-80	Doubtful	Nil		Nil		
> 80	Unsuitable	Nil		Nil		

Table 6. Classification of water based on SAR values.

SAR (value)	Class	Samples falling in dissimilar seasons			
		Pre-monsoon		Post-monsoon	
		%	No.of samples and Samples No.	%	No.of samples and Samples No.
<10	Excellent	100	All samples	100	All samples
10-18	Good	Nil		Nil	
18-26	Fair	Nil		Nil	
>26	Poor	Nil		Nil	

Table 7. Classification of water based on RSC values.

RSC (meq/L)	Class	Samples falling in dissimilar seasons			
		Pre-monsoon		Post-monsoon	
		%	No. of samples and Samples No.	%	No. of samples and Samples No.
<1.25	Safe	100	All samples	100	All samples
1.25-2.5	Marginal	Nil		Nil	
>2.5	Unsuitable	Nil		Nil	

Table 8. Classification of water based on KR values.

KR meq/L	Class	Samples falling in dissimilar seasons			
		Pre-monsoon		Post-monsoon	
		%	No. of samples and Samples No.	%	No. of samples and Samples No.
<1	Safe	100	All samples	100	All samples
>1	Unsuitable	Nil		Nil	

hence, the groundwater quality suitable for irrigation.

Conclusions

Evaluation of groundwater quality for irrigation were carried out using different index methods like SP, SAR, RSC, PI, KR and EC; among these, majority of index results were similar to SP, SAR, RSC, PI, and KR implying that the 100% of the groundwater samples fall under excellent and were excellent to good category in pre and post monsoon seasons. But, only based on EC, 80% of the samples fall under the high salinity category (751 to 2250 µS/cm); it is suitable for horticultural crops. Therefore, the results were concluded, that the study area groundwater quality was in general suitable for irrigation. Observed from the analyzed results of groundwater quality was diminutive and changed due to monsoon impacts of a lesser amount of rain fall, runoff, infiltration and rock water interaction (geogenic reaction) in the study area.

REFERENCES

APHA (1995). Standard Methods for the Examination of Water and Waste Water. 19th edition, Washington DC, American Public Health Association.

Bai Y (1988). Pollution of Irrigation Water and its Effects, Beijing, China, Beijing Agriculture University Press.

Biswas SN, Mohabey H, Malik ML (2002). Assessment Of The Irrigation Water Quality Of River Ganga In Haridwar District. Asian J. Chem., 16.

Chang Y, Hans MS, Haakon V (2001). The Environmental Cost of Water Pollution in Chongqing, China. Environ. Dev. Econ. 6:313-333.

Datta KK, Dayal B (2000). Irrigation with Poor Quality Water: An Empirical Study of Input Use, Economic Loss and Cropping Strategies. Indian J. Agric. Econ. 55(1):26-37.

Domenico PA, Schwartz FW (1990). Physical and Chemical Hydrology. John Wiley and Sons, New York, pp. 410-420.

Doneen LD (1964). Notes on Water Quality in Agriculture, Department of Water Science and Engineering, University of California, Water Science and Engineering, p. 400.

Eaton FM (1950). Significance of Carbonate in Irrigation Water. Soil Sci. 67:112-133.

Handa BK (1969). Description and Classification of Media for Hydrogeochemichal Investigation. Symposium on Groundwater Studies in Arid and Semi-arid Regions.

Joshi DM, Kumar A, Agrawal N (2009). Assessment of the Irrigation Water Quality of River Ganga in Haridwar District India. J. Chem. 2(2):285-292.

Kelly WP (1963). Use of Saline Irrigation Water. Soil Sci. 95(4):355-39.

Lindhjem H (2007). Environmental Economic Impact Assessment in China: Problems and Prospects. Environ. Impact Assess. Rev. 27(1):1-25.

Michael AM (1978). Irrigation Theory and Practice. Vikas Publishing House Pvt.Ltd, New Delhi, pp. 713-713.

Raghunath HM (1987). Groundwater, 2nd Ed. Wiley Eastern Ltd. New Delhi, India, pp. 344-369.

Richards LA (1954). Diagnosis and Improvement of Saline and Alkali Soils. USDA and IBH Pub, Coy Ltd., New Delhi, India. Agric. Handbook 60:98-99.

Todd DK (1995). Groundwater Hydrology. John Wiley and Sons Publications, 3rd Ed, New York.

Talukder MSU, Shirazi SM, Paul UK (1998). Suitability of Groundwater for Irrigation at Kirimganj Upazila Kishoreganj. Progress Agric. 9:107-112.

Wilcox LV (1955). Classification and Use of Irrigation Water, Washington: US Department of Agriculture. Circular No.969, p. 19.

16

Water quality of domestic wells in typical African communities: Case studies from Nigeria

B. A. Adelekan

Department of Agricultural and Mechanical Engineering, College of Engineering and Technology, Olabisi Onabanjo University, P. M. B. 5026, Ibogun, Ogun State, Nigeria. E-mail: jideadelekan@yahoo.com.

Several African communities obtain their domestic water supplies from dug wells and this study takes Ibogun, Pakoto and Ifo communities in Ogun State, Nigeria as being typical. The objective is to assess the quality of the water supply from these dug wells, ascertain the contamination problems that may confront the consumers, and provide appropriate remedies. Twenty dug wells were randomly selected in each community. Water samples from the selected dug wells were collected during July - August 2009. The samples were checked for odour, colour and taste; and through standard methods, they were analyzed for pH, total solids, total hardness, chlorides, sulphate, nitrates, magnesium, calcium, manganese, sodium, copper, zinc, iron and lead; total viable count, total coliform count, faecal coliform count and faecal streptococci count. Most of the pH values of the samples were outside the recommended range of 6.5 – 8.5 for drinking water. Predominantly, the ionic dominance pattern observed were Na > Ca > Mg and HCO_3 > Cl > SO_4, indicating typical cationic characteristics and anionic characteristics of groundwater. For total solids and total hardness, guideline values were largely met. Levels of iron did not exceed the WHO guideline value of 0.2 mg/l for Fe in drinking water. Mean levels of Mn measured were far in excess of the average of 0.01 mg/l for fresh water, while in relation to the WHO guideline value of 0.4 mg/l for Mn in drinking water, the levels measured were low. Generally, the levels of nitrates, sulphate, chlorides, magnesium, manganese and iron were moderately high, but the WHO guidelines were not exceeded. The WHO guidelines for microbiological quality of water were met in several cases. Matching of non-technical and techno-social remedial measures is recommended. These include sensitization of the populace on merits of qualitative domestic hygiene and environmental protection practices such as cleaner compounds and strict enforcement of environmental protection laws.

Key words: Dug wells, water, physico-chemical analyses, microbial analyses.

INTRODUCTION

In many parts of Nigeria and several other African countries, piped water supply is either unavailable or irregular especially in the small-sized communities and towns. Even in most Nigerian cities, the supply of water for domestic purposes has several accompanying inadequacies. According to Sangodoyin (1993), reasons given for these inadequacies include enormous socio economic rate of development, a growing industrial base, poor planning, insufficient funding, haphazard implementation of programs, lack of maintenance culture as well as technically deficient personnel. Adedeji (2001) surveyed the water supply pattern in Lagos, Nigeria and

found that for households connected to the public water supply, just 75.2 L per capita per day were assured. Those not connected to the public supply had far less. Furthermore, the national average for access to drinking water was estimated at 32% of households. Even at that, wide variations existed among the states of the federation, with low values of 6 and 8% recorded for Taraba and Benue states respectively and the highest values of 58 and 78% recorded in Ogun and Lagos states respectively. Given such a grim situation, residents are left with no other choice than to seek sources of freshwater from streams, rainfall, and groundwater (by digging wells). A dug well is a large diameter well constructed by excavating with hand tools or power machinery instead of drilling and driving and has the basic purpose of supplying water for domestic needs. It is distinct from a drilled well (commonly called borehole) which is constructed by cable tool or rotary drilling methods. In the absence of perennial streams, groundwater remains the most accessible source.

The assessment of water quality in dug wells is essential because these are often the main sources of water for human consumption in typical African communities. The well-being of people is dependent on the quality of water which they ingest or otherwise make use of.   It has become imperative to assess the quality of the water supply from these dug wells and identify the various sources of contaminants in order to ascertain the contamination problems that may confront the consumers. From results obtained, a scientific basis will then be provided for finding appropriate remedies to the situations and the inherent impacts on residents in communities that depend on the wells. It is with this objective in mind that the present study was undertaken in Ibogun, Pakoto and Ifo communities in Ogun State, Nigeria. Populations reported for Ibogun and Pakoto are less than 5,000 persons in each case. Ifo is bigger although its population is still about 20,000 persons (National Population Commission, 2009). The major occupation in Ibogun, Pakoto and Ifo is farming. Ibogun hosts a campus of a university and the student population is included in its total population. In addition to farming, some trading goes on in Ifo, making it less agrarian than the other two communities. All the communities have no industry of note. Essentially the 3 of them are typical African communities and findings regarding them can be of relevance in such small-sized communities scattered all around the continent.

The quality of ground water depends on the various chemical constituents and their concentrations in the water, and they are mostly derived from the geology of the particular location. Although ground water is the purest naturally occurring water, it is still readily contaminated by industrial waste and municipal waste in their various forms. In most parts of the world, the most common form of contamination of raw water sources is from human sewage and in particular human fecal pathogens and parasites. In 2006, waterborne diseases were estimated to cause 1.8 million deaths each year while about 1.1 billion people lacked safe drinking water (Clansen et al., 2007).

To live in good health, people need to have access to good quality water in adequate quantity. Parameters for drinking water quality typically fall under three categories namely physical, chemical and microbiological. Physical quality involves such para-meters as odour, colour and taste. Chemical parameters include pH, total solids, nitrates, sulphates, chlorides, hardness, metals generally

as well as some other elements. Microbiological parameters include coliform bacteria, streptococci, *E. coli,* and parasites. This study evaluated the concentrations of these parameters in the samples of water sourced from various wells in these communities, compared those values to International Drinking Water Quality Standards of the World Health Organization, determined the level of pollution of the water and made recommendations as to their safe use.

MATERIALS AND METHODS

Sampling

Water samples from the selected dug wells were collected during July - August 2009, with a sample collected from each well. 20 dug wells were randomly selected in each community, giving a total of 60 water samples in all. Two sets of samples, each of 150 ml volume were collected for each well; one set for physico-chemical analysis and the other set for microbial analysis. pH and temperature were measured on site. Each sample was collected in a new factory-fresh plastic bottle with the cap securely tightened. After collection, the samples were immediately placed in ice coolers for transportation to the laboratory where they were then transferred to the refrigerator. Laboratory analyses commenced the same day and within 30 min of arrival at the laboratory in every case. The analyses were conducted in the laboratories of the Institute of Agricultural Research and Training (IAR&T), and Federal College of Agriculture (FCA), both located in Ibadan, Nigeria.

Methods for physico-chemical analyses

The samples were first checked for odour, colour and taste. Afterwards, they were analyzed for major physical and chemical water quality parameters namely pH, total solids (TS), total hardness (TH), chlorides, sulphate, nitrates, magnesium, calcium, manganese, sodium, copper, zinc, iron and lead. Samples were subjected to filtration prior to chemical analysis. The determination of TS was done by a gravimetric process, while the total hardness was carried out by EDTA complexometric titration method. Nitrate was determined by colorimetric procedure. The methods of analyses were those detailed in APHA et al. (1998).

Methods for bacteriological analyses

For the preparation of sterile water, 9 ml distilled water was pipetted into a clean dry test tube plugged with clean cotton wool and wrapped with aluminium foil. The test tube was placed inside an autoclave and sterilized by autoclaving at 121°C for 15 min. The media were afterwards prepared as follows:
For preparation of nutrient agar (NA), 28 g of powdered commercially prepared nutrient agar was weighed on a Mettler analytical balance into a clean dry 1 L conical flask and 100 ml of distilled water was added. The flask was then placed inside a water bath set at 90°C to allow the agar dissolve. The solution was then measured into Macconkey bottles and the bottles placed was for 15 min inside autoclave set at 121°C.
For preparation of Macconkey agar (MCCA), 55 g of commercially prepared agar was weighed into a 1 L conical flask and gradually heated to dissolve the agar. It was then distributed

into Macconkey bottles and autoclaved as for nutrient agar.

For preparation of potato dextrose agar (PDA), 39 g of commercially prepared agar was weighed into a 1 L conical flask and gradually heated to dissolve the agar. The solution was then distributed into Macconkey bottles and autoclaved as for nutrient agar.

In all the cases, after autoclaving, the media were placed inside a water bath set at $45^{\circ}C$ to maintain the media in a molten state. One millilitre of each water sample was weighed into a test tube containing 9 ml of sterile distilled water in Macconkey bottles was shaken vigorously on a vortex mixer and serially diluted. From it, 1 ml of $10^4 - 10^6$ dilutions were plated on to nutrient agar (NA) and potato dextrose agar (PDA). The media were individually poured into separate plates and were duplicated. After solidifying, the plates were incubated at $37^{\circ}C$ for nutrient agar and Macconkey agar while potato dextrose agar was incubated at $30^{\circ}C$. All the plates were incubated invertedly. The plates were counted at 48 h for Nutrient Agar and Macconkey Agar while it was read for potato Detrose Agar at 72 h. The pathogens were identified using cultural and morphological features. Isolation and identification of bacteria in the water samples were done using methods detailed in Collins and Lyne (1984), Adegoke et al. (1993) and APHA et al. (1998). The procedures for the biochemical characterization of the isolates which includes (1) cultural profiles on different culture media; (2) growth in air; (3) gram staining; (4) coagulase, catalase and oxidase tests, and (5) carbohydrate fermentation were strictly followed.

RESULTS AND DISCUSSION

Results of the physico-chemical and bacteriological analyses are presented in the Figures 1 – 9.

Since communities within the study areas use the untreated water for domestic purposes, results stated above are compared with both the World Health Organization (WHO) guidelines for drinking water (WHO, 1984, 1993, 1996, 2004) and the international average for fresh water.

Physical characteristics of water samples studied

Water meant for human consumption should be completely free of odour, colour and taste. The presence of any of these in a water sample makes it aesthetically unpleasant and typically reduces its acceptability among consumers. Odour test was performed but odour was absent in all of the 60 water samples analysed. Colour is vital as most water users, be it domestic or industrial, usually prefer colourless water. Determination of colour can help in estimating costs related to elimination of discolouration of the water samples. In Ifo, most of the water samples were colourless.

However in 3 samples, light brown, brown and milky discolourations were observed. In Ibogun, light brown colour was observed in 1 sample while milky colour was observed in another and all the rest were colourless. Light brown colouration was present in 1 sample in Pakoto, while all the rest were colourless. Therefore

ordinarily, most of the water samples from the wells in these communities would be aesthetically acceptable to the consumers. This situation in itself may pose some danger and suggests considerable caution among the consumers since aesthetically pleasant water does not necessarily imply that the water is hygienically safe for consumption.

Chemical characteristics of water samples

Though pH has no direct effect on human health, nevertheless, high pH causes a bitter taste, makes pipes and appliances to become encrusted and depresses the effectiveness of chlorine disinfection. Water having low pH will corrode and dissolve metals. pH is also an indicator of biological life since most of them thrive in a quite narrow and critical pH range. For human beings, pH value of 7.0 is considered as best or ideal although a range of 6.5 - 8.5 is permissible (WHO, 1993). In the present study the majority of the water samples were typically acidic having pH ranges of 2.6 to 6.0 in Ifo; 4.0 to 5.9 in Ibogun; and 3.8 to 7.1 in Pakoto (Figure 1). These pH ranges elicited further interest to later obtain soil samples around the study areas in an effort to find out what is responsible for general acidity of those samples from the dug wells. In general, the lower the value of pH, the higher the level of corrosion. Gupta et al (2009) observed that in some cases decrease in pH is accompanied by the increase in bicarbonate, carbonate and hydroxyl ions.

This observation was partly correlated in this study as well. In Ifo for example, the highest value of 1600 mg/l of total solids (Figure 2) and the second highest value of 158 mg/l of total hardness (Figure 3) were measured for a water sample having a pH of 4.0. In Ibogun, the second highest value of 96 mg/l of total hardness (Figure 3) was from a sample which had a pH of 4.0 (Figure 1) while in Pakoto, the highest value of 1900 mg/l of total solids (Figure 2) was measured in a sample having a pH of 4.4 (Figure 1). For these com-munities studied, it is likely that the low pH of the waters is traceable to the presence of carbonates. The geology of the area studied is known to be rich in limestone.

WHO (2004) stated that water meant for human consumption should have a maximum total solids (TS) content of 1500 mg/l. In Figure 2 as shown for Ifo, wells 7 and 20 have values of 1600 and 2200 mg/l respectively, effectively placing these above the WHO guideline. It is also seen from the same figure that well 6 in Ibogun and well 14 in Pakoto have values of 1700 and 1900 mg/l respectively for total solids. All the rest are below the guideline, with the majority being below 500 mg/l. Therefore water from most of the wells where within the guideline of WHO (2004) as far as total solids content is concerned. Total solids (TS), is really a sum of two terms

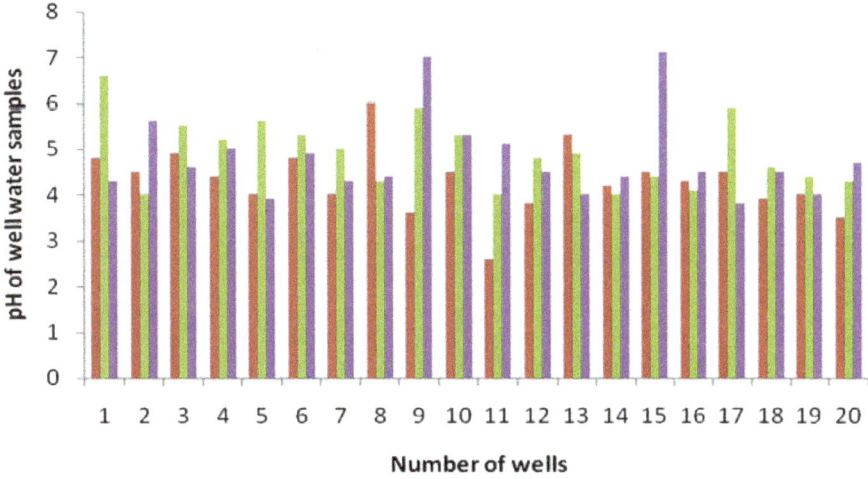

Figure 1. pH of water samples in Ifo, Ibogun and Pakoto.

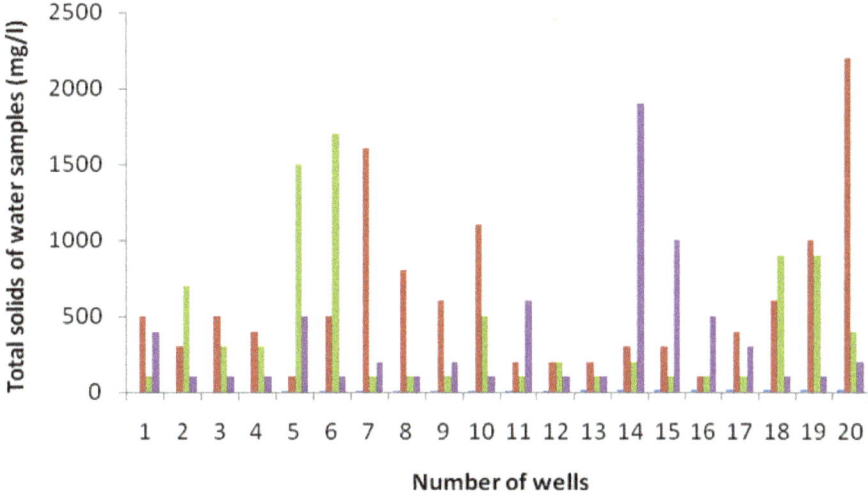

Figure 2. Total solids of water samples in Ifo, Ibogun and Pakoto.

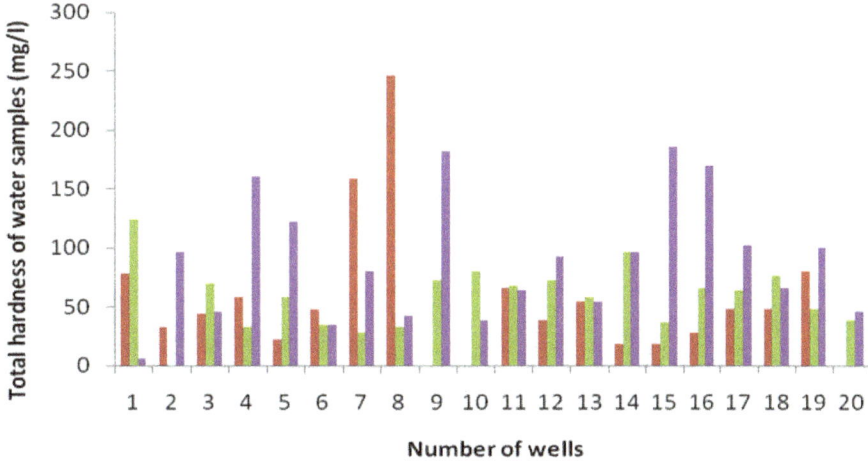

Figure 3. Total hardness of water samples in Ifo, Ibogun and Pakoto.

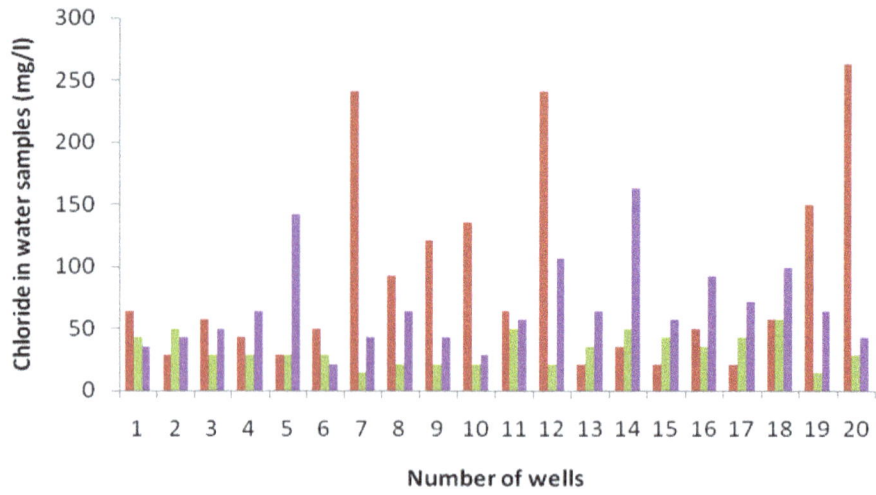

Figure 4. Chloride in water samples of Ifo, Ibogun and Pakoto.

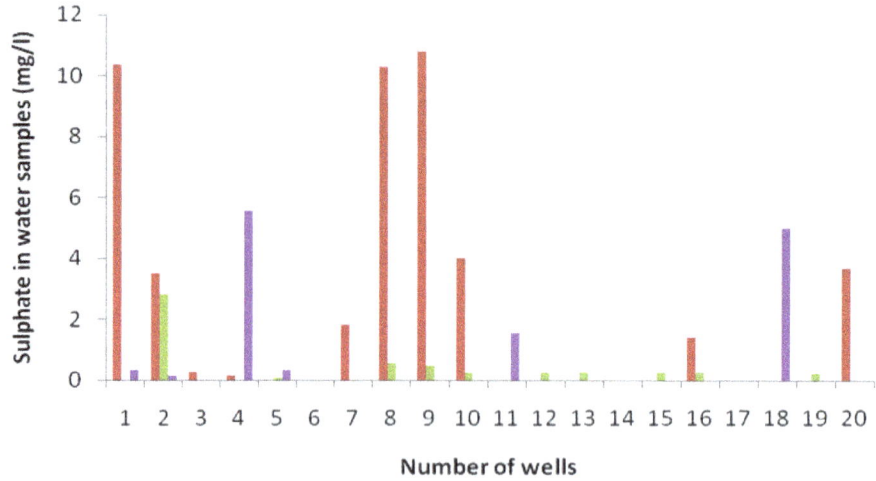

Figure 5. Sulphate in water samples of Ifo, Ibogun and Pakoto.

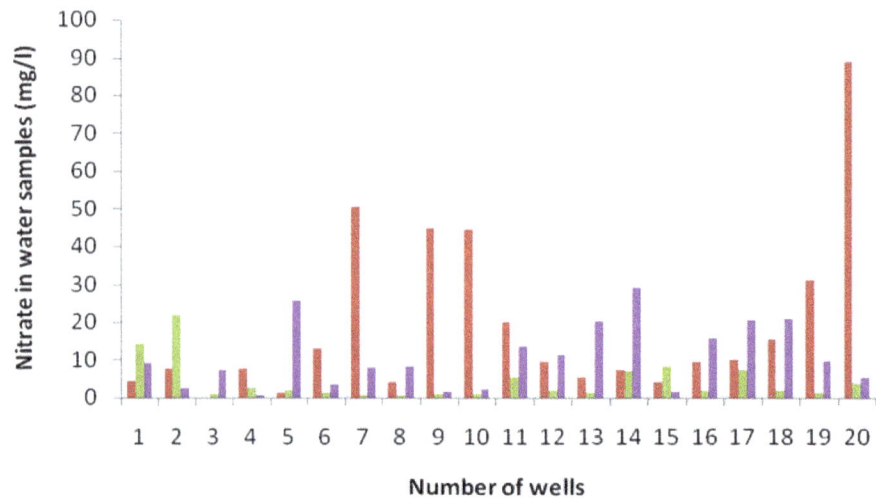

Figure 6. Nitrate in water samples of Ifo, Ibogun and Pakoto.

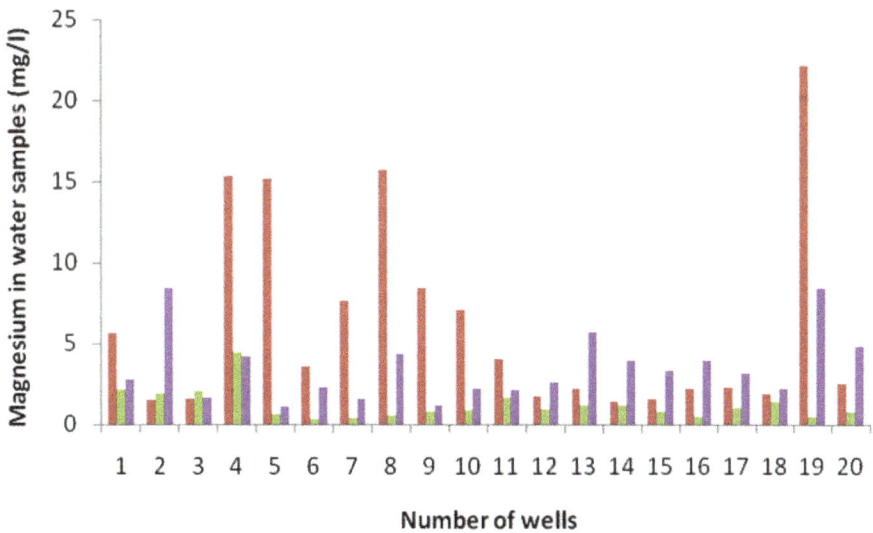

Figure 7. Magnesium in water samples of Ifo, Ibogun and Pakoto.

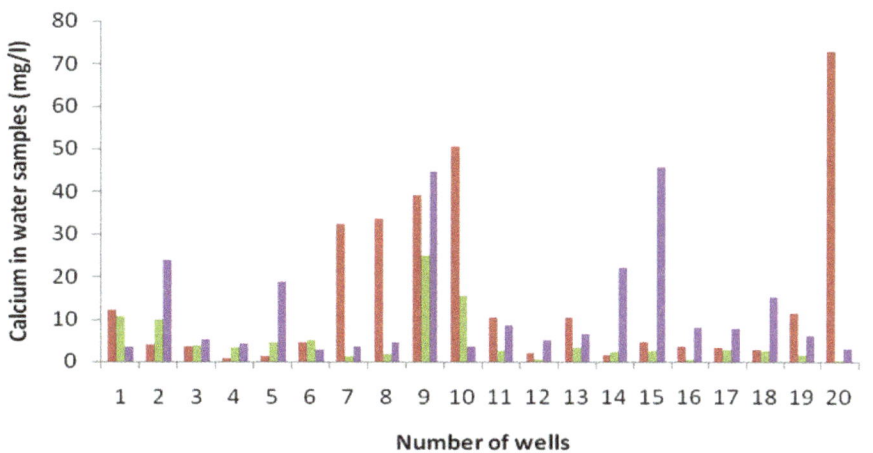

Figure 8. Calcium in water samples of Ifo, Ibogun and Pakoto.

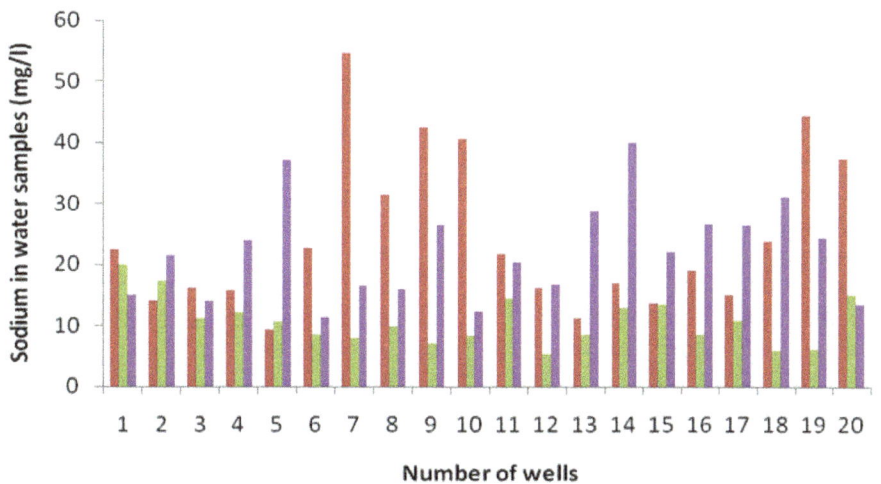

Figure 9. Sodium in water samples of Ifo, Ibogun and Pakoto.

namely total suspended solids (TSS) and total dissolved solids (TDS). Total suspended solids in freshwater is an indication of the amount of erosion that took place nearby or upstream. This parameter would be a most significant measurement as it would depict the effectiveness of and indicate the level of compliance with control measures, for example prevention of direct contact between well water and surface erosion. An internationally accepted maximum value of TDS in fresh water is 100 mg/l (Meybeck and Helmer, 1989). As far as the studied communities are concerned, the values of TDS were less than 100 mg/l in 50% of water samples from Ibogun and Pakoto. This correlates to an accepted common practice of building a 1 m high protective concrete ring on top of each well and fixing a tight cover over it, thus preventing contamination with dust and other solids from outside the system. In Ifo, the measured TDS was lower than 100 mg/l in just 40% of the wells. Perhaps the reason for this is that the number of people depending on each well was higher in Ifo than in the other 2 communities. Therefore more people had direct access to the well, increasing the risk of contamination of the water with foreign particles borne on containers being used to draw water from the well.

Water hardness is the traditional measure of the capacity of water to react with soap; hard water requiring considerably more soap to produce lather. Hardness is one of the very important properties of ground water from utility point of view particularly for domestic purposes. As reported in the literature, the concentration of total hardness in drinking water sources ranged between 75 and 1110 mg/l (Nawlakhe, 1995). Shastri et al. (1996) also reported a study where water samples from ponds and wells were very hard ranging from 222.8 - 1094.4 mg/l. However, WHO (2004) specified a maximum guide-line value of 400 mg/l of total hardness for water meant for human consumption. In the present study, no sample exceeded that maximum limit. The highest values of total hardness measured in Ifo were in wells 8 and 7 which showed 246 and 158 mg/l respectively, while the rest were 80 mg/l (well 19) or less (Figure 3). Similarly in Ibogun, the highest value was 124 mg/l found in well number 1, while the rest were 96 mg/l (well 14) or less as seen in the same figure. In Pakoto, as shown in Figure 3, the highest and least values were 186 and 6 mg/l measured in wells 15 and 1 respectively. Whenever freshwater has a high concentration of total hardness, this may be due to dissolution of polyvalent metallic ions from sedimentary rocks, seepage and run off from soil. However high concentration was not measured in any water sample analysed. It is well known that hardness is not caused by a single substance but by a variety of dissolved polyvalent metallic ions, predominantly calcium and magnesium cation, although other cations like barium, iron, manganese, strontium and zinc also contribute. In the present study, some trace metals

namely copper, iron and zinc were analysed but found to be 0 mg/l in all the samples studied. WHO (2004) guidelines for these are 2, 0.2 and 3 mg/l respectively. All the samples therefore did not exceed these limits. As regards magnesium and calcium, the measured values were remarkably low. As seen in Figure 7, values of magnesium in water samples ranged from 1.5 - 22.2 mg/l; 0.35 - 4.48 mg/l; and 1.1 - 8.5 mg/l in Ifo, Ibogun and Pakoto, respectively. For calcium the ranges were 1.2 - 72.9; 0.5 - 24.9 and 2.9 - 44.6 mg/l respectively as shown in Figure 8. These values were below the maximum guidelines given by WHO (2004). These comparatively low values for these cations definitely contributed to the low values measured for total hardness in the water samples.

The WHO guideline value of 0.2 mg/l for iron (Fe) in drinking water (WHO, 2004) was not exceeded by water from dug wells in the 3 communities since iron was analysed but was found to be 0 mg/l in all of the samples. By implication, the levels also did not exceed the iron average of 0.01 mg/l for fresh water (Meybeck and Helmer, 1989). It may thus be inferred that iron is minimally present in the soil in those communities. Manganese (Mn) is a common problem element in natural waters. In drinking waters, this element may cause unsightly stains and produce a brown/black precipitate. Although it is an essential element, the chronic ingestion of Mn in drinking water is associated with neurological damage (Kondakis et al., 1989). In the present study, Mn was analysed but was found to be 0 mg/l in about 50% of water samples from Ifo and Pakoto and in almost 100% of the samples from Ibogun. Among the measured values, ranges of 0.02 -0.62 and 0.02 - 0.08 mg/l were recorded in Ifo and Pakoto respectively. All these values were higher than the average of 0.01 mg/l of Mn for fresh water (Meybeck and Helmer, 1989). However, when compared to the guideline value of 0.4 mg/l for Mn in drinking water (WHO, 2004), almost 100% of the levels measured in this study were lower.

As shown in Figure 4, chloride was measured in all the samples with values ranging from 21.3 - 262.7; 14.2 - 56.8, and 21.3 - 99.4 mg/l in Ifo, Ibogun and Pakoto respectively. The WHO (2004) guideline for chlorides in drinking water is 250 mg/l. Only well number 20 in Ifo which measured 262.7 mg/l exceeded this limit, while no sample in Ibogun and Pakoto exceeded it. Remarkable differences were noticed in the average readings of chloride in the water samples of the wells in the 3 communities and these were 89.1, 39.4 and 67.4 mg/l for Ifo, Ibogun and Pakoto respectively. These differences may perhaps be due to urbanization. Ifo is by far the most populated and most urbanized of the three and generates wastes of different kinds at a higher quantity which are amassed in open dumps. Incident rainfall leaches salts from these dumps and from running effluent from homes which later infiltrate into the ground to dissolve in the water. Ibogun is the most agrarian among the three and the use of fertilizers is not commonly practiced on their farms. As a result, there appears to be fewer natural sources of chlorides to pollute the water there. Sulphate was analysed and found to be 0 mg/l in 50% of water samples in Ifo and Ibogun as well as in 70% of samples in Pakoto (Figure 5). Among the measured samples, sulphate ranged from 0.2 - 10.8 mg/l in Ifo; 0.1 - 2.8 mg/l in Ibogun and 0.2 - 5.6 mg/l in Pakoto. Although the concentrations of sulphate in some samples from wells 1, 8 and 9 in Ifo and wells 4 and 18 in Pakoto exceeded the average of 4.8 mg/l reported for fresh water (Meybeck and Helmer, 1989), yet the WHO (2004) guideline for sulphate is 500 mg/l. Therefore it may be inferred that sulphate does not have a strong presence in the groundwater of any of the communities studied. Sodium (Na) was detected in all the water samples. As shown in Figure 9, values ranged from 9.3 - 54.7; 5.3 - 17.3, and 11.5 - 40.0 mg/l in Ifo, Ibogun and Pakoto respectively. However, these values were well within the limits of the guideline value of 200 mg/l set by WHO (2004). The ionic dominance for water bodies according to Stumm and Morgan (1981) are: $Ca > Mg > Na > K$ and $HCO_3 > SO_4 > Cl$ for fresh waters. Generally, the ionic dominance pattern observed in this study for groundwater was $Na > Ca > Mg$ and $HCO_3 > Cl > SO_4$. Some of these cations were in relatively high concentrations.

Health considerations regarding infants point in the direction of having domestic water entirely free of nitrates. Nitrates are strongly linked to the occurrence of blue baby syndrome otherwise known as infant methaemoglobinaemia (Powlson et al., 2008). However, water which has zero nitrates hardly occurs in nature. WHO (2004) set an upper limit of 50 mg/l of nitrates for water meant for domestic consumption. In this study, concentrations of nitrates ranged from 1.4 - 88.1; 0.8 - 22.0, and 0.8 - 29.2 mg/l for Ifo, Ibogun and Pakoto respectively as shown in Figure 6. It was observed that the WHO (2004) guideline was exceeded only by 10% of the samples in Ifo and no samples in Ibogun and Pakoto. However, considering the global average of 0.1 mg/l for nitrates in fresh water (Meybeck and Helmer, 1989), it is seen that the levels of nitrates measured in the samples are much higher than this value. This could be as a result of infiltration from domestic wastes and fertilized farm lands. Nitrate is made in the human body (Green et al., 1981), the rate of production being influenced by factors such as exercise (Allen et al., 2005).

Therefore presence of nitrates in groundwater in domestic settings is one of the indicators of contact with human wastes. Also the use of highly soluble fertilizers which contain nitrates causes water pollution problems as rainwater leaches out these nitrates and carries them into nearby streams while some portion infiltrates into the ground (Stoddard et al., 2005). Various health concerns have been expressed regarding the issue of drinking water containing high levels of nitrates (WHO, 1996; L'hirondel, 2002; Ward et al., 2006). Agricultural activities are reported to contribute about 50% of the total pollution source of fresh water by means of the higher nutrient enrichment, mainly ammonium ion (NH_4) and nitrate (NO_3), derived from fertilizers (Islam and Tanaka, 2004; Benson et al., 2006). Water used for drinking and other domestic purpose should also be free from toxic elements. In the present study, the heavy metal lead (Pb) was analysed and found to be absent in all the samples. By implication, no sample exceeded the guideline value of 0.01 mg/l for drinking water (WHO, 2004). This is understandable since the 3 communities are largely agrarian and they contain no heavy industries which may be the source of heavy metals in the environment.

Microbiological analyses

Total viable count is the sum of the total coliform count, faecal coliform count, faecal streptococci count and other pollutants in the water. Although high viable counts are usually indicative of the contamination of the water and the presence of bacteria other than aquatic bacteria, they do not necessarily indicate pollution by faeces and or sewage. Furthermore, the bacteria may be mostly soil saprophytes. Thus a high viable count alone is not evidence that a source of water is potentially dangerous due to the possible presence of intestinal pathogens. Nevertheless, water supplies with high viable counts are undesirable since they still carry the associated risk of possible pollution. As seen in Table 1, none of the water samples analysed met the guideline for total viable count which is 0/250 ml for all waters intended for drinking. Total coliform count is the microbiological test used to detect the level of pollutions caused by living things especially human beings who reside in a location. According to WHO (1993), coliform bacteria must not be detectable in any 100 ml sample of all water intended for drinking. The guideline value for total coliform is therefore 0/100 ml (WHO, 1993) for water intended for such a purpose. This guideline value was met in 55, 30 and 100% of samples in Ifo, Ibogun and Pakoto, respectively (Table 2). Also, for the test conducted at 37°C, the maximum colony count allowed is 20 cfu/ml (WHO, 1993). By implication, more than half of total samples analysed did not meet the WHO guidelines for total coliform count for water intended for drinking. This indicates that the ground water has been polluted by microbes sourced from living organisms. The ready explanation for this is that there may be some form of interaction between these wells and nearby septic tanks or waste dumps. However this could not be confirmed. WHO (1993) points out that on its own, total coliform count is not a definite indicator of the sanitary quality of rural water supplies, particularly in tropical areas where many bacteria of no sanitary

Table 1. Total viable count (x 10^{5}cfu/ml)

Well No.	Ifo	Ibogun	Pakoto
1	1.8	1	0.6
2	1.8	1.4	1.1
3	2	1.2	1
4	1.4	1	0.9
5	1.6	1.8	1.2
6	1.2	1	1.1
7	2.2	0.6	0.8
8	0.8	0.6	0.6
9	1.2	1.4	0.4
10	0.8	1.6	0.8
11	0.6	1.2	1.4
12	1.2	0.8	1.6
13	0.8	0.6	1.1
14	2	1.6	1
15	1.9	1.4	1.1
16	1.6	1.6	0.6
17	1.4	0.8	0.6
18	1.2	0.6	0.4
19	1	0.6	0.8
20	1.4	0.4	1

Table 2. Total coliform count (×10^{5}cfu/ml).

Well No.	Ifo	Ibogun	Pakoto
1	0	0.2	0
2	0	0.2	0
3	0	0.3	0
4	0	0.1	0
5	0.1	0	0.1
6	0.1	0	0
7	0.2	0.1	0
8	0	0.1	0
9	0	0	0
10	0	0	0.2
11	0.2	0.2	0
12	0.1	0.6	0
13	0.1	0.3	0
14	0	0.1	0.1
15	0	0.2	0.1
16	0.2	0.4	0
17	0.2	0	0
18	0.1	0.1	0
19	0	0.2	0

significance occur in almost all untreated supplies. Therefore the certainty of pollution of all these waters in which coliform bacteria was detected may not be concluded. Faecal coliform count is a test based on coliform bacteria as the indicator organism. The presence of these indicative organisms is evidence that the water has been polluted with faeces of humans or other warm-blooded animals. The WHO (1993) guideline for faecal coliform count in water intended for drinking is 0/100 ml. This guideline was met in 80, 70 and 100% of samples in Ifo, Ibogun and Pakoto respectively (Table 3). Therefore 20, 30 and 0% of water samples in these communities had contamination which originated from human beings or other warm blooded animals.

The occurrence of faecal streptococci as a supplement to faecal coliforms reinforces the incidence of faecal pollution by warm blooded animals. Since faecal strepto-cocci have a limited survival time outside the animal intestine and are not capable of multiplying in the environment, their presence indicates a very recent pollution. With respect to faecal streptococci count, 80, 85 and 75% of wells in Ifo, Ibogun and Pakoto (Table 4) respectively did not meet up with the guideline of 0/100 ml set by WHO (1993). Therefore the quality of the majority of those samples can definitely be improved with adequate treatment.

Conclusions

Most of the pH values of the samples were outside the recommended range of 6.5 – 8.5 for drinking and other form of ingestion. Predominantly, the ionic dominance pattern observed were Na > Ca > Mg and HCO_3 > Cl > SO_4, indicating typical cationic characteristics and anionic characteristics of groundwater. Levels of iron did not exceed the WHO guideline value of 0.30 mg/l for Fe in drinking water in the 3 communities studied. Mean levels of Mn measured were far in excess of the average of 0.01 mg/l for fresh water, while in relation to the WHO guideline value of 0.40 mg/l for Mn in drinking water, the levels measured in this study were low. Generally, the levels of nitrate, sulphate, chloride, trace elements (e.g. manganese and iron) were moderately high, indicating organic contamination of the groundwater but the WHO guidelines were not exceeded. Guidelines for microbio-logical quality were met in several cases and the physical guidelines for colour, odour and taste were met in most cases.

Therefore aesthetically pleasant water that is sourced from groundwater through these dug wells appeared safe for consumption. Most of the wells did not have any readily noticeable source of contamination in their environments. The use of such dug wells in domestic settings in typical African communities is recommended. However there is a need for the enforcement of existing laws against improper disposal of domestic wastes as well as those guiding the location of septic tanks in relation to the location of dug wells in domestic settings.

Matching of non-technical and techno-social remedial

Table 3. Faecal coliform count ($\times 10^5$ cfu/ml).

Well No.	Ifo	Ibogun	Pakoto
1	0	0	0
2	0	0	0
3	0	0	0
4	0	0	0
5	0	0	0
6	0	0	0
7	0	0.1	0
8	0	0	0.1
9	0	0	0
10	0	0	0.1
11	0.1	0.2	0
12	0.1	0.1	0
13	0.1	0	0
14	0	0	0
15	0	0.2	0.1
16	0.2	0.1	0.1
17	0	0	0
18	0	0	0
19	0	0.1	0
20	0	0	0

Table 4. Faecal streptococci count ($\times 10^5$ cfu/ml).

Well No.	Ifo	Ibogun	Pakoto
1	0.2	0.2	0.1
2	0.4	0.1	0
3	0.4	0.3	0.1
4	0.3	0.1	0.1
5	0.2	0	0.2
6	0	0.4	0.1
7	0.6	0.1	0.3
8	0.2	0.1	0.1
9	0.4	0.3	0.1
10	0.1	0.2	0.2
11	0	0.4	0
12	0	0.2	0.1
13	0.2	0.1	0.1
14	0.6	0	0.1
15	0.8	0	0.2
16	0.6	0.3	0
17	0.7	0.2	0
18	0.2	0.2	0
19	0.2	0.1	0.2
20	0.1	0	0.1

measures is recommended. These include sensitization of the populace on merits of qualitative domestic hygiene

and environmental protection practices such as cleaner compounds and strict enforcement of environmental protection laws. Another important issue for pollution mitigation measures pertains to the need to educate all stakeholders such as the local population, the wards, local governments, policy and legal authorities, and other interested parties, and include them in planning and decision making regarding sourcing water for use in domestic settings and the protection of the environments.

REFERENCES

Adedeji AO (2001). Quality of Drinking Water in Lagos State. Discovery innovation, 7: 56.

Adegoke GA, Ashaye OA, Shridar MKC (1993). Microbiological and Physico-chemical Characteristics of Water Used by Some Brewery, Bakery and Soft Drinks Plants in Oyo state, Nigeria. J. Agric. Sci. Technol., 3(1):92-95.

Allen JD, Cobb FR, Gow AJ (2005). Regional and Whole-body Markers of Nitric Oxide Production following Hyperemic Stimuli. Free Radical Biol. Med., 38: 1164-1169.

APHA, AWWA, WPCF (1998). Standard methods for the examination of water and wastewater, 20th edn. Washington, D.C.

Benson VS, VanLeeuwen JA, Sanchez J, Dohoo IR, and Somers GH (2006) Spatial Analysis of Land Use Impact on Ground water Nitrate Concentrations. J. Environ. Qual., 35:421-432.

Clansen T, Schmidt W, Rabie T, Roberts I, Caincross S (2007). Interventions to improve water quality for preventing diarrahoea: A Systematic Review and Meta-analysis. Brit. Med. J. doi:10.1136/bmj.39118.489931.BE

Collins CH, Lyne PM (1984). Microbiological Methods. 5th Ed. Butterworth and Co. Publ. Ltd., London. pp. 331-345.

Green LC, Ruiz de Luzuriaga K, Wagner DA, Rand W, Isfan N, Young VR, Tanneenbaum SR (1981). Nitrate Biosynthesis in Man. Proc. Natl. Acad.Sci. USA, 78: 7764-7768.

Gupta DP, Sunita A, Saharan JP (2009). Physiochemical Analysis of Ground Water of Selected Area of Kaithal City (Haryana) India. Researcher 1(2), pp1-5. http://www.sciencepub.net

Islam SM, Tanaka M (2004). Impacts of Pollution on Coastal and Marine Ecosystems including Coastal and Marine Fisheries and Approach for Management: A Review and Synthesis. Mar. Pollut. Bull., 48: 624-649.

Kondakis XG, Makris N, Leotsinidis M, Prino M, Papapetropoulos T (1989) Possible Health Effects of High Manganese Concentration in Drinking Water. Arch. Environ. Health, 44(3): 175-178.

L'hirondel JL (2002). Nitrate and Man: Toxic, Harmless, or Beneficial? CABI Publ., Wallingford, Oxfordshire, UK.

Meybeck M, Helmer R (1989). The Quality of Rivers: from Prestine Stage to Global Pollution. Palaeogeography,Palaeoclimatology, Palaeoe-cology (Global Planet Change Section), 75: 283-309.

Nawlakhe WG, Lutade SL, Patni PM, Deshpande LS (1995). Total Hardness in Drinking Water Sources. Indian J. Env. Prot., 37(4): 278-284.

National Population Commission (2009). Population Figures of Nigeria's Census Conducted in March 2006. Federal Government Official Gazzette. Published February 2.

Powlson DS, Addiscott TM, Benjamin N, Cassman KG, de Kok TM, Grinsven HV, L'hirondel JL, Avery AA, Kessel CV (2008). When Does Nitrate Become a Risk for Humans. America Society of Agronomy, Soil Science Society of America, Crop Science Society of America. Published in J. Environ. Qual., 37: 291-295

Sangodoyin AY (1993). Consideration on Contamination of Groundwater by Waste Disposal Systems in Nigeria. Environ. Technol., 14: 957-964.

Shastri SC, Bakra PP, Khan, JI (1996). Industry Environment and the Law, R 13SA Publishers, Jaipur.

Stoddard CS, Groove JH, Coyne MS, Thom WO (2005). Fertilizer, Tillage and Dairy Manure Contributions to Nitrate and Herbicide Leaching. Published in J. Environmental Quality, 34: 1354-1362.

Stumm W, Morgan JJ (1981). Aquatic Chemistry. Wiley. New York. 780 pp.

Ward MH, de Kok TM, Levallois P, Brender J, Gulis G, VanDerslice J, and Nolan BT (2006). Response to Dietary Nitrate: Where is the Risk? Environ. Health Perspect., 114: A459-A460.

World Health Organization, WHO (1984). Guidelines for Drinking Water Quality, vol. 1 and 2. World Health Organization, Geneva, Switzerland.

World Health Organization, WHO (1993). Guidelines for Drinking Wate Quality. World Health Organization, Geneva, Switzerland.

World Health Organization, WHO (1996). Guidelines for Drinking Wate Quality, vol. 2, 2nd edn. Health criteria and other supporting information World Health Organization, Geneva, Switzerland.

World Health Organization, WHO (2004). Guidelines for Drinking Wate Quality, vol. 1, 3rd edn. World Health Organization, Geneva Switzerland.

17

Evaluation of ground water quality in Lucknow, Uttar Pradesh using remote sensing and geographic information systems (GIS)

Anju Verma[1], Biswajeet Thakur[2], Shashwat Katiyar[3], Dharam Singh[1] and Madhu Rai[4]

[1]Department of Environmental Sciences, IBSBT, CSJM University, Kanpur.
[2]Birbal Sahni Institute of Palaeobotany, Lucknow.
[3]Department of Biochemistry, IBSBT, CSJM University, Kanpur.
[4]Department of Geology, University of Lucknow, Lucknow.

Lucknow being capital city of most poapulated state of India is facing tremendous population pressure. This has led to overexploitation of natural resources and among them water is most valuable natural resource essential for human survival and ecosystem. This study monitors ground water quality, relating it to land use/land cover and habitation mask of different water quality parameters are prepared by using geographic information systems (GIS) and remote sensing technique. Base map was prepared by Survey of India toposheets on 1:50.000 scale. The land use / land cover map was made from satellite imagery and GIS software like ERDAS Imagine and ARC GIS 9.3. The ground water samples were collected from the selected locations and were analyzed for different physico-chemical analysis and a water quality index was prepared. Water quality index (WQI) was then calculated on the basis of WHO standards to classify suitability for drinking water. The WQI map was interpolated using inverse distance weight (IDW) method on GIS for spatial variation and suitability of quality assessment.

Key words: Remote sensing, ground water, water quality index, urban sprawl, inverse distance weight (IDW) method, WHO standard.

INTRODUCTION

Urbanization is characterized by clustering of people in relatively small areas and is recognized as an inevitable historical process (UN, 2004). The urbanization leads to many changes which have adverse impacts on Environment, including ecology, especially hydro-geomorphology, water resources and vegetation. Rapid growth of urban areas has further affected the ground water quality due to over exploitation of resources and improper practices (Mohrir et al., 2002). Lucknow is the capital city of the most populous state Uttar Pradesh and is one of the fastest developing urban centers of India. Lucknow district is a part of Central Ganga Plain covering an area of 2, 528 km[2]. and lies between North latitudes 26°30' and 27°10' and East longitudes 80°30'and 81°13' with total population of 34 lakhs as per 2011 (Anonymous,

2011).

The city is facing a rapid change in environmental quality. Rapid urbanization leads to many problems as it places huge demand on land, water, housing, transport, health, Education etc (Gyananath et al., 2001). The city has an alarming increase in population it increased from 0.497 million in 1951 to 2.267 million in 2001 and 2.714 million in 2006 to 3.306 in 2011 increased 4.56 times (456 per cent) during the last fifty years. The growth rate of population Lucknow (UA) was at 7.12% per annum (Lucknow Master Plan, 2010). This rising population density has major impact on natural resources of the area especially on water quality and quantity.

Fresh water is most important natural resources for the life but overexploitation and unjustified use of water has led to deterioration of quality of water. In the last few decades, ground water has become an essential resource due to its use for drinking, industrial and irrigation purposes. Remote sensing and geographic information

---

*Corresponding author. E-mail: anjuverma4@rediffmail.com.

systems (GIS) are effective tools for land use/ land cover, water quality mapping for monitoring and detection of change in environment (Ferry et al., 2003). One of the greatest advantages of using remotely sensing data for hydrological investigations and monitoring its ability to generate information in spatial and temporal domain, which is very crucial for successful analysis, prediction and validation (Saraf, 1999; Epstein et al., 2002). GIS can be a powerful tool for dealing with water resource problems and developing solutions (Skidmore et al., 1997). In this paper we have integrated remote sensing, GIS and field studies for evaluating impacts of land use change on water quality of Lucknow City.

METHODOLOGY

Study area

Lucknow district is a part of Central Ganga Plain in the state of Uttar Pradesh covering an area of 2,528 km^2 and lies between North latitudes 26°30' and 27°10' and East longitudes 80°30'and 81°13' (Figure 1) with total population of 34 lakhs as per 2011 (Anonymous, 2011]. General elevation of the district varies between 103 and 130 (Anonymous, 2009-10) meters above mean sea level showing southeasterly slope. Irrigation in the district takes place through Sharda Canal and Sharda Sahayak network systems and tube-wells. The Gomati River, the chief geographical feature, meanders through the city, dividing it into the Trans-Gomati and Cis-Gomati regions. The climate of Lucknow city is of subtropical type with three distinct seasons namely summer, monsoon and winter. The maximum temperature remains 45°C during month of May and minimum temperature remains 5°C during January. The average annual rainfall of the city is 1014.7 mm.

Data used

Different data products required for the study include Survey of India toposheets 63B/9, 10, 13, 14, 63F/1, 2, 63 A/ 12, 16, 63 E/ 4 on 1:50,000 scale. Besides this guide map of Lucknow on 1:20,000 scale have also been referred during interpretation as well as for base map preparation. IRS- P6 LISS III of 2005 have also been utilized for land use and urban settlement map.

The study area was demarcated using Toposheets of 1:50,000 by identifying the district boundaries by Scanning, projecting, geo-referencing and digitizing toposheets of the area manually using ERDAS. Various land use / land cover features were studied and a base map was prepared by visual interpretation using toposheets and false colour composite image of LISS III sensor. Water samples were taken from the area using random sampling techniques and with the help of GPS (Global Positioning System) the co-ordinates were noted down.

Spatial database

The spatial database is prepared by using different thematic Thematic layers like base map of the study area, land use / land cover, drainage network from SOI toposheets on 1:50,000 scale using ERDAS, Arc/Info GIS software to obtain baseline data. All maps are digitized to convert data into vector format. Landuse / land cover maps are prepared by using GIS software through supervised classification. SOI toposheets, satellite data and GPS (global Positioning data) together used with ground truth data.

Attribute database

Ground water samples were collected from predetermined locations that is, Urban and Suburban areas selected from the satellite imagery (Figure 2). The water samples taken from these locations and the area were then analyzed for eight water quality physico-chemical parameters adopting standard protocols (APHA, 1998). The water quality data thus obtained is used as database for present study (Table 1).

The standards prescribed by BIS were used foe the calculation of water quality indices (BIS, 1991)

GIS based spatial modeling for pollutant distribution

GIS can act as a powerful tool for modeling water quality. Various Thematic maps which are helpful understanding and managing water resources can be prepared with the use of GIS. In this study spatial interpolation technique through inverse distance weighted (IDW) approach of GIS has been used. In this technique sample points on different locations are selected for estimating output grid value. It determines cell value using a linearly weighted combination of sample points and controls the significance of known points upon the interpolated values based upon their distance from the output point thus a surface grid is generated as thematic isolines (Asadi et al., 2007).

Calculation

Water quality index (WQI)

Water quality index is regarded as one of the most effective way to communicate water quality (Sinha et al., 2004; Srivastava et al. 1994) WQI of water collected from 52 (22 + 30) locations of urban and sub-urban areas of Lucknow district were calculated. It is very useful method for assessing water quality of drinking water. In this a rating scale is fixed on the basis of importance and incidence on the overall quality of drinking water I terms of different physic-chemical parameters (Anonymous 2009-10; Horton, 1965). For calculating WQI different formulas given below are used (Sinha et al., 2006 Singh et al., 1999).

(i) Water Quality Rating, $Q_n = [(V_a - V_i) / (V_s - V_i)] \times 100$ (1)

Q_n = Quality rating for total water quality parameter.
V_a = Actual value of parameter obtained from Laboratory analysis.
V_i = Ideal value of the parameter obtained from the standards. (For pH it is 7 and for others it is zero).
V_s = Value recommended by BIS India of water quality.

(ii) Unit weight (Wn) = K / Sn   (2)

Sn is accepted drinking water quality standards by ISO
K = Proportionality Constant
Calculated by $K = [1 / (\Sigma^n_{n=i} 1/S_i]$
Sn = Standard values of the water quality.

Based on the above water quality values, the water samples quality is categorized as Excellent, Good, poor, Very Poor, Unfit for Drinking (Tiwari et al., 1985) (Table 2).

Land use land cover distribution

Land use change with time has great impacts on the environmental quality of the area. The change in land use is highly associated with ground water quality (Dasgupta et al., 2001). In the present study the land use has been classified into ten classes. In the overall

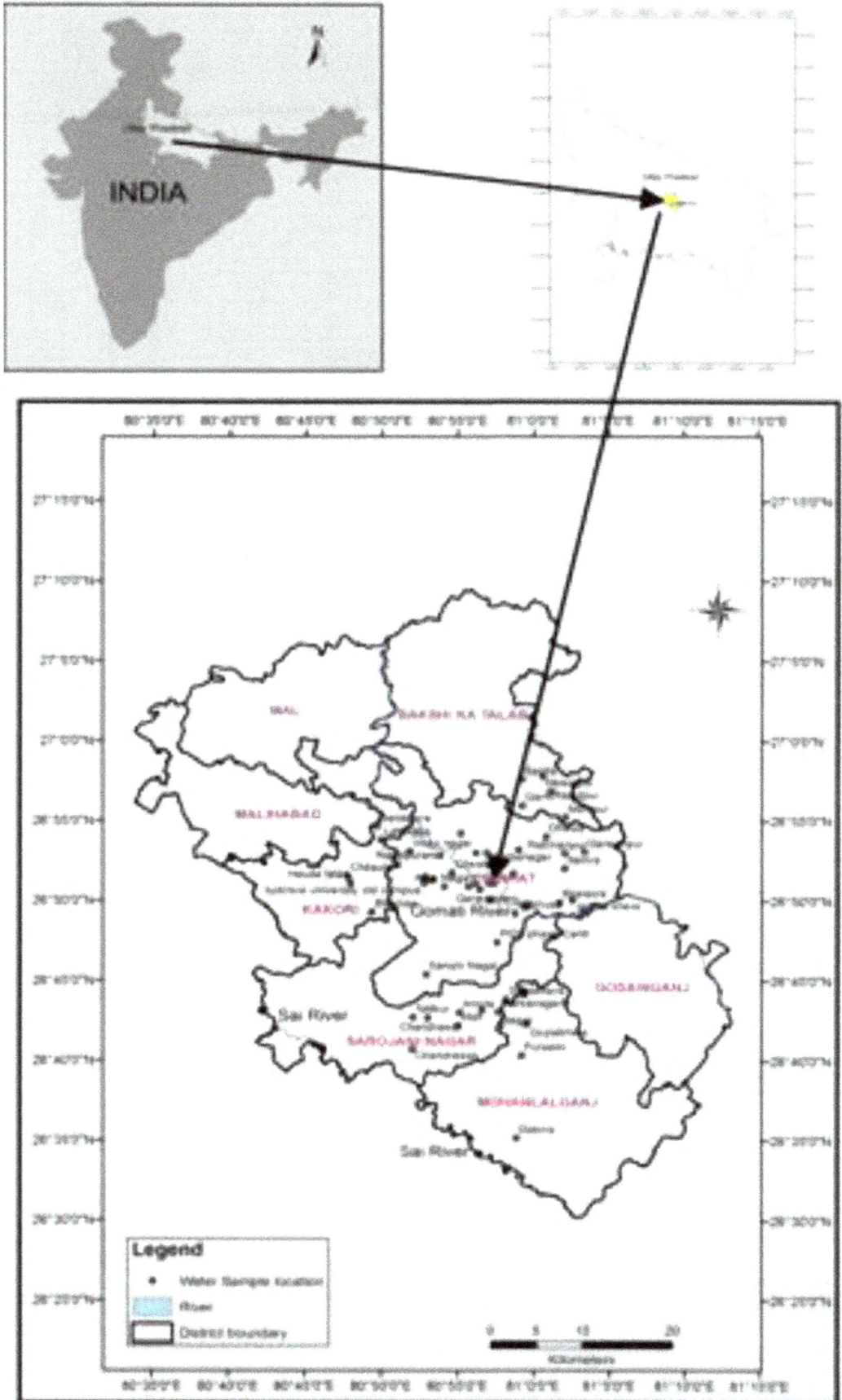

Figure 1. Location map of the study area -Lucknow city, Uttar Pradesh, India.

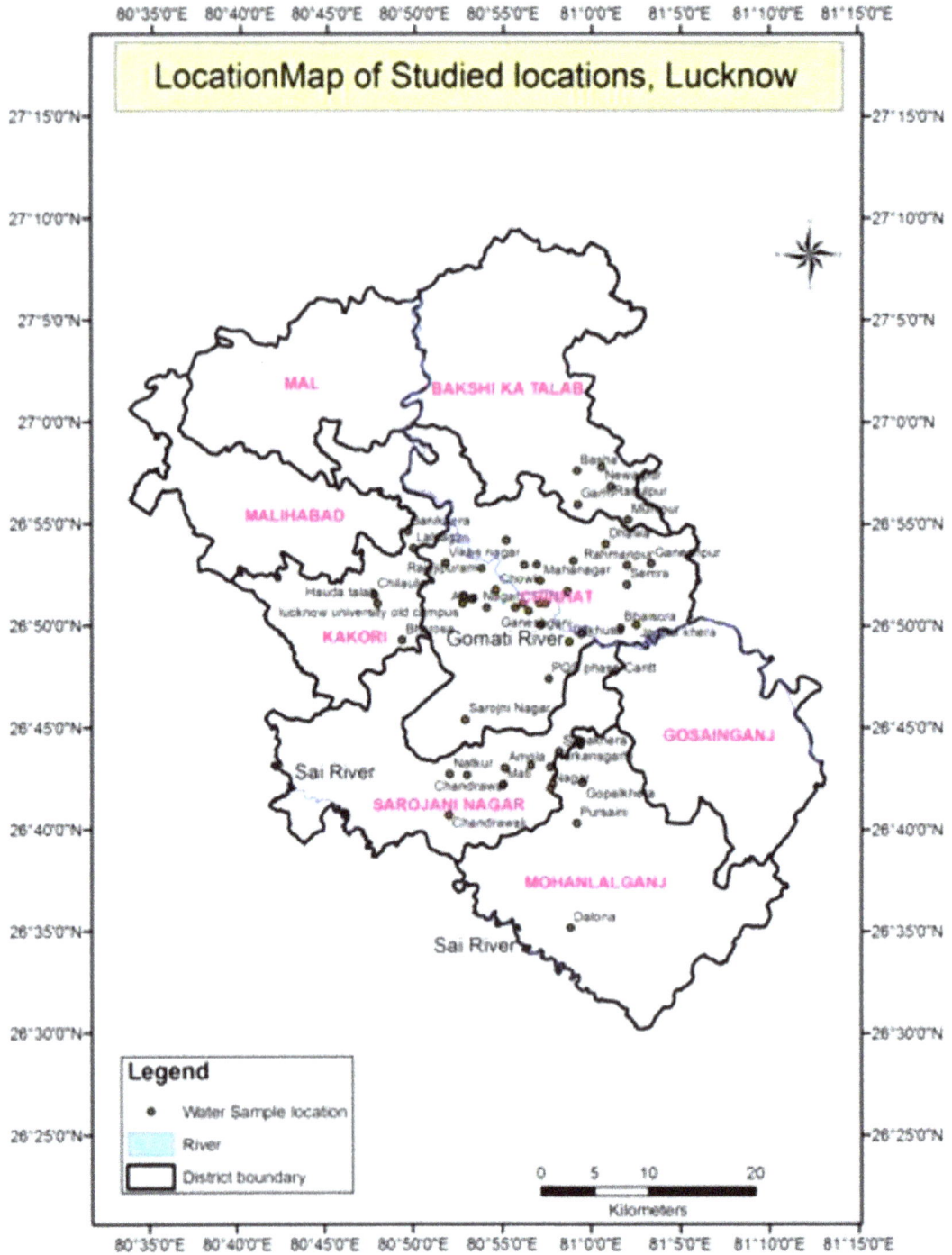

Figure 2. Sample location of studied water samples.

view of Lucknow district, the land use represents very few classes and hence the area is well suited for development projects. The expansion of the city's spatial limit has lead to a surmounting pressure on both natural and built drainage systems (Krishna et al., 2001). A measurable amount of settlement can be observed in most of the regions showing that the agricultural region has been converted into residential or urban purposes. In the overall view of Lucknow district, the land use represents very few classes and hence the area is well suited for development projects (RSACUP, 2000). The expansion of the city's spatial limit has lead to a surmounting pressure on both natural and built drainage systems. Drainage modifications and land use changes have lead to alterations in the regional hydrology, that is, the natural pattern of drainage has been modified by the changed land use. The build up area has increased at a faster rate and most categories are getting converted in buildup area. Table 3 shows that different categories of land have been transformed into build up area. The total land use change in the district has been shown in Figure 3.

Table 1. Physico-chemical analysis of water of Lucknow district.

S. No	Locations	pH	HCO₃	Cl	NO₃	SO₄	F	Ca	Mg	Na	Ec	WQI	Category of water
	Physico-chemical analysis of urban area of Lucknow district												
1	Bhujal Bhawan	8	183	7	3	Nil	0.6	16	22	9	325	30.54	Good
2	Indira Nagar	8.2	317	71	118	29	0.4	24	75	53	1000	70.17	Very poor
3	Gomti Nagar	8.1	403	28	19	43	0.3	40	68	26	900	62.32	poor
4	Narahi	8.2	51	78	2	29	0.4	24	63	115	1160	52.3	poor
5	Gulistan Colony	8.2	500	35	19	24	1.2	40	54	83	1040	70.11	Very poor
6	PQS Phase Catt.	8	310	21	3	10	1.1	32	46	12	600	51.47	Poor
7	Cariappa Road Catt.	8	290	14	3	Nil	0.7	32	19	42	550	38.63	Good
8	Dilkusha	8.1	409	35	2	14	0.7	48	34	64	840	101	UBD
9	Sarojini Nagar	8	177	57	12	10	0.3	40	15	35	530	35.28	Good
10	Mahanagar	8.2	366	185	155	43	0.6	96	78	80	1500	118.3	UFD
11	Vikas Nagar	8	293	57	12	38	0.1	56	32	46	785	48.84	poor
12	Lucknow University(New Campus)	8	232	28	13	10	0.8	16	29	42	530	39.26	Good
13	Lucknow University(Old Campus)	8.2	457	142	23	65	0.5	56	63	400	1360	88.49	Very poor
14	Aminabad	8.2	427	142	56	38	0.7	38	51	113	1300	82.05	Very Poor
15	River Bank Colony	8.2	457	57	10	14	0.4	48	49	67	1000	63.21	Poor
16	Chowk	8.1	256	71	10	24	0.4	40	39	37	714	49.94	Good
17	Campbell Road	8.2	275	99	70	28	0.8	80	49	16	926	59.20	Poor
18	Rajajipuram	8	171	28	12	10	0.6	32	15	18	400	33.86	Good
19	Arya Nagar	8	226	71	12	24	1.1	48	20	40	658	48	Good
20	Ganesh Ganj	8.1	409	99	4	24	0.4	64	34	90	996	59.86	Poor
21	New Hyderabad	8.2	659	64	2	67	0.3	8	141	51	1440	98.75	Very Poor
22	Nirala Nagar	8.2	494	64	50	48	0.5	8	88	92	1200	86.27	Very Poor
	Physico-Chemical analysis of Sub-urban area of Lucknow district												
1	Ahirankhera,Thakur ganj	7.85	329	14	2.4	20	0.78	25	42	23	320	46.91	Good
2	Lalnagar, Kakori	7.5	214	11	4	10	0.25	30	26	22	480	30.65	Good
3	Baniyakhera,Malihabad, Kakori	7.7	207	14	3	8	0.26	24	20	25	490	28.77	Good
4	Hauda Talab, Kakori	7.9	244	14	2.7	4	0.27	27	15	20	350	29.81	Good
5	Chilauli,kakori	7.9	245	15	2.1	5	0.58	26	25	17	440	35.42	Good
6	Bharosa,Kakori	8	216	7.1	2.7	5	0.52	20	30	25	350	35.82	Good
7	Chandraval, Sarojini nagar.	8.1	366	18	4.8	12	0.42	30	24	8	550	41.08	Good
8	Chandraval	8	299	14	2.8	14	0.62	32	28	10	410	40.66	Good
9	Natkur, SN	7.8	287	28	3.5	10	0.48	31	23	8	480	35.74	Good
10	Bhaisora, GN	8.1	159	14	2.1	8	0.22	24	15	12	260	26.93	Good
11	Jagpalkhera,GN	8	178	14	2.5	8	0.24	27	17	14	270	28.38	Good
12	Semra, Chinhat	8.1	189	7.1	1.2	2	0.25	22	20	24	280	29.63	Good
13	Rehmanpur, Chinhat	8	175	8	2.5	1	0.29	25	22	25	240	30.21	Good
14	Ganashpur, Chinhat	8.05	287	7.5	0.7	4	0.29	22	24	23	230	34.64	Good
15	Dhawa, Chinhat	7.8	195	10	2.5	4	0.25	24	24	22	250	29.84	Good
16	Murlipur, Chinhat	7.7	232	9	1.3	2	0.16	25	21	24	260	28.48	Good
17	Rasoolpur ,Chinhat	8	145	7.1	2.7	5	0.65	26	25	26	240	29.54	Good
18	Gadahi, Chinhat	8.1	234	7.5	2.1	2	0.46	22	21	25	230	33.47	Good
19	Basha, Kursi road	8.1	268	11	0.9	7	0.43	44	30	5	410	40.18	Good
20	Newajpur, Kursi road	8	287	14	1.2	5	0.79	30	35	3.5	420	42.49	Good
21	Sabhakhera, Mohanlalganj	7.8	366	18	2.7	5	0.69	20	30	35	580	42.24	Good
22	Kalli Pachim, Mohanlalganj	7.9	305	11	2.5	2	0.32	25	28	32	570	37.32	Good
23	Gareniankhera, Mohanlalganj	7.7	217	14	1.1	4	0.44	20	24	30	460	31	Good
24	Amola,Mohanlalganj	7.9	220	11	3.2	5	0.55	22	20	28	380	32.53	Good
25	Matee, Mohanlalganj	7.8	240	12	2.1	4	0.52	28	25	30	460	31.21	Good
26	Harkansgadhi, Mohanlalganj	7.6	250	14	2.3	5	0.44	20	20	33	450	30.55	Good

Table 1. Contd.

27	Dalona, Mohanlalganj	7.5	230	11	2.5	4	0.78	25	22	35	370	34.45	Good
28	Nagar, Mohanlalganj	7.6	219	13	1.4	2	0.05	28	30	34	430	31.06	Good
29	Gopalkhera, Mohanlalganj	7.8	250	14	2.7	3	0.45	30	25	36	520	31.86	Good
30	Pursaini, Mohanlalganj	7.8	230	11	2.3	2	0.69	28	26	31	490	36.11	Good

Table 2. Water quality index categories.

Water quality Index	Category
0-25	Excellent
26-50	Good
51-75	Poor
76-100	Very Poor
> 100	Unfit for Drinking (UFD)

Table 3. Parameter wise standards and their assigned weight.

S. No. (Wn)	Parameter	BIS standard	Assigned unit Wt.
1	pH Value	8.5	0.09818
2	Hardness	300.00	0.0027818
3	Chloride	200.00	0.0041727
4	Nitrates	50.00	0.00441347
5	Sulfate	250.00	0.0033381
6	Fluoride	1.00	0.8345365
7	Calcium	100.00	0.0083454
8	Magnesium	30.00	0.02738179
9	Sodium	200.00	0.0041727

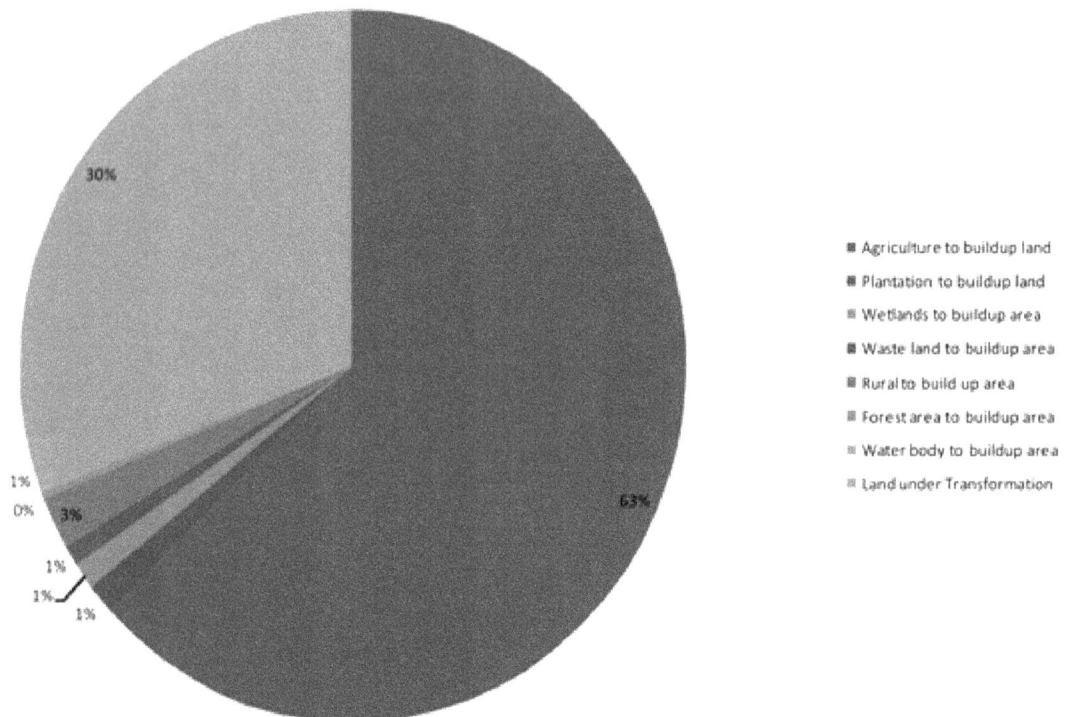

Legend:
- Agriculture to buildup land
- Plantation to buildup land
- Wetlands to buildup area
- Waste land to buildup area
- Rural to build up area
- Forest area to buildup area
- Water body to buildup area
- Land under Transformation

Figure 3. Pie diagram of Land transformation of Lucknow in the last 25 years.

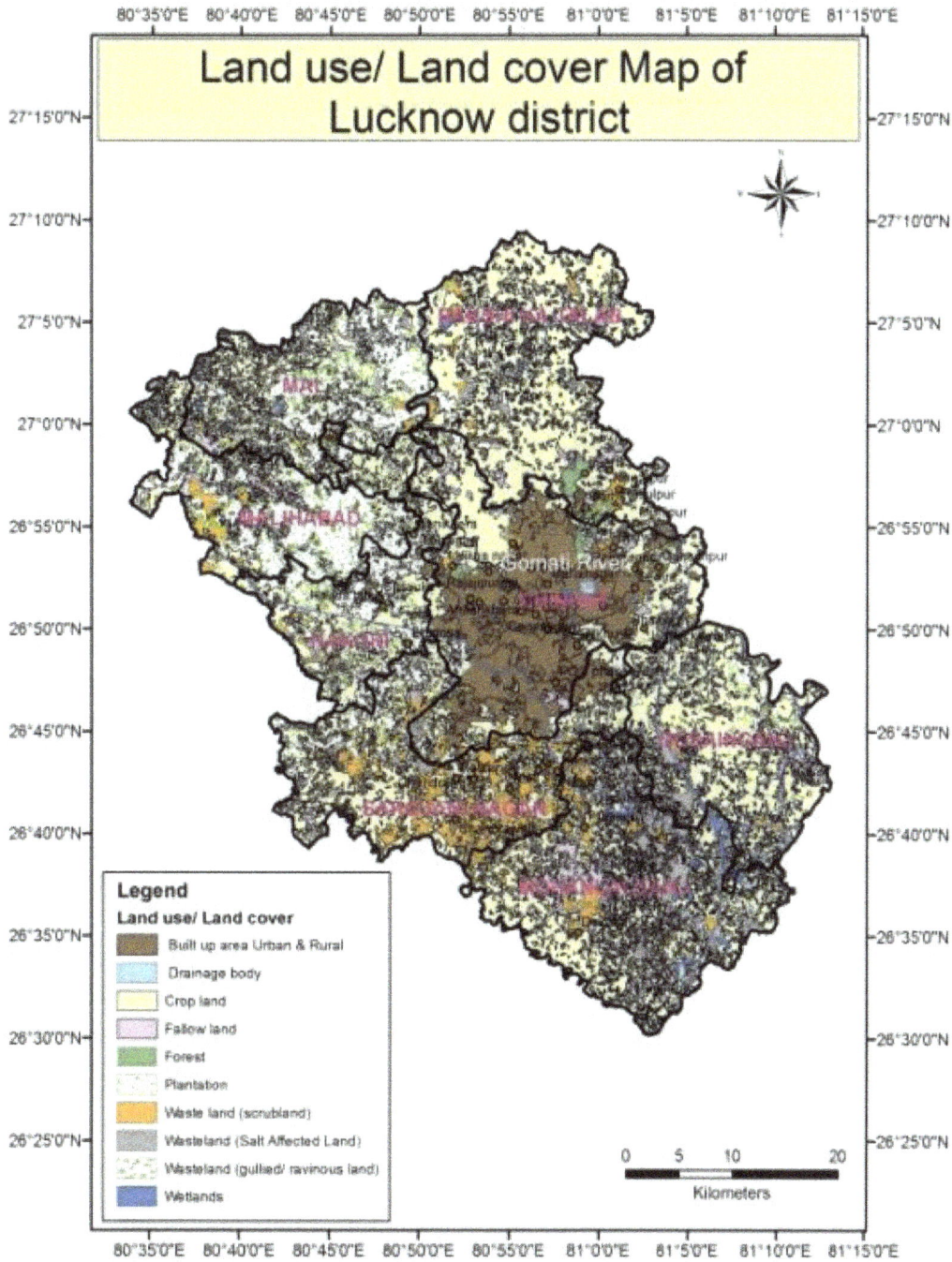

Figure 4. Landuse / Land cover map of luck now district.

RESULTS AND DISCUSSION

The land use/land cover data of the study area is showing rapid urbanization. Table 3 data shows that in last 25 years most of the land categories have changed in to buildup area. The water quality Index calculated for drinking purpose. Water quality index values shows that ground water in urban areas (Table 1) Like Dilkusha and Mahanagar is unfit for drinking, while areas like Indira Nagar, Gomti Nagar, Narhi, Gulistan colony, PQSphase cantt, Lucknow University old campus, Aminabad, New Hyderabad, Nirala Nagar, Campbell Road, Ganeshganj have very poor water quality and it is not good for human consumption. The rest of the sampling locations have comparatively good quality of water. WQI map is shown in (Figure 4) since water samples were collected from urban and sub urban areas, the water quality of the samples from urban areas was very poor but away from

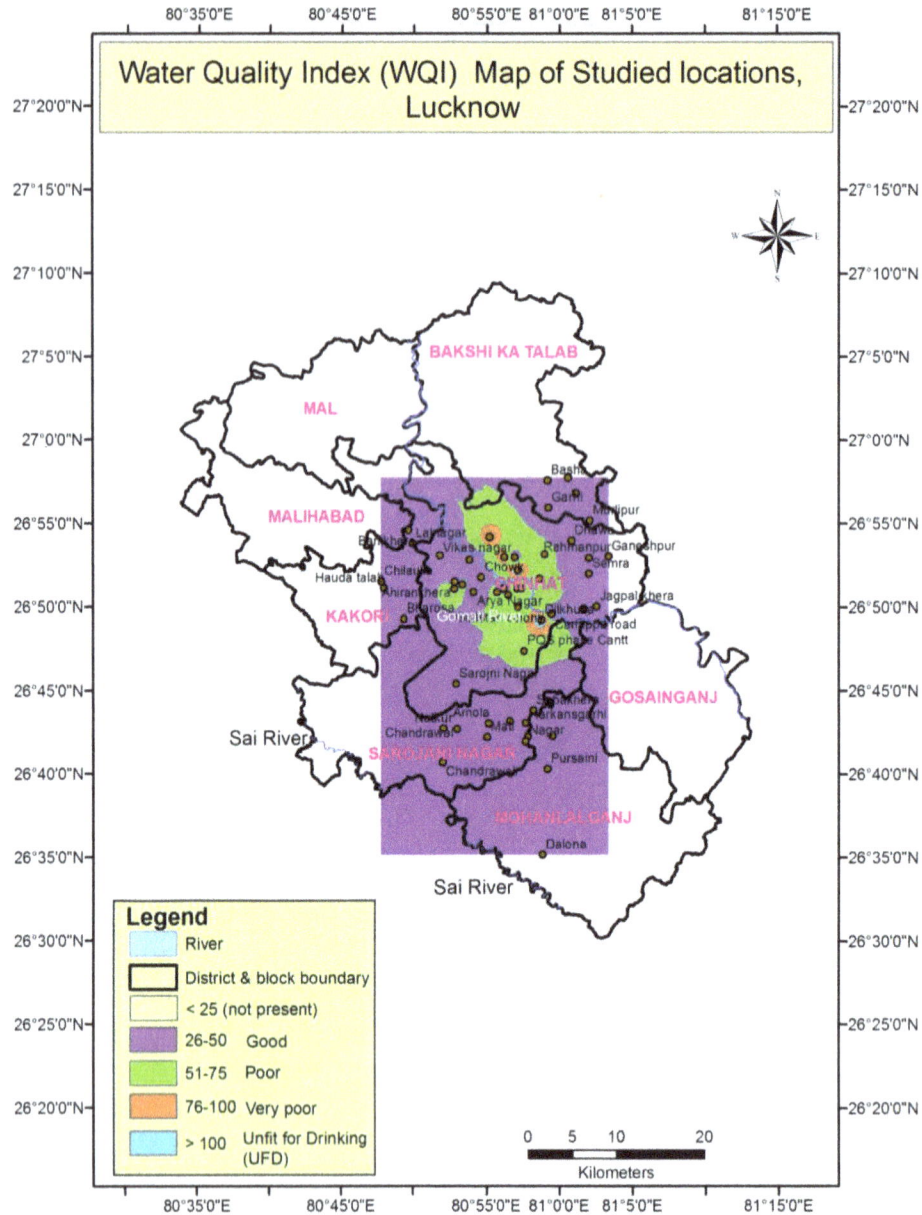

Figure 5. Water Quality Index (WQI) map.

the urban center water quality was only fairly good. The areas like Thakurganj, Newajpur and Sabha khera which are in close contact with the city areas are also showing deterioration in water quality.

The correlation of land use and water quality is depicted in Table 4. Figure 5 indicates that the extent of water quality deterioration has a linear correlation with urbanization. The water quality zones can be identified on the basis of water quality and land use. The old areas like Mahanagar, Dilkusha etc need special attention so that ground water quality can be improved. The analysis of the results drawn at various stages of work has integrated remote sensing and GIS as effective tools for the preparation of various thematic layers and final water

quality zonation map and correlating it with land use map of the study area. The WQI contour map in the studied sections (Figure 6) shows that the severity of the quality deterioration is in city vicinity that is, areas like Mahanagar, Indira nagar and Dilkusha region. The close compactness of the contour shows the intense deterioration increasing at an alarming rate in the city vicinity.

Conclusion

On the basis of above discussion, it may be concluded that underground water quality in the study area is

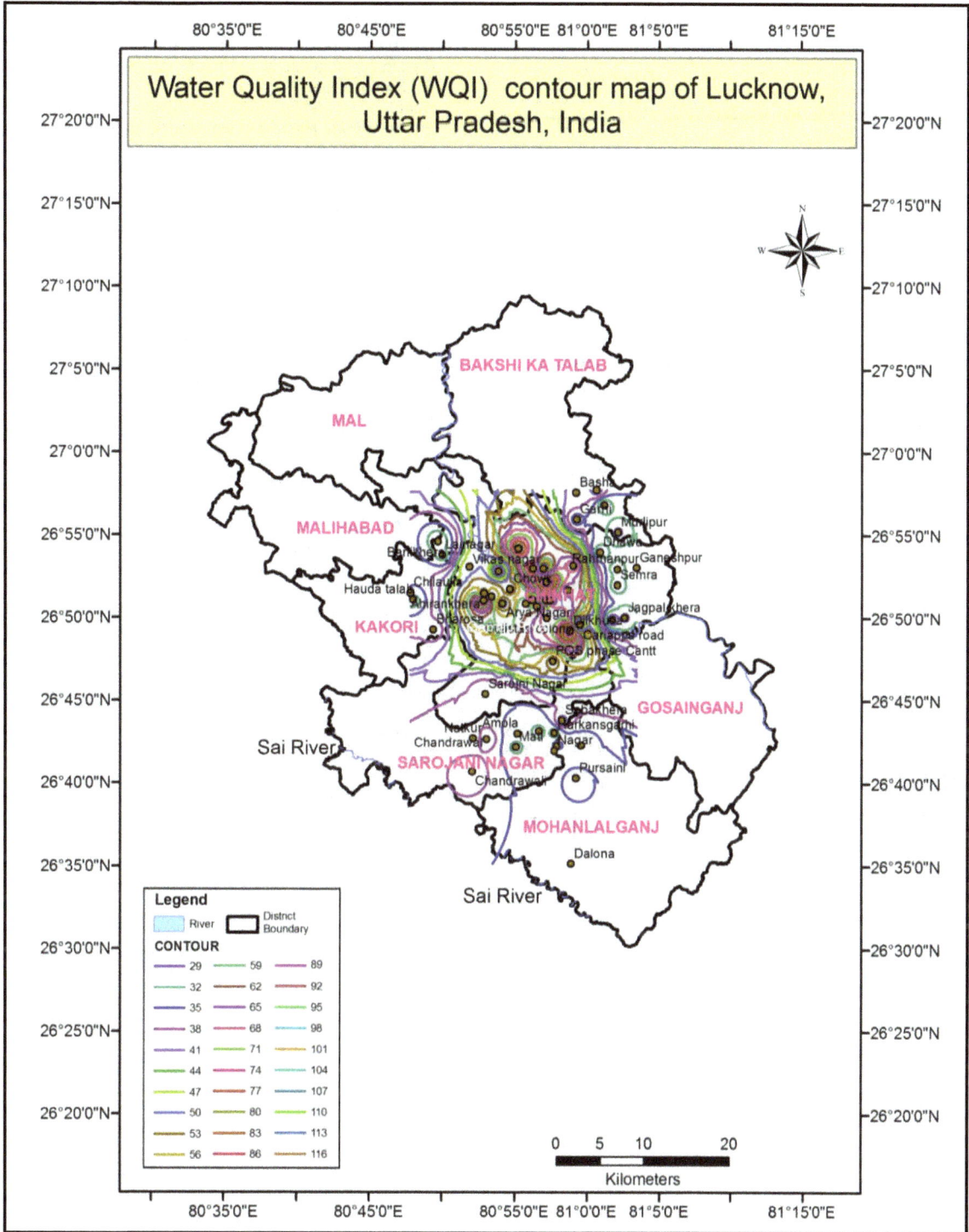

Figure 6. Contour map of the Water Quality Index (WQI) of studied location.

showing high correlation with the land use. The drinking water is highly polluted in the residential areas with high population density. The water quality of Sub urban areas having less population density and build up area are having comparatively better quality of water. Samples of suburban areas which are in close vicinity of the city like Thakurganj, Newajpur, Sabahpur are showing poor quality of water. Therefore, some effective measures are urgently required to enhance the drinking water quality by an effective management plan. the domain of the study focuses that if proper planning and measures are not taken then in the near future this will engulf the outskirts

Table 4. Land transformation of Lucknow in the last 25 years.

Categories of land transformation	Area (%)	
	(Km Sq)	**%**
Agriculture to buildup land	76.73	62.98
Plantation to buildup land	1.69	1.4
Wetlands to buildup area	1.45	1.19
Waste land to buildup area	1.17	0.96
Rural to build up area	2.93	2.42
Forest area to buildup area	0.39	0.32
Water body to buildup area	0.74	0.61
Land under Transformation	36.52	30.12
Total	**121.26**	**100.00**

of the city where at present the water quality is still good.

REFERENCES

Anonymous (2009-10). Ground water brochure of Lucknow district, Uttar Pradesh. p. 8.

Anonymous (2011). Ground water brochure of Lucknow district, Uttar Pradesh. p. 8.

APHA, AWWA, WPCF (1998). Standard Methods for the Examination of Water and Waste water. 20th edition, American Public Health Association, Washington, DC, New York, USA.

Asadi SS, Vuppala P, Reddy AM (2007). Remote Sensing and GIS Techniques for Evaluation of Groundwater Quality in Municipal Corporation of Hyderabad (Zone-V), India. Int. J. Environ. Res. Public Health 4(1):45-52.

BIS (1991). Indian Standards Specification for drinking Water, B.S. 10500. Government of India.

Das Gupta M, Purohit KM, Jayita D (2001). Assessment of drinking water quality of River Brahmani. J. Environ. Pollut. 8:285-291.

Epstein J, Payne K, Kramer E (2002). Techniques for mapping suburban sprawl. Photogram. Eng. Rem. Sens. 63(9):913-918.

Ferry Ledi T, Mohammed AK, Aslam MA (2003). A Conceptual Database Design For Hydrology Using GIS. Proceedings of Asia Pacific Association of Hydrology and Water Resources. March, 13-15, Kyoto, Japan.

Gyananath G, Islam SR, Shewdikar SV (2001). Assessment of Environmental Parameter on ground water quality. Indian J. Environ. Prot. 21:289-294.

Horton RK (1965). An index number system for rating water quality. J Water Pollut. Control Fed. 37:300.

Krishna NDR, Maji AK, Krishna YVN, Rao BPS (2001). Remote sensing and Geographical Information System for canopy cover mapping. Indian Soc. Rem. Sens. 29(3):108-113.

Lucknow Master Plan (2010). Town and Country Planning and Lucknow Development Authority, Lucknow, Uttar Pradesh.

Mohrir A, Ramteke DS, Moghe CA, Wate SR, Sarin R (2002). Surface and ground water quality assessment in Bina region. Indian J Environ. Prot. 22(9):961-969.

RSACUP (2000). Project Report on Geographic Information System Application Using High Resolution IRS-IC data for Urban Land Mapping, using Modeling and Monitoring Urban Sprawl of the city of Lucknow.

Saraf AK (1999). A report on Land use Modelling in GIS for Bankura District, Project sponsored by DST, NRDMS division, Government of India.

Singh AP, Ghosh SK (1999). Water Quality Index for River Yamuna Pollut. Res. 18:435-439.

Sinha DK, Saxena S, Saxena R (2004). Water Quality Index for Ram Ganga River at Moradabad, Pollut. Res. 23(3):527-531.

Sinha DK, Saxena R (2006). Statistical Assessment of Underground Drinking Water Contamination and Effect of Monsoon at Hasanpur J.P. Nagar (Uttar Pradesh, India). J. Environ. Sci. Eng. 48(3):157-164.

Skidmore AK, Witske B, Karin S, Lalit K (1997). Use of Remote sensing and GIS for sustainable land management. ITC J. 3(4):302-315.

Srivastava AK, Sinha DK (1994). Water Quality Index for river Sai at Rae Bareli for the pre-monsoon period and after the onset of monsoon. Indian J. Environ. Prot. 14:340-345.

Tiwari TN, Mishra M (1985). A preliminary assignment of water quality index of major Indian rivers. Indian J. Environ. Prot. 5(4):276-279.

United Nations (UN) (2004). World Urbanization Prospects: The 2003 Revision Database. Department of Economic and Social Affairs Population Division. New York: United Nations.

Permissions

The contributors of this book come from diverse backgrounds, making this book a truly international effort. This book will bring forth new frontiers with its revolutionizing research information and detailed analysis of the nascent developments around the world.

We would like to thank all the contributing authors for lending their expertise to make the book truly unique. They have played a crucial role in the development of this book. Without their invaluable contributions this book wouldn't have been possible. They have made vital efforts to compile up to date information on the varied aspects of this subject to make this book a valuable addition to the collection of many professionals and students.

This book was conceptualized with the vision of imparting up-to-date information and advanced data in this field. To ensure the same, a matchless editorial board was set up. Every individual on the board went through rigorous rounds of assessment to prove their worth. After which they invested a large part of their time researching and compiling the most relevant data for our readers.

The editorial board has been involved in producing this book since its inception. They have spent rigorous hours researching and exploring the diverse topics which have resulted in the successful publishing of this book. They have passed on their knowledge of decades through this book. To expedite this challenging task, the publisher supported the team at every step. A small team of assistant editors was also appointed to further simplify the editing procedure and attain best results for the readers.

Apart from the editorial board, the designing team has also invested a significant amount of their time in understanding the subject and creating the most relevant covers. They scrutinized every image to scout for the most suitable representation of the subject and create an appropriate cover for the book.

The publishing team has been an ardent support to the editorial, designing and production team. Their endless efforts to recruit the best for this project, has resulted in the accomplishment of this book. They are a veteran in the field of academics and their pool of knowledge is as vast as their experience in printing. Their expertise and guidance has proved useful at every step. Their uncompromising quality standards have made this book an exceptional effort. Their encouragement from time to time has been an inspiration for everyone.

The publisher and the editorial board hope that this book will prove to be a valuable piece of knowledge for researchers, students, practitioners and scholars across the globe.

List of Contributors

B. A. Adelekan
Department of Agricultural Engineering, Federal College of Agriculture, Moor Plantation, Ibadan, Oyo State, Nigeria

N. Oguntoso
Department of Agricultural Engineering, Federal College of Agriculture, Moor Plantation, Ibadan, Oyo State, Nigeria

Naima EL Hammoumi
Laboratory of Catalysis and Environment, Faculty of Science, University Hassan II Ain Chock- km 8 Road of El Jadida P. O. BOX 5366, Casablanca Morocco
Research Team of Hydrogeology, Treatment and Purification of Water and Climate Change, Department of Hydraulics,Hassania School of Public Works, km 8 Road of El Jadida P. O. BOX 8108, Casablanca Morocco

M. Sinan
Research Team of Hydrogeology, Treatment and Purification of Water and Climate Change, Department of Hydraulics,Hassania School of Public Works, km 8 Road of El Jadida P. O. BOX 8108, Casablanca Morocco

B. Lekhlif
Research Team of Hydrogeology, Treatment and Purification of Water and Climate Change, Department of Hydraulics,Hassania School of Public Works, km 8 Road of El Jadida P. O. BOX 8108, Casablanca Morocco

M. Lakhdar
Laboratory of Catalysis and Environment, Faculty of Science, University Hassan II Ain Chock- km 8 Road of El Jadida P. O. BOX 5366, Casablanca Morocco

M. D. T Gnazou
Laboratoire de Chimie de l'Eau, Faculté Des Sciences, Université de Lomé, B.P 1515, Lomé, Togo.

L. M Bawa
Laboratoire de Chimie de l'Eau, Faculté Des Sciences, Université de Lomé, B.P 1515, Lomé, Togo.

O Banton
Laboratoire d'Hydrogéologie, Université d'Avignon et des Pays du Vaucluse, France.

G Djanéyé-Boundjou
Laboratoire de Chimie de l'Eau, Faculté Des Sciences, Université de Lomé, B.P 1515, Lomé, Togo.

O. J Akintorinwa
Department of Applied Geophysics, Federal University of Technology, Akure, Nigeria

T. S. Olowolafe
Department of Applied Geophysics, Federal University of Technology, Akure, Nigeria

Michael A. Nwachukwu
Department of Environmental Technology, Federal University of Technology Owerri Nigeria

Aslan Aslan
Earth and Environmental Studies, Montclair State University New Jersey, U.S.A

Maureen I. Nwachukwu
Department of Geosciences, Federal University of Technology, Owerri Nigeria

Ijeh Boniface Ikechukwu
Department of Physics, College of Natural and Applied Sciences, Michael Okpara University of Agriculture, P. M. B. 7267, Umuahia, Abia State, Nigeria

Onu Nathaniel Nze
Department of Geosciences, Federal University of Technology, Owerri, Imo State, Nigeria

S. I. Ibeneme
Department of Geosciences, Federal University of Technology, Owerri Imo State, Nigeria

L. N. Ukiwe
Department of Chemistry, Federal University of Technology, Owerri Imo State, Nigeria

A. O. Selemo
Department of Geosciences, Federal University of Technology, Owerri Imo State, Nigeria

C. N. Okereke
Department of Geosciences, Federal University of Technology, Owerri Imo State, Nigeria

J. O. Nwagbara
Department of Geosciences, Federal University of Technology, Owerri Imo State, Nigeria

Y. E. Obioha
Department of Geosciences, Federal University of Technology, Owerri Imo State, Nigeria

A.G. Essien
Department of Geosciences, Federal University of Technology, Owerri Imo State, Nigeria

B. O. Ubechu
Department of Geosciences, Federal University of Technology, Owerri Imo State, Nigeria

E. S. Chinemelu
Department of Geosciences, Federal University of Technology, Owerri Imo State, Nigeria

E. A. Ewelike
Department of Geosciences, Federal University of Technology, Owerri Imo State, Nigeria

R. N. Okechi
Department of Biotechnology, Federal University of Technology, Owerri Imo State, Nigeria

GUHA Hillol
Bechtel Power Corporation, Geotechnical and Hydraulic Engineering Services, U.S.A

DAY Garrett
Bechtel Power Corporation, Geotechnical and Hydraulic Engineering Services, U.S.A

CLEMENTE José
Bechtel Power Corporation, Geotechnical and Hydraulic Engineering Services, U.S.A

H.O. Nwankwoala
Department of Geology, College of Natural and Applied Sciences, University of Port Harcourt, Nigeria

S. A. Ngah
Institute of Geosciences and Space Technology, Rivers State University of Science and Technology, P. M. B. 5080, Nkpolu-Oroworukwo, Port Harcourt, Nigeria

K. Madi
Geology Department, University of Fort Hare, Private Bag, X1314, alice, 5700, Eastern Cape, South Africa

B. Zhao
TWP Projects (PTY) LTD, P.O Box 61232, Marshalltown 2107, Johannesburg, South Africa

Hamdy A. El-Ghandour
Department of Irrigation and Hydraulics, Faculty of Engineering, Mansoura University, Egypt

Ahmed Elsaid
Department of Mathematics and Engineering Physics, Faculty of Engineering, Mansoura University, Egypt

R. N. Tiwari
Department of Geology, Government P.G. Science College Rewa, Madhya Pradesh, India

Shankar Mishra
Department of Chemistry, P.G. College Semariya, Rewa, Madhya Pradesh, India

Prabhat Pandey
Department of Physics, Government P.G. Science College Rewa, Madhya Pradesh India

Abdur Rahman
Department of Civil Engineering, Stamford University Bangladesh, Dhaka, Bangladesh

M. A. Zafor
Department of Civil Engineering, Leading University, Sylhet, Bangladesh

Mursheda Rahman
Department of Civil Engineering, Stamford University Bangladesh, Dhaka, Bangladesh

Shivakumar J. Nyamathi
Civil Engineering Department, UVCE, Bangalore University, Bangalore, India

R Bharath
Civil Engineering Department, UVCE, Bangalore University, Bangalore, India

A. V. Hegde
Applied Mechanics Department, NITK, Surathkal, Mangalore, India.

K. Srinivasa Reddy
Central Research Institute for Dryland Agriculture (CRIDA), Santoshnagar, Hyderabad-500 059, India

B. A. Adelekan
Department of Agricultural and Mechanical Engineering, College of Engineering and Technology, Olabisi Onabanjo University, P. M. B. 5026, Ibogun, Ogun State, Nigeria

Anju Verma
Department of Environmental Sciences, IBSBT, CSJM University, Kanpur

Biswajeet Thakur
Birbal Sahni Institute of Palaeobotany, Lucknow

Shashwat Katiyar
Department of Biochemistry, IBSBT, CSJM University, Kanpur

Dharam Singh
Department of Environmental Sciences, IBSBT, CSJM University, Kanpur

Madhu Rai
Department of Geology, University of Lucknow, Lucknow